Energy-smart farming

Efficiency, renewable energy and sustainability

It is widely recognised that agriculture is a significant contributor to global warming and climate change. Agriculture needs to reduce its environmental impact and adapt to current climate change whilst still feeding a growing population, i.e. become more 'climate-smart'. Burleigh Dodds Science Publishing is playing its part in achieving this by bringing together key research on making the production of the world's most important crops and livestock products more sustainable.

Based on extensive research, our publications specifically target the challenge of climate-smart agriculture. In this way we are using 'smart publishing' to help achieve climate-smart agriculture.

Burleigh Dodds Science Publishing is an independent and innovative publisher delivering high quality customer-focused agricultural science content in both print and online formats for the academic and research communities. Our aim is to build a foundation of knowledge on which researchers can build to meet the challenge of climate-smart agriculture.

For more information about Burleigh Dodds Science Publishing simply call us on +44 (0) 1223 839365, email info@bdspublishing.com or alternatively please visit our website at www.bdspublishing.com.

Related titles:

Water management
Print (ISBN 978-1-78676-176-7) Online (ISBN 978-1-78676-178-1; 978-1-78676-179-8)

Climate change and agriculture
Print (ISBN 978-1-78676-320-4) Online (ISBN 978-1-78676-322-8; 978-1-78676-323-5)

Reducing greenhouse gas emissions from livestock production
Print (ISBN 978-1-78676-192-7); Online (ISBN 978-1-78676-195-8, 978-1-78676-196-1)

Measuring soil health
Print (ISBN 978-1-78676-439-3) Online (ISBN 978-1-78676-441-6; 978-1-78676-442-3)

Chapters are available individually from our online bookshop:
https://shop.bdspublishing.com

climate•SMART•publishing
IN AGRICULTURAL SCIENCE

BURLEIGH DODDS SERIES IN AGRICULTURAL SCIENCE
NUMBER 115

Energy-smart farming

Efficiency, renewable energy and sustainability

Edited by Emeritus Professor Ralph E. H. Sims, Massey University, New Zealand

burleigh dodds
SCIENCE PUBLISHING

Published by Burleigh Dodds Science Publishing Limited
82 High Street, Sawston, Cambridge CB22 3HJ, UK
www.bdspublishing.com

Burleigh Dodds Science Publishing, 1518 Walnut Street, Suite 900, Philadelphia, PA 19102-3406, USA

First published 2022 by Burleigh Dodds Science Publishing Limited

Library of Congress Control Number: 2022934571

British Library Cataloguing in Publication Data
A catalogue record for this book is available from the British Library

ISBN 978-1-78676-835-3 (Print)
ISBN 978-1-78676-838-4 (PDF)
ISBN 978-1-78676-837-7 (ePub)
ISSN 2059-6936 (print)
ISSN 2059-6944 (online)

DOI 10.19103/AS.2022.0100

Typeset by Deanta Global Publishing Services, Dublin, Ireland

Contents

Series list

Title	Series number
Achieving sustainable cultivation of maize - Vol 1 From improved varieties to local applications Edited by: Dr Dave Watson, CGIAR Maize Research Program Manager, CIMMYT, Mexico	001
Achieving sustainable cultivation of maize - Vol 2 Cultivation techniques, pest and disease control Edited by: Dr Dave Watson, CGIAR Maize Research Program Manager, CIMMYT, Mexico	002
Achieving sustainable cultivation of rice - Vol 1 Breeding for higher yield and quality Edited by: Prof. Takuji Sasaki, Tokyo University of Agriculture, Japan	003
Achieving sustainable cultivation of rice - Vol 2 Cultivation, pest and disease management Edited by: Prof. Takuji Sasaki, Tokyo University of Agriculture, Japan	004
Achieving sustainable cultivation of wheat - Vol 1 Breeding, quality traits, pests and diseases Edited by: Prof. Peter Langridge, The University of Adelaide, Australia	005
Achieving sustainable cultivation of wheat - Vol 2 Cultivation techniques Edited by: Prof. Peter Langridge, The University of Adelaide, Australia	006
Achieving sustainable cultivation of tomatoes Edited by: Dr Autar Mattoo, USDA-ARS, USA and Prof. Avtar Handa, Purdue University, USA	007
Achieving sustainable production of milk - Vol 1 Milk composition, genetics and breeding Edited by: Dr Nico van Belzen, International Dairy Federation (IDF), Belgium	008
Achieving sustainable production of milk - Vol 2 Safety, quality and sustainability Edited by: Dr Nico van Belzen, International Dairy Federation (IDF), Belgium	009
Achieving sustainable production of milk - Vol 3 Dairy herd management and welfare Edited by: Prof. John Webster, University of Bristol, UK	010
Ensuring safety and quality in the production of beef - Vol 1 Safety Edited by: Prof. Gary Acuff, Texas A&M University, USA and Prof. James Dickson, Iowa State University, USA	011
Ensuring safety and quality in the production of beef - Vol 2 Quality Edited by: Prof. Michael Dikeman, Kansas State University, USA	012
Achieving sustainable production of poultry meat - Vol 1 Safety, quality and sustainability Edited by: Prof. Steven C. Ricke, University of Arkansas, USA	013
Achieving sustainable production of poultry meat - Vol 2 Breeding and nutrition Edited by: Prof. Todd Applegate, University of Georgia, USA	014

Introduction

The global food supply system, including production, processing and transport, accounts for around one third of global end-use energy demand and one quarter of total annual greenhouse gas emissions. Around 15% of these food supply emissions result from heat, electricity and fuel used directly for on-farm production activities. Since 1970, these have increased to now exceed 1000 Mt CO_2-eq in 2019.[1]

Around 15% of global agricultural production costs on-farms are energy-related, mainly for the purchase of fossil fuels. With heightened concerns around the sustainability of crop and livestock production, the agricultural sector must develop and invest in alternative methods of farming that recoup the same results but have reduced costs and environmental impacts.

This volume reviews recent research and practical applications on smart ways of reducing the costs and environmental impacts of on-farm energy use. The chapters are broken down into three parts.

- Part 1 focuses on the efficient use of electricity, heat fuel and water;
- Part 2 examines the use of on-farm renewable energy heat and power generation;
- Part 3 provides case studies of energy-smart farming, with chapters specifically dedicated to discussing dairy, pig and poultry farming.

Part 1 Efficient use of electricity, heat, fuel and water

Part 1 opens with a chapter that focuses on measuring and auditing on-farm energy use. Chapter 1 begins by reviewing energy analysis in agriculture, focusing on how energy analysis methods can help to identify potential issues in conventional energy use so that farmers can consider utilising alternative lower-carbon energy resources. The chapter discusses how agriculture is an energy consumer, before assessing aspects such as energy conservation, energy saving, on-farm energy inputs and methods for assessing energy use in irrigation. A case study on energy consumption in wheat production using neural networks is also provided before the chapter concludes with an overview of why developing new methods to analyse energy inputs in all agricultural sectors is important.

Chapter 2 focuses on advances in energy efficient lighting and ventilation for food production systems. The chapter first examines the benefits and shortcomings of using various light sources including sunlight, incandescent

1 Flammini et al., 2022. *Emissions of greenhouse gases from energy use in agriculture, forestry and fisheries: 1970-2019*. Earth Systems Science Data, 14, 811-821, https://essd.copernicus.org/articles/14/811/2022/.

light, compact fluorescent light and light emitting diodes (LEDs). The chapter discusses how switching to more energy-efficient light resources can help reduce energy costs, then goes on to review the recent shift from using incandescent lights to LED lamps. A section on issues arising from using light dimmers is also included, followed by an examination of novel ultraviolet (UV) technologies. Greenhouse and controlled environment lighting applications are also discussed, as well as ventilation advances in livestock housing. The chapter concludes by emphasising the potential energy saving by switching to LED lamps.

The next chapter of Part 1 reviews improving farm machinery operation and maintenance to optimise fuel use efficiency. Chapter 3 begins by discussing the current techniques for optimising tractor fuel consumption, such as adaptive driving, optimising tyre inflation pressure, wheel slip and tractor ballasting, estimating tractor power needs and purchasing a fuel-efficient tractor. The chapter moves on to examine the developments in fuel-efficient tractor operation and how tractor performance can be measured. It also reviews the energy effects of farming systems, focusing specifically on no-tillage and controlled traffic farming. Energy savings and other benefits are addressed and a section on controlled farming operations in complex topographies is also included. A section on the embodied energy from the manufacture of machinery, fertilisers and herbicides is provided. The chapter concludes with an analysis of potential incentives to encourage more sustainable energy use in farm machinery.

Chapter 4 closes Part 1 with an analysis of efficient water management and irrigation on farms. The chapter begins by providing an overview of irrigation in agricultural production, focusing on supply-side and demand-side water management. It then reviews water use efficiency, specifically looking at water productivity, irrigation practices, water efficient irrigation technologies and the adoption of efficient water-use irrigation innovations. The chapter also highlights the challenges of water and energy trade-offs, the integrated analysis of trade-offs, the challenges of sustainable intensification and the development of integrated evidence-based policies relating to agricultural water and energy.

Part 2 On-farm renewable energy heat and power generation

The first chapter of Part 2 focuses on assessing and modelling the costs of on-farm distributed renewable energy systems. Chapter 5 first examines the private costs of on-farm distributed renewable energy systems, which is then followed by a review of the social costs of energy, addressing the use of cost-benefit analysis and life cycle assessment techniques. The chapter also highlights the various risk analysis methods and provides a case study from the Duero Valley in Spain describing a cost assessment for the potential use of a solar photovoltaic system to power water pumps. It concludes by providing a summary of the potential for decarbonisation of the agrarian sector as a result of using renewable energy systems, supported by an annex that includes methodologies for encouraging the deployment of distributed renewable energy systems on farms.

The next chapter addresses energy-smart innovation and renewable energy systems on farms. Chapter 6 begins by highlighting how energy efficiency can help decarbonise food production systems, then goes on to address the current trends in renewable energy. A discussion of the current policies and measures that are in place to encourage the deployment of renewable energy on farms is also provided, followed by an overview of the range of farm-scale renewable energy technologies and systems that are commercially available. The chapter concludes by emphasising the importance of switching from fossil fuel dependence to renewable energy systems to help reduce agriculture's greenhouse gas emissions.

Chapter 7 examines the use of on-farm biomass heat technologies for providing heat and power. The chapter first discusses the primary energy conversion technologies that are available such as combustion, gasification and pyrolysis. It moves on to discuss secondary conversion technologies such as internal combustion engines, steam turbines, steam engines, Stirling engines, indirect-fired gas turbines, direct-fired pressurised gas turbines, micro turbines and hydrogen. The chapter also examines cogeneration system analysis at the small scale, before providing a case study on the conversion of coal boilers to biomass in greenhouse production enterprises. It concludes by highlighting the potential of using sustainably-produced biomass to provide heat at the farm scale in the future.

The final chapter of Part 2 discusses developments in agrivoltaics and how synergies can be achieved through combining crop and pasture plant production with solar photovoltaic power systems. Chapter 8 begins by reviewing the design of agrivoltaics systems to maximise land use and the synergy between energy and agricultural production. It then moves on to discuss the physiological and agronomical aspects of crop cultivation under agrivoltaic systems. The chapter also describes crop modelling applications for the management of agrivoltaic systems, focusing on simulation aspects and design and optimisation. A section on the benefits of agrivoltaic systems is also provided.

Part 3 Energy-smart farming: case studies

The first chapter of Part 3 is a case study that focuses on the tools and technologies that can be used to reduce fossil energy use on dairy farms. Chapter 9 first examines how total energy demand can be monitored through the analysis of direct energy and indirect energy inputs on dairy farms. It moves on to discuss energy modelling methods such as mechanistic modelling, regression models and machine learning algorithms. A section on smart energy strategies is also provided, focusing on energy mitigation strategies and renewable energy technologies. The chapter includes a case study of a Dairy Energy Decision Support Tool, before concluding with a section on how developing energy-smart farming methods on dairy farms is crucial to improving energy savings.

Chapter 10 focuses on energy-smart pig farming. The chapter starts by reviewing energy generation on pig farms, focusing on electricity and heating fuels which provide the thermo-neutral environment required for pigs. It addresses feed energy related to pig production and how buildings can be designed to improve energy efficiency. The chapter also highlights the practices available to improve energy efficiency of existing buildings, such as pig barns, by means of optimal operation of temperature control and ventilation systems and identifying structural deficiencies. It concludes by providing an overview of how the trend towards energy-smart pig farming will continue due to the growing demand for more environmentally-friendly farming methods to help reduce demands for fossil fuels and their effects on climate change.

The final chapter of the book describes how energy is used in poultry production buildings today and discusses the factors that have driven the growth of energy use. Chapter 11 identifies opportunities toward energy-smart production including existing technologies and management practices that can lead to significant energy conservation and improving energy efficiency. It presents technologies that are emerging in certain regions with high potential for wider adoption. It also reviews the current status of research and development of alternative energy technologies with potential for application and development in poultry production.

Acknowledgement

We wish to acknowledge the following for their help in reviewing a particular chapter:

- Chapter 6: Associate Research Professor Daniel Ciolkosz, Pennsylvania State University, USA

Part 1

Efficient use of electricity, heat and fuel

Chapter 1

Measuring and auditing on-farm energy use

Majeed Safa, Lincoln University, New Zealand

1 Introduction

Food production and supply systems can be classified into primary agriculture, including cropping, livestock, horticulture and fisheries; food processing and packaging; trade and transport; and household preparation and cooking (Wallgren and Höjer, 2009). This extensive range covers all activities from farm to kitchen. Crop yields and food supplies to markets are directly linked to energy inputs (Ilyas et al., 2020; Lin et al., 2021; Stout, 1990; USDA, 2008). Agricultural operations include all farming operations that occur after the land has been cleared of native vegetation and developed, such as tillage, planting, fertilising, pest control, harvest, post-harvest, and transport behind the farm gate.

As an energy conversion process, agriculture converts two naturally abundant materials, water and carbon dioxide, to carbohydrates and other complex organic materials through the photosynthetic process that occurs when growing plants use solar energy. Agriculture also conserves and recycles mineral resources (Fluck and Baird, 1980; Odum and Odum, 1976; Pimentel and Pimentel, 2008; Stout, 1990; Tester, 2005). Food production, processing and packaging, and the distribution of food products from farms to consumers, need around 42 GJ/person/year (Pimentel et al., 2007).

In the early 1900s, on-farm energy sources around the world were mostly in the form of biomass derived from agriculture or forests. In rural areas of many

http://dx.doi.org/10.19103/AS.2022.0100.01

developing countries, the use of traditional biomass in the form of fuelwood, charcoal and animal dung remains the main local energy source to provide heat for cooking and comfort. Historically, many industrial products were also mainly made from plant materials, and early transport fuels also came from agriculture. Once cheap coal and oil became readily available, there was a rapid rise of industrialised agriculture that depended upon these fossil fuel inputs.

The growing risk of climate change impacts has led to increased interest in replacing the combustion of fossil fuels to provide heat, power and mobility. Producing carbon-neutral energy biomass such as wood chips, biogas, crop residues from agricultural by-products or from purpose-grown energy crops has grown in popularity, as has the generation of renewable electricity.

During the last two centuries, the amount of energy consumed in the agriculture sector per kilogram of protein produced has increased more in developed countries than in developing countries. The total energy demand for primary production in farms and fisheries is about 3% of the national energy consumption in developed countries and about 3.6% in developing countries (Karkacier et al., 2006; Sauerbeck, 2001; Stout, 1990). The average energy input is about 7.7 GJ/ha for subsistence farming in developing countries, mainly for human labour and animal power, and about 37.9 GJ/ha in industrialised agricultural systems in developed countries, mainly provided by fossil fuels for mechanisation and fertiliser manufacture inputs (Pimentel and Pimentel, 2008).

The entire global food system consumes around one-third of total end-use energy, emits around one-quarter of total greenhouse gases, and when production, processing, packaging and transport are included, comprises on average about 15% to 20% of a nation's total energy consumption (Pimentel and Pimentel, 2008; Stout, 1990; Ziesemer, 2007). In the United States, for example, it was estimated that around 19% of fossil energy consumed is used in the food supply system (Pimentel and Pimentel, 2008).

National statisticians often pay little attention to the energy consumption of the agriculture sector. In some countries, only fuel purchased by farmers at subsidised prices is considered when analysing fuel consumption. The diesel and petrol purchased by farmers from public service stations are included in the transport sector. Only a small percentage of the total electricity consumed in the food supply system is used on farms, while most indirect energy inputs such as for fertiliser and machinery manufacture are included in the industrial sector (Pellizzi, 1992).

2 Energy analysis in agriculture

Energy analysis serves a range of economic, management and technical purposes. Research into energy use in the food supply chain began after the first oil shock in the 1970s. Improving agricultural methods and finding new energy resources were advocated to reducing dependency on fossil fuel energy resources (Fluck and Baird, 1980; Kitani, 1999; Smil, 1991; Stout, 1990).

In 1974, Odum established the first energy analysis method (SSSA, 1997). At the same time, another evaluation method was suggested by the International Federation of Institutes for Advanced Study (Fluck, 1992). Since then, several other methods have been used to determine and analyse energy consumption in agricultural production (Fluck and Baird, 1980). These include statistical analysis, process analysis and input–output analysis when energy inputs and outputs are identified and evaluated (Kitani, 1999).

One of the main problems when comparing energy studies is the use of different data collection methods. Four analytical methods provide rational information on which to base energy-related decisions: life-cycle assessment, energy consumption, cost-effectiveness and environmental assessments (Randolph and Masters, 2008; Safa et al., 2011). Conversely, some energy-use studies focus on gaining measurements or responses to questionnaires from a number of sample farms.

Weather patterns, soil physicochemical factors, management conditions, pests, diseases, weeds, field size, degree of mechanisation, energy prices and the interaction of many other factors affect energy demands on farms. Crop growth models usually include carbon, nitrogen and water as well as energy balances (CIGR, 1999; Kuesters and Lammel, 1999; Liu, 2009; USDA, 2008; Vlek et al., 2004). Analyses can include the identification, estimation and measurement of energy use in agricultural systems (Fluck and Baird, 1980).

Concern about the rising reliance of agricultural production systems on fossil fuel inputs prompted the use of energy analysis techniques to study the level of energy dependence and comparative energy efficiency of farming systems (Stout, 1990). Odum (1994) attempted to understand the principle of general systems theory in relation to environmental systems. He discussed the relationship between energy inputs and outputs in ecological systems using mathematics. He also stated that energy analyses in agriculture have much wider error margins than energy analyses in industry.

The results of energy studies depend on the set of assumptions used, such as defining outputs and inputs, and the energy equivalent of inputs (Conforti and Giampietro, 1997). Local results may not be representative of other areas (Liu, 2009).

Because there are different methods to estimate energy consumption, any comparison and evaluation of results from a number of studies are difficult. For example, human labour has been considered as an energy input in some studies, but not in others (Conforti and Giampietro, 1997; Fluck, 1992; Hülsbergen and Kalk, 2001; Sartori et al., 2005; Saunders et al., 2006). Furthermore, a general international agreement on how to estimate energy input has been difficult to achieve. In addition, a lack of reliable data for each country and region often forces researchers to take values from other countries, without making adjustments for the different circumstances in those countries compared with their own (Conforti and Giampietro, 1997; Kitani, 1999). Moreover, in some studies, the latest references of energy coefficients are mentioned instead of the original sources.

One of the most important problems in energy analysis is the non-homogeneity of different sources (Fluck and Baird, 1980) and the different norms and coefficients that have been used in various studies. For example, the same amount of fertiliser can have a different energetic cost, depending on the technical level of the local manufacturing industry. Energy contents also depend on the transport distance, which is variable, but an average value can be assumed for a region (Kitani, 1999). There are also problems with assigning shares of energy inputs where there are multiple outputs from a system – such as sheep production for meat and wool, or cereal production for grain and straw. In this instance, it is difficult to divide the energy inputs between the outputs.

Comparing one set of energy consumption data with other published assessments in different countries is a challenge and requires preliminary checks to be made on the primary data, the definitions of inputs and outputs, and the conversion factors used in the calculation (Conforti and Giampietro, 1997; Kitani, 1999; Safa et al., 2011). This is explained in the following section.

Due to the varying farm conditions and the effects of several direct and indirect factors on the system, an appropriate management tool for energy use on farms is essential. Mathematical modelling, for example, has been successfully applied to tillage, spraying and crop harvesting machines (Alvarez, 2009; Bertocco et al., 2008; Fang et al., 2000; Safa et al., 2015). Before designing a usable mathematical model, obtaining sufficient relevant data and information is necessary. A survey can be designed as a data collection process to identify the main inputs and outputs of a specific farm enterprise. The methods adopted by earlier researchers should be investigated carefully and the most appropriate ones chosen to determine the most important farming parameters. Direct energy sources and consumption should be analysed on the basis of specific farm operations and indirect energy inputs. Those used in the manufacture of fertiliser, machinery, maintenance, animal feed brought into the farm and seed production should be included.

3 Agriculture as an energy consumer

Land, labour, seed and water were the most interdependent factors in early agricultural societies that were first established around 3000 BC in Mesopotamia. Over the following centuries, humans have slowly improved the techniques and tools to increase productivity and reduce labour intensity. Oxen and horses were domesticated to help farmers cultivate more land. However, around 10% of farm land had to be devoted to grow feed for those work animals.

Prior to the nineteenth century, farmers had lived in a subsistence economy (as many still do, in developing economies). After the industrial revolution, populations increased in some countries, and a large proportion migrated from rural areas to industrial cities to find better employment opportunities. To reduce this labour gap, farming efficiency was improved by introducing larger and more powerful breeds of horses, artificial fertilisers and farm mechanisation (Boyle et al., 2003; Pimentel and Pimentel, 2008). Energy consumption in agriculture has become more intensive as after World War II, the 'Green Revolution' led to the use of high-yielding seeds, fertilisers and chemicals, as well as tractors with modern diesel engines and electricity use on farms (Hatirli et al., 2006).

Although today's modern farming systems are highly dependent on energy inputs, farmers aim to minimise production costs (Nguyen and Hignett, 1995). Energy consumption has been increasing in response to the limited supply of arable land, increasing population, technological changes, and a desire by the growing middle classes in developing countries for increased protein intake and improved standard of living (Hatirli et al., 2006; Kizilaslan, 2008; Manaloor and Sen, 2009). Over the last century, the global land area that was cultivated increased by 80–100% and food harvested on farms grew six-fold. However, in the same period, energy consumption increased by around 85 times (Smil, 2008).

A trend in agricultural production called 'from farm to the last consumer' is to utilise different forms of energy sources for producing, transporting, processing, packaging and retailing. The proportion of energy demand for each part of the food supply chain depends on a number of factors such as the properties of the crops, the distance between farm and market, the method of processing and the retail system.

The energy requirements for on-farm cereal or vegetable crop production can be divided into crop protection, nutrition, cultivation and culture (Tzilivakis et al., 2005). These can be further sub-divided into:[1]

1. energy for the manufacture of chemicals, fertilisers and other farm inputs (including packaging and transport to the farm);
2. energy required for carrying out field operations with the value assigned based on the type and working width of the machine and, in the case of tillage operations, the operating depth and soil type; and
3. indirect energy required for the manufacture of machinery and its maintenance, which varies with the operational lifetime and depreciation period.

1 The categories would be different for horticultural crops, meat production and dairy.

There is a correlation between energy consumption in agriculture and the global rise of urbanisation (Safa and Samarasinghe, 2011; Smil, 2008). Furthermore, energy has an important and unique role in economic and social development, especially in developing countries. However, there is a general lack of rural energy development policies that focus on agriculture. This is mainly due to the low levels of government attention being given to agricultural production, especially in developing countries. Another reason might be the 'follower' characteristics of developing countries that place less value on agricultural production as more industrialisation is reached by the developed countries. In addition, a poorer educated and less organised rural population in developing countries has not significantly influenced many politicians (Karkacier et al., 2006).

3.1 Energy conservation in the agriculture sector

Overall, energy conservation in the food supply sector is very complex. It comprises a wide range of activities that can be applied on farms, during food processing and transport, and in homes and restaurants.

On farms, there is a positive relationship between energy usage and productivity (Baruah and Bora, 2008; Hatirli et al., 2006; Karkacier and Gokalp Goktolga, 2005; Karkacier et al., 2006; Outlaw et al., 2005; Singh et al., 2004; Smil, 2008). There is also a significant relationship between energy inputs and local climate, price, yield and technology (Ozkan et al., 2004). A study of energy consumption in agricultural operations showed which operations are more important for energy saving (Pellizzi et al., 1988). Changes that could reduce energy consumption across the food supply chain include using more efficient technologies, converting to organic farming methods, eating more local and seasonal products to reduce 'food miles,' and reducing consumption of meat and dairy products (Sauerbeck, 2001; Wallgren and Höjer, 2009). Producing and consuming more vegetable protein instead of animal protein can reduce the demand for farm land and water as well as the use of energy (Pimentel and Pimentel, 2008; Wallgren and Höjer, 2009). Overall, there is potential to conserve energy demand in the agriculture sector by between 10% and 40%.

Organic farming can improve energy efficiency by avoiding on-farm use of mineral fertilisers and pesticides because conventional farming requires higher energy inputs per hectare or per tonne of product (Alfoldi et al., 1994; Dalgaard et al., 2001; Grastina et al., 1995). For example, energy inputs per hectare for organic corn production are around 30% less than for conventional systems (Pimentel et al., 2005). However, the average crop yields of cereal crops are 20% to 34% lower from organic farming than for conventional farming (Joseph et al., 2019). Therefore, organic products often have a price premium. However, the demand for them continues to increase.

Intensive livestock production for meat, dairy, eggs etc. depends on providing on-farm feed or buying concentrated feed products (Wallgren and Höjer, 2009). Where cattle graze all year-round on pastures, the energy inputs are less per kilogram of protein produced than for intensive cattle reared on feedlots that depend totally on conserved and concentrated feed.

In dairy production and in intensive livestock and poultry farming, a large share of the energy used is in the form of electricity for producing heat, including for hot water, and in ventilation systems. Electricity is also consumed for running milking shed equipment and chilling milk (Kitani, 1999). Therefore, energy-efficient technologies for water heating and chilling milk, such as insulated water cylinders and milk vats, can significantly reduce energy use (Centre for Advanced Engineering, 1996).

Agricultural mechanisation has improved yields and reduced costs over recent decades but has replaced inefficient human and animal energy with energy from fossil fuels. Reducing tractor and harvester fuel use in crop and pasture production is possible by improved operation by the driver, which can improve overall energy efficiency.

Crop yields and energy inputs are highly influenced by any variability in soil and climatic conditions (Bertocco et al., 2008). There is also potential for energy efficiency gains by improving irrigation pumping and storage systems (that can also save water) and by the more precise and efficient application of fertilisers and agrichemicals that can also reduce environmental impacts (Centre for Advanced Engineering, 1996; Nemecek et al., 2008; Outlaw et al., 2005).

Transport is one of the more energy-intensive components of the food supply chain. Shipping food products from farms to markets and then to consumers is more complex than in many other sectors, especially for fresh products. Most crops, milk and animal products, however, are processed and packaged by food industries and the food products are then shipped to wholesale and distributing centres. From there, they are transported to groceries and supermarkets near population centres. Individual customers then buy and transport fresh and packaged foods to their homes or restaurants.

Large amounts of goods, supplies and machinery are also transported to farms. In the United States, for example, around 600 kg/ha of different materials are transported to each farm annually (Pimentel and Pimentel, 2008). The energy for transportation is mainly from petroleum products, and can be reduced by improving the energy efficiency of vehicles, improving the logistics, using alternative fuels, changing modes (e.g. from road to rail or coastal shipping), supporting local and regional food production to reduce the delivery distance of food products and using e-commerce to decrease dependence on private cars for food collection (Pimentel et al., 2007; Wallgren and Höjer, 2009).

3.2 Energy saving in agricultural operations

There is a high energy demand in the arable farms and market gardens growing vegetables for tractor fuel - for cultivation, planting, fertiliser distribution, spraying, harvesting etc. Each specific farm operation can use a variety of methods and machinery. For example, cultivation can vary from no-tillage[2] to conventional ploughing followed by secondary cultivation operations. In each operation, different factors may affect energy use, such as speed and depth of operation, soil type and moisture content, and width of machinery, all of which govern the number of passes needed across a field (McLaughlin et al., 2008). Additionally, the total energy requirement for each operation has different components. For example, energy inputs for tillage depend on soil-tool interactions; interactions between tilled and fixed soil masses; soil deformation; and the acceleration of the tilled soil (Ashrafi Zadeh and Kushwaha, 2006). Consequently, the choice of the most suitable tillage system differs with the specific farm conditions, the farmer's knowledge, financial constraints and energy factors, if considered (Bertocco et al., 2008). For example, fuel consumption for conservation tillage (such as shallow discing then planting) can be two-thirds less than for conventional ploughing, harrowing and sowing (Smil, 1991). The selection of the appropriate method and machine in order to reduce energy can be complex, as crop yields may also be affected. Saving energy should normally not result in lost productivity.

New designs of agricultural machinery are continually being developed to improve timeliness in the field and also reduce fuel inputs. For example, a combination of disc harrows and cultivator sweeps gives a seedbed in a single field operation.

The number of tractors in use worldwide has increased from 11 million in 1961 (FAO, 2008) to around 30 million. In some regions, older tractors have been replaced with fewer but more powerful new ones (Smil, 2008; Stout, 1990). There are a considerable variety of makes, models, power outputs, designs, engine types and transmission and tyre types available, with many innovations happening during the twentieth century (Smil, 2008). Selecting the right tractor and equipment can make a significant difference to on-farm efficiency. Fuel consumption can be reduced by as much as 30% when the tractor is selected to match the field machine, set up correctly and driven with maximum efficiency (Centre for Advanced Engineering, 1996; Pellizzi et al., 1988).

2 No tillage, sometimes called zero tillage, leaves residues from the previous crop on the field as a way of growing crops from year to year without soil preparation through tillage. No-tillage is an emerging technique which can increase soil C content, conserve the amount of water in the soil, decrease erosion and reduce fuel consumption. It may also increase the amount and variety of natural life in, and on, the soil, such as the number of earth worms. However it may require increased herbicide usage. The risk of yield reduction is higher than for conventional tillage (Baker et al., 2007; Gajri et al., 2002).

To achieve optimum fuel consumption, the tractor and equipment should be adjusted for each specific task and good driving practices must be followed. Regular maintenance, optimum wheel slip, correct tyre size and pressure, the use of four-wheel drives when advantageous, and operating in higher gears at lower engine speeds can all help to reduce fuel consumption as well as save operator time. Reducing transport distances, creating larger and longer fields, selecting appropriate speeds and depths of operations and choosing the right weather conditions for field activities are the other components (Ashrafi Zadeh and Kushwaha, 2006; Barber, 2004; Centre for Advanced Engineering, 1996; Conforti and Giampietro, 1997; Kitani, 1999; Pellizzi et al., 1988; Smil, 2008; Stout, 1990).

The mismatch of tractor size with equipment capacity is common on farms. In heavy-load operations, such as primary and secondary tillage, tractor size can significantly influence fuel consumption per hectare. Usually, farmers have a limited number of tractors on which to attach different field machines to match the correct load application. Typically, more powerful tractors are used for heavy-load applications such as tillage, and lower power tractors for lighter load applications such as mowing and drilling. Rather than purchase an expensive over-powered tractor, contractors could be employed for specific heavy-duty tasks.

Many indirect factors, such as cultural practices, the availability of capital, personal opinions and the availability of machinery dealers also influence the selection of tractors and equipment (Barber, 2004; McLaughlin et al., 2008).

3.3 On-farm energy inputs

Since the mid-twentieth century, the quantity of fossil fuels, the electricity consumption in global agriculture and the use of fertilisers and pesticides have increased by about 20- to 50-fold, resulting in increased productivity. However, the rate of input increase was significantly higher than the resulting yield increases. For example, between 1950 and 1980, fertiliser used in corn production in the US increased from 5 kg/ha to about 150 kg/ha (30 times) (Pimentel and Pimentel, 2008), but yields only doubled (Fig. 1).

Energy use on farms can be categorised into direct (fuel and electricity), indirect (fertilisers, agrichemicals, seeds and animal feeds) and capital or embedded (energy used to manufacture equipment (farm vehicles, machinery and irrigation) and buildings, fences etc. (Wells, 2001). However, in most research analyses, only direct and indirect inputs have been included.

On-farm direct energy (operational energy) involves the consumption of fuels such as diesel, petrol, furnace oil, electricity and woody biomass (Mohtasebi et al., 2008; Safa et al., 2011). In one typical example of a cropping

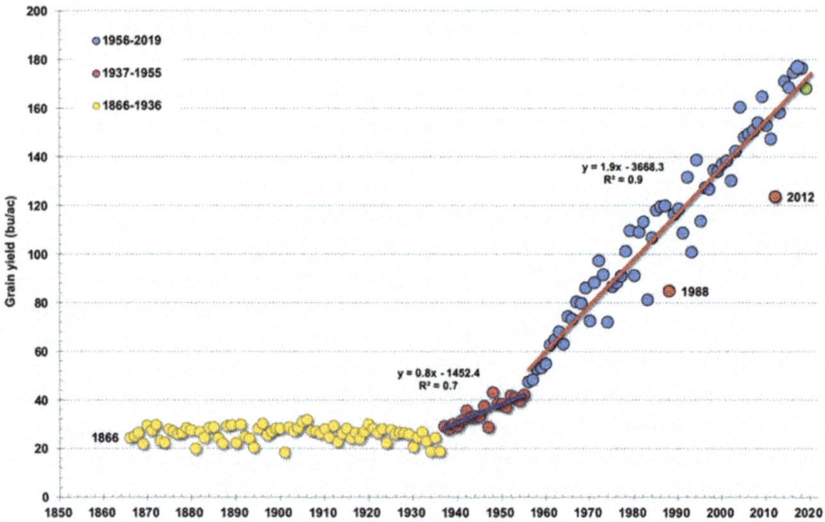

Figure 1 US average corn yields from 1866 until 2019. The impact of the 'Green Revolution' after World War 2 is evident. Departures from the trends in 1901, 1988 and 2012 are due to droughts (Nielson, 2020).

farm, the use of diesel in tractors and other engines to power harvesters, water pumps, etc. contributed around 27% to the total energy input under irrigated conditions, with electricity used for irrigation consuming 12.7% of total on-farm energy (Singh and Mittal, 1992).

Indirect energy accounts for approximately 70% of total energy use on dairy and pig farms and about 50% on arable farms (Bailey et al., 2008; Meul et al., 2007; Wells, 2001).

The overall energy inputs also include the energy expended during extraction, conversion, refining and transport of fossil fuels (Barber, 2004; Centre for Advanced Engineering, 1996; Kitani, 1999; Mohtasebi et al., 2008). This upstream energy is not always included in on-farm energy analyses.

For each energy source, there is a corresponding norm called a conversion coefficient or energy equivalent. Conversion coefficients help to standardise the unit of all inputs to MJ/ha. However, different coefficients have been used in various studies for the same energy input. Therefore, some form of standardisation is needed to select a suitable energy coefficient for each energy input.

Energy used in the agriculture sector can also be classified in other sectors. For example, fuel consumed by farm trucks may be classified in the transport sector, or, indirect energy for fertiliser manufacture may be included in the industrial sector. Consequently, the official national statistics do not usually show accurate data for energy used in agriculture (Pellizzi, 1992; Safa and Samarasinghe, 2013). The total on-farm energy input (E) is determined from the sum of the input factors (A_i) multiplied by the appropriate energy conversion coefficient for each factor (C_i):

$$E = \Sigma(A_i C_i) \quad (1)$$

When assessing farm inputs and outputs for energy inputs, energy conversion coefficients are used (Table 1). The differences between analysts arise from the technology and estimation methods used. Therefore, selection of a consistent energy conversion coefficient is important when comparing energy equivalents across different studies.

3.3.1 Human labour

Before the invention of the tractor, humans and draught domestic animals were the only energy input choices for agricultural operations. The introduction of machinery reduced manual labour requirements, although it still plays a role in many field activities (Smil, 2008). In subsistence farming in many developing countries, human labour remains the main source (73%) of energy in agricultural operations (Stout, 1990). In 1992, around 48% of the global labour force

Table 1 Coefficients used for various energy inputs and outputs in cereal crop production by selected analysts

	Energy coefficients (MJ/unit)					
	Mani et al. 2007 (India)	Saunders et al. 2006 (NZ)	Pellizzi 1992 (Italy)	McChesney et al. 1982 (NZ)	Ozkan 2007 (Turkey)	Kitani 1999 (US)
Inputs (unit)						
Human (h)	1.96	#	#	#	2.3	#
Diesel (L)	56.31	43.6	#	46.7	56.31	47.8
Petrol (L)	#	39.9	#	42.3	#	#
Nitrogen (kg)	60	65	73.75	65	64.4	78.1
Phosphorus (kg)	11.1	15	13.14	15	11.96	17.4
Potassium (kg)	6.7	10	9.10	10	6.7	13.7
Sulphur (kg)	#	5	#	5	#	#
Herbicides (kg)	#	310	85.95	270	#	*
Insecticides (kg)	#	315	50.55	#	#	*
Fungicides (kg)	#	210	#	#	#	*
Seed (kg)	14.7	#	9.12	2.5	25	13
Outputs						
Grain (kg)	14.7	#	9.12	#	14.7	13
Straw (kg)	12.5	#	10.5		12.5	

* For different crop varieties, different equivalents were used.
# The reference did not indicate any value for the item.

worked in the agriculture sector (CIGR, 1999), which was dropped to 26.7% in 2019 (Worldbank, 2021) However, the percentage of farm labourers of the total workforce in developed countries is significantly higher than in developed countries (Christiaensen et al., 2020).

If all crop production operations were to be undertaken manually, at least 1200 hours per hectare would be required. This means that each person can manage around just one hectare during a growing season (Pimentel and Pimentel, 2008). Using machinery, human input is still needed for almost every task including driving, repairing machinery, managing irrigation, spraying, and fertiliser distribution, but only 11 hours per hectare is needed (Pimentel, 2009).

In future, human labour on fully mechanised farms using mechatronics and robotics could be reduced to almost zero. However, organic farming may need some manual work for harvesting and weeding (Pimentel et al., 2005; Wallgren and Höjer, 2009; WCED, 1987), although robotic weeders are also being developed.

Human energy can be analysed by measuring heart rates and recording oxygen consumption (Stout, 1990). It varies with gender, body size, weight, age, activity and climate (Smil, 1994). In wheat production, depending on the technology used, labour inputs range from 684 h/ha in Kenya to only 7.8 h/ha in the US (Pimentel and Pimentel, 2008). The average hourly input per tonne of wheat has reduced from 30 hours in 1800 to just two hours in 1970 (Coley, 2008).

In modern agriculture, human energy used is much less than from other energy inputs, and can be down to 1%. Therefore, it is often not calculated in recent energy studies. Where needed, the energy output of a male worker is typically 1.96 MJ/h and it is 0.98 MJ/h for a female worker (Mani et al., 2007; Singh and Mittal, 1992). In developed countries, human labour is the most expensive source of energy in field operations, which encourages farmers to invest in more machinery.

3.3.2 Fuel

Until the 1890s, the availability of nutrients and the amount of suitable land for growing food were the major limits to agriculture. Fossil fuels solved these problems (Coley, 2008).

Official statistics pay little attention to fuel consumption in agriculture. Only fuel purchased by farmers at subsidised prices is considered when analysing fuel consumption, and not petrol and diesel when purchased from service stations - which are classified in the transport sector (Pellizzi, 1992). Depending on the type of fuel and the amount of time a tractor or machine is used for, fuel and lubricant costs will usually represent between 16% and 45% of total machine costs (Siemens and Bowers, (1999). Due to subsidies, the percentage of fuel and lubrication costs is lower in some countries than in others.

In warm, dry climatic areas, more fuel is used for irrigation than for other operations such as tillage and seeding (Centre for Advanced Engineering, 1996; Safa and Tabatabaeefar, 2002). Fuel consumption in specific operations depends on soil conditions, crop type, ground-speed and rolling resistance (Smil, 1991), but better-educated farmers tend to consume less fuel (Safa et al., 2010).

Fuel consumption expressed as litres per hectare (L/ha) is a more useful measurement than litres per hour (L/h) because it uses the same basis to compare different inputs and operations (McLaughlin et al., 2008). Specific volumetric fuel consumption (SVFC, measured as L/kW/h) is the most common method used to estimate the energy efficiency of a tractor, although specific volumetric fuel efficiency (SVFE measured as kW/h/L) is also used (Grisso et al., 2004).

Diesel engines are robust, have a higher fuel efficiency and a longer life than petrol engines; therefore they are mostly used on farms (Kitani, 1999). Petrol engines are used mainly for utilities, light trucks, chainsaws, portable sprayers etc. There are several methods to estimate the fuel consumption of tractors based on their engine power. However, due to the influence of height above sea level, soil conditions, residue cover, air pressure, humidity, temperature and operator skills on tractor fuel consumption, most of these methods work only for specific situations (Bertocco et al., 2008; McLaughlin et al., 2002; Serrano et al., 2007). These methods are useful to predict the fuel consumption of diesel engines working under full load, but when under partial loads, and when engine speeds are reduced from full throttle, they are less accurate (Siemens and Bowers, 1999).

According to the ASAE EP496.2 (2006), most agricultural tractors tested over the last 25 years have diesel engines. Their fuel consumption assessment is:

$$Q_{avg} = 0.223 \times P_{pto} \quad (2)$$

where, Q_{avg} is the average diesel consumption (L/h) and P_{pto} is the maximum PTO power (kW). Siemens and Bowers suggested that results be increased by approximately 15% to account for farming conditions that are dissimilar to the original study. This equation was modified due to technology improvements leading to more efficient fuel consumption, which has decreased by about 4.8% annually over the last 20 years (Grisso et al., 2004). Bowers (1985) (Equation 3) and Riethmuller (1989) (Equation 4) suggested linear relationships between the draught per unit of equipment width and fuel consumption per hour.

$$Q_{avg} = 1.2774 \times P_{drf} \quad (3)$$

$$Q_{avg} = 1.1306 \times P_{drf} \quad (4)$$

where Q_{avg} is the average diesel consumption (L/h) and P_{drf} is the specific draught (kN/m).

Serrano et al. (2007) introduced a nonlinear relationship between PTO power utilisation and fuel consumption per hour:

$$Q_{avg} = 266.4 + 884.5e^{(-Ppto/124)} \quad (5)$$

where, Q_{avg} is the average diesel consumption (L/h) and P_{pto} is the PTO power utilisation (kN/m).

The best way to measure diesel fuel consumption is by measurements in the field by filling the tractor fuel tank to the top before and after each operation. For operations such as harvesting, this method is less practical. Fuel consumption is related to many factors, and it is incorrect to use the same rate of fuel consumption for different operations (Table 2).

The efficiency of tractors and self-propelled machines affects fuel consumption and can be analysed with respect to engine, power transmission and traction (Pellizzi et al., 1988; Serrano et al., 2007). Matching of tractor and implement, using hydraulic three-point linkage equipment correctly, using power-take-off (PTO) equipment, selecting the optimum travel pattern in the field, having large paddocks, regular machinery servicing, adjusting tyre inflation pressure, matching engine speed and gear selection, improving

Table 2 Average fuel consumption rates for a range of agricultural operations in different studies (L/h)

	Wells NZ (2001)	Dalgaard Denmark (2001)	Witney US (1988)	Kitani (1999)
Mouldboard plough	18	22	21	25
Chisel plough	#	#	#	13
Heavy-duty discs	12	#	13	9
Field cultivator	6	6.2	8	8
Spring tine harrow	4	4	#	#
Rotary hoe cultivator	#	#	13	4
Combined tillage	#	#	#	24
Air seeder	5	#	#	5
Grain drill	10	3.2	4	5
Fertiliser spreader	3	1.9	3	2
Boom-type sprayer	3	1.2	1	1.5
Combine harvester	#	14	11	18

The reference did not indicate the value of the item.

traction efficiency, using turbochargers, and improving farmers' awareness are all methods that could lead to fuel savings and improved field efficiency (Barber, 2004; Grisso et al., 2004).

The average fuel consumption rates for agricultural operations vary in different studies (Table 2). One reason is that in many developing countries, diesel is used to fuel water pumps, whereas more efficient electric pumps are commonly used in developed countries.

3.3.3 Fertiliser

Insufficient soil nutrients are the most important barriers to crop productivity, other than water supply (Pimentel and Pimentel, 2008). For better growth, farmers apply extra nutrients, either as natural manure, compost, or more commonly as artificial mineral fertilisers. Around 60% of world fertiliser demand comes from developing countries, where it is used mainly in cereal production (55-58% of the total) (FAO, 2000a). Mineral fertilisers have increased crop and pasture yields more than plant breeding or other innovations (Smil, 1991, 2008) have done. Traditionally, soil fertility was maintained and improved by adding livestock manure, planting nitrogen-fixing legumes and leaving plant residues for incorporation into the soil to improve organic matter and soil carbon contents (CIGR, 1999).

During and after the Green Revolution, fertiliser use increased dramatically (Manaloor and Sen, 2009). Without using chemical fertilisers, more land would have been needed to be converted from native forests and more grasslands to arable and pasture farms. Land use changes such as deforestation can produce more greenhouse gas (GHG) emissions per hectare than fertiliser manufacture and use (Vlek et al., 2004). Nowadays, one-third of the protein in global food supplies is derived as a result of the Haber-Bosch manufacturing process (Pimentel and Pimentel, 2008; Smil, 1994). By 2000, the average annual global consumptions of N, P and K were 53 kg/ha, 9 kg/ha, and 12 kg/ha, respectively (Smil, 2008).

Nitrogen and phosphorus have been applied more than other nutrients, and so have had greater environmental impacts (Meisterling et al., 2009). New chemical components, more accurate application methods, and better agricultural management, including appropriate crop rotations, timely sowing and improved water management can significantly enhance the efficiency of fertiliser applications and minimise potential environmental degradation, including nitrates affecting local fresh water quality (Ashrafi Zadeh and Kushwaha, 2006; Centre for Advanced Engineering, 1996; Kitani, 1999; McLaughlin et al., 2002; Murphy and Power, 2008; Nemecek et al., 2008; Pellizzi et al., 1988; Stout, 1990).

Most chemical fertilisers are produced using fossil fuel resources (Fluck and Baird, 1980; Kitani, 1999), especially natural gas. Around 5% of total annual gas consumption is used by the Haber-Bosch synthesis (Smil, 2008). About 30–70% of the total energy input in crop production is based on nitrogen fertiliser, varying with crop and location (Kuesters and Lammel, 1999; Mohtasebi et al., 2008; Pimentel et al., 2007; Safa and Samarasinghe, 2011; Vlek et al., 2004). There are similar trends in global fertiliser, wheat and oil prices. Thus, any fluctuations in the oil market can lead to increased food production costs.

Environmental impacts include eutrophication, poisoning of waterways, biodiversity depletion and greenhouse gas emissions, particularly carbon dioxide from natural gas when used in their manufacture, and nitrous oxide from applying nitrogenous fertilisers (Bertocco et al., 2008; Boyle et al., 2003; Nemecek et al., 2008). Additional increase in the amount of fertiliser applied results in a lower rate of increase in yield, until a peak is reached, after which further application can reduce the yield (Fluck and Baird, 1980). Growing leguminous clover crops in pasture can provide 100–200 kg/ha nitrogen, and together with crop rotation, can reduce fertiliser requirements by about 40% (Pimentel, 2009; Pimentel et al., 2007). Crop rotations can also help control pests and diseases. Improved timing and application management can also reduce nitrogen fertiliser inputs without yield reductions (Pimentel et al., 2007) which can also be achieved using controlled-release fertilisers (Pimentel et al., 2005).

Nitrogenous fertilisers (such as urea, ammonia, phosphate ammonium and ammonium sulphate) are energy-intensive, whereas super phosphate (P_2O_5) and potash (K_2O) require less energy to extract and refine. Only around 10–20% of nitrogen applied is absorbed by the plants, influenced by soil type, temperature and rainfall (Pimentel et al., 2005; Witney, 1988). Fertiliser losses include from run-off, soil erosion to rivers, leaching to ground water and nitrous oxide emissions (Nemecek et al., 2008).

One kilogram of nitrogen is needed to produce one kilogram of protein (Da Rosa, 2005). Ammonia is the basic source of nitrogen fertilisers, with the hydrogen input currently taken from the natural gas methane in a process that emits much CO_2. Green hydrogen from the electrolysis of water using renewable electricity can be used as a low-carbon alternative.

The global demand for nitrogen fertiliser has increased at the rate of 1.8% annually (FAO, 2000a). It is, however, possible to recapture some valuable nutrient resources by recycling crop residues and livestock manure. Global crop residues are estimated to be 430 Mt/y and contain about 4.3 Mt N, 0.4 Mt P and 4.0 Mt K, together with many tonnes of other useful elements. Most of the nitrogen volatilises as ammonia when the manure is left on the land surface; therefore it should be incorporated into the soil to maximise recovery, based on present technologies (Pimentel and Pimentel, 2008).

3.3.4 Pesticides

The average worldwide growth in the use of a wide range of agrichemicals is around 4.4% per year (Vlek et al., 2004). Insects destroy 13%, plant pathogens 12% and weeds 12% of all potential agricultural production prior to harvest every year. When the post-harvest losses are added to these pre-harvest losses, total losses due to pests increase to 52% (Pimentel and Pimentel, 2008). Chemical, mechanical and biological methods of pest control can be applied to control or eliminate fungus, insects and weeds. Mechanical methods are used mainly on small farms, organic farms, orchards and vineyards and in areas with cheap labour, although new technologies based on robotics are now commercial (Carley, 2020). However, most farmers choose chemical methods because they are faster and cheaper; they are also more effective (Pimentel and Peshin, 2014). Nevertheless, using pesticides is much more energy-intensive than mechanical pest control. For example, using cultivators on organic farms uses less than half the energy in manufacturing and applying herbicides (Pimentel, 2009).

The energy component in agrichemicals, when formulated in powder, emulsive oils or granules, comes mainly from their manufacture, packaging and transport (CIGR, 1999; Kitani, 1999; Stout, 1990). Most raw materials used in agrichemical production come from petrochemical industries and are therefore are energy-intensive (Stout, 1990).

The potential hazards from pesticide use, including human and animal poisoning, cancer, other chronic effects, reduced biological diversity and water pollution, should be balanced against the benefits from their use. Through appropriate management, it may be possible to reduce pesticide use without reducing crop yields (Pimentel and Pimentel, 2008).

3.3.5 Tractors, equipment and vehicles

Most commercial energy in agriculture is used in agricultural machinery manufacture, maintenance, repair and operation. The number of tractors and machines have increased during the last century; the number of tractors worldwide has risen from 11 million in 1961 to 28 million, in 2006 (FAO, 2008). Estimating the total energy inputs for machinery is complex (Smil, 2008) because manufacturing companies use a range of processes. It is also difficult to share the proportion of embedded energy between different agricultural products.

To calculate the annual energy input from tractors and other equipment, it is necessary to know the weight, working life and typical surface on which the machine is to be used. Due to different technologies and different components, energy input per kilogram is not an ideal estimation index to compare energy consumption when manufacturing machinery. As a result, there are large differences between different estimations, such as Roller et al.

(1975) 75 MJ/kg; McChesney et al. (1978) 90 MJ/kg; Hornacek (1979) 80.23 MJ/kg; Fluck and Baird (1980) 27 MJ/kg; Stout (1990) 85 MJ/kg; Wells (2001) 80 MJ/kg for implements and 160 MJ/kg for tractors; and also Kitani, 1999 (Table 3).

It appears that improving technology does not change the energy demand for producing agricultural machinery. Kitani (1999) considered several steps when calculating these energy coefficients:

1. the energy required for producing the raw materials;
2. the energy used in the manufacturing process; and
3. the energy consumption for transporting the machine to the consumer and during repairs and maintenance.

An attempt was made to check the power and weight of tractors, combine harvesters and other machinery using manufacturer catalogues and websites (Table 4). There has been a general trend towards higher power tractors to save time and overall costs (Centre for Advanced Engineering., 1996; Outlaw et al., 2005; Serrano et al., 2007; Witney, 1988).

For estimating the weight of machinery, it is possible to use studies that find correlations between different machinery properties. For example, Serrano et al. (2007) presented the relationship between disc harrow mass and width.

$$m = -965.71 + 1041.9w \tag{6}$$

where m is disc harrow mass (kg) and w is the implement width (m).

Table 3 Energy coefficients for producing and repairing different types of agricultural machinery over their lifetime (Kitani, 1999)

Equipment	Energy (MJ/kg)
Tractor	138
Mouldboard plough	180
Chisel plough	149
Heavy-duty disc and field cultivator	149
Spring tine harrow	149
Rotary cultivator	148
Combined tillage	180
Air seeder and grain drill	133
Fertiliser spreader	129
Boom-type sprayer	129
Combine harvester	116

Table 4 Average mass and estimated life of tractors and farm implements

	Estimated life (h) (ASAE, 2009)	Field efficiency (%) (ASAE, 2009)	Estimated mass (kg) (Wells, 2001)
Tractor, two-wheel drive	12 000	#	*
Tractor, four-wheel drive	14 000	#	*
Mouldboard plough	2000	85	1500
Chisel plough	2000	85	1500
Heavy-duty discs	2000	85	1000
Field cultivator	2000	85	1000
Spring tine harrow	2000	85	200
Rotary cultivator	2000	80	200
Air seeder	1500	70	2000
Grain drill	1500	70	2000
Fertiliser spreader	1200	70	200
Boom-type sprayer	1500	65	100
Combine harvester	3000	70	*

* Mass estimated by Equation 7 and using manufacturers' catalogues.

Wells (2001) also showed that there was a correlation between tractor mass and related power (kW), with tractor power in New Zealand at that time ranging from approximately 20 kW to 300 kW.

$$\text{Mass}(\text{kg}) = 54.7 \times \text{power}(\text{kW}) + 190 \tag{7}$$

The energy used to produce and repair agricultural machinery can be calculated:

$$\text{ME} = (\text{G} \times \text{E}) / (\text{T} \times \text{Ca}), \tag{8}$$

where ME is machine energy (MJ/ha); G is the weight of the implement (kg); E is the embedded energy sequestered (MJ/kg); T is the economic life of the machine (h) (Table 4); and Ca is effective field capacity (ha/h) calculated by:

$$\text{Ca} = (s \times w) \times \text{FE} / 10 \tag{9}$$

where s is ground speed (km/h); w is the width of the machine (m); and FE is field efficiency (%) as taken from the ASAE Standard D497.5 (2009).

3.3.6 Electricity

Electricity is mainly used in arable farms for pumping water from wells and rivers to irrigate farms. Also in pig, poultry and dairy farms, it is mostly used for heating, cooling, ventilation and operation of fixed machinery in buildings. The energy conversion factor for electricity is 3.6 MJ/ per kWh, being the unit commonly used. This conversion factor does not take into account the efficiency of converting primary energy (oil, gas, coal and renewables) to electricity in power generation plants. Thus, depending on power plant efficiency and the mix of generation in an electricity grid, the primary energy inputs can exceed 10 MJ/kWh (Saunders et al., 2006).

3.3.7 Seed

Agricultural crops can be propagated by seeds, tubers or bulbs. There is little information about energy requirements for seed production but many crops are similar to growing other arable crops. Some farmers grow, dry and store their own seed ready for next season but most sell to commercial seed companies that process and package it.

3.4 Energy usage in irrigation

Water resources are crucial for agricultural production and are becoming more limited in many farming areas (Pimentel and Pimentel, 2008). Agriculture consumes about 70% of freshwater withdrawals per year (FAO, 2000b; UNESCO, 2001). The water demand by crops ranges from 600 to 3000 litres of water per dry kilogram of crop yield. Producing 9 t/ha of corn requires around 7 million litres of water, either rain-fed or from irrigation. Wheat requires about 2.4 million L/ha of water for a yield of 2.7 t/ha (Pimentel et al., 2009).

Energy use in irrigation consists of energy for pumping and energy for distribution. Irrigation can be one of the most expensive operations in a farm enterprise, whether extracted from local bores, wells or waterways (Pimentel and Pimentel, 2008). Typically, crop yields on irrigated farms are greater than on dryland farms, with around 17% of total agricultural land being irrigated to produce around 40% of global food (FAO, 2002). Therefore, increasing the amount of land irrigated would increase agricultural production, especially where climate change projections are for less annual precipitation.

Improving the efficiency of irrigation systems is the best way to increase the area of land irrigated. Until the second half of the twentieth century, irrigation systems depended on gravity-fed systems from storage dams. Since then, more efficient engines, pumps, and sprinklers enable farmers to use rainfall storage systems. Furrow irrigation may be around 20–40% efficient whereas some modern irrigation systems can be more than 65% or up to 95% using drip irrigation. Diesel pumps are used in smaller systems, especially in developing countries where grid electricity may not be available (Centre for Advanced Engineering, 1996).

Due to the increased precision of application and reductions of unnecessary water usage, water can be better conserved and energy saved. The energy consumption of pressure irrigation distribution systems is influenced by both the performance of the pumping station and the spatial distribution and layout of the networks (Abadia et al., 2008; CIGR, 1999).

In areas close to surface water supply or with high water tables, the energy use for pumping is less than in other areas needing bores. It is difficult to estimate a standard figure for the power requirements for any system as there are many variables (Abadia et al., 2008; Mukherji, 2007). The required energy varies with the depth of water table, type of system, crop water requirements, frequency of application, water resources (rainfall, dam storage, groundwater and waterways), pump efficiency, delivery system and distance, and energy sources used (Smil, 2008; Stout, 1990; Vlek et al., 2004). The total energy consumption is related to the frequency and duration of irrigation, the quantity and source of the water applied, the distance from the source and the type of irrigation system. Energy is also needed to construct the water supply and storage system, provide the conveyance works and maintain and operate the system (Stout, 1990). The highest energy demand on some farms is for irrigation (Devi et al., 2009; Safa and Samarasinghe, 2011; Safa and Tabatabaeefar, 2002), especially in dry areas. Irrigated wheat needs three times more energy than rain-fed wheat for producing the same amount of grain per hectare with similar water volumes received (Pimentel et al., 2009).

4 Case study: Energy consumption in wheat production using neural networks

A study was conducted over 35 300 hectares of irrigated and dryland wheat fields in Canterbury, New Zealand over the 2007–2008 harvest year (Safa and Samarasinghe, 2013). Around 160 direct and indirect factors were identified and used to create a model of artificial neural networks (NN) to predict on-farm energy demands based on-farm conditions, crop area, social considerations (such as the level of farmer education), direct and indirect energy inputs for N and P applications and irrigation depending on frequency. The model predicted energy use for these arable farms with an error margin of ±12% (± 2.9 GJ/ha).

4.1 Methodology

Energy modelling is a useful tool for engineers and scientists concerned with energy production and consumption and environmental impacts (Al-Ghandoor et al., 2009; Tester, 2005). A range of models have been used, from geological models for research on natural resources to modelling future energy demand (Tester, 2005). Regression analysis was the common modelling technique used in energy studies, however, NN have been increasingly used for solving various problems due to their ability to model complex nonlinear systems in a flexible and adaptive manner (Jebaraj and Iniyan, 2006).

NN have been applied by researchers in a wide range of applications such as mathematics, engineering, medicine, economics, environment and agriculture, as well as for classification, prediction and problem-solving in the energy field (Sözen, 2009). Modelling of scenarios by NN can help solve problems for which no explicit formulations are available (Fang et al., 2000). The main advantage of NN is that they are able to use historical underlying data to develop an accurate representation of a process or relationship of interest.

Energy consumption in wheat production was analysed based on direct and indirect energy sources including human labour and embedded energy in supplies of fuel, fertilisers, pesticides, machinery and seed. The total energy input (E) was determined as the sum of the input factors (A_i) multiplied by the appropriate energy conversion coefficient for each factor (C_i) as:

$$E = \Sigma\left(A_i C_i\right) \tag{10}$$

4.2 Energy consumption

Total energy consumption by wheat production in the Canterbury region was about 22.57 GJ/ha (Table 5). Of this total, 36% was for direct energy in the form of diesel, at 3.12 GJ/ha, and electricity accounted for 4.87 GJ/ha. Indirect energy for fertiliser supply, especially urea, was 10.65 GJ/ha and accounted for 47% of total energy inputs.

For irrigated wheat production alone, the total energy consumption was 25.60 GJ/ha. This was around 50% more when compared with rain-fed dryland wheat farms (17.46 GJ/ha) due to the high consumption of electricity used for water pumping (Table 6). In both irrigated and non-irrigated farming systems, fertiliser was ranked as the highest energy consumer, making up 40% and 66% of total energy use respectively. Electricity was the second highest on irrigated farms and diesel fuel inputs were also high. Therefore, energy conservation efforts should focus on fertiliser, electricity and diesel fuel consumption.

Table 5 Average shares of direct, indirect and embedded energy inputs in wheat production systems in Canterbury, New Zealand (GJ/ha)

	Human	Seed	Fertiliser	Pesticides	Electricity	Machinery manufacture	Diesel fuel	Total
GJ/ha	0.006	1.27	10.65	0.91	4.87	1.74	3.12	22.57
%	0.03	6	47	4	22	8	14	100

Table 6 Comparisons of direct, indirect and embedded energy inputs in irrigated and dryland wheat-farming systems in Canterbury, New Zealand (GJ/ha)

	Indirect and embedded Energy				Direct energy			
	Seed	Fertiliser	Pesticide	Machinery manufacture	Human	Electricity	Fuel	Total
Irrigated	1.29 (5%)	10.19 (40%)	1.04 (4%)	2.17 (9%)	0.008 (0.03%)	7.76 (30%)	3.10 (12%)	25.60
Dry land	1.16 (7%)	11.43 (66%)	0.69 (4%)	1.02 (6%)	0.004 (0.02%)	0 (0%)	3.16 (18%)	17.46

5 Conclusions and future trends in research

Energy is an important input across all agricultural sectors. A range of methodologies can be used to estimate energy inputs and energy usage on farms. Without a standard technique, it is not easy to compare published data and articles covering a single type of farm enterprise. Energy demand reduction is not a primary target for farmers; however, it is a good indicator when used to manage and compare farms and farming systems.

Every farm is unique, and when comparing farm systems, energy usage is only one factor. For example, energy usage in an area with high precipitation is less on a per hectare basis than in dryer areas that need to employ irrigation systems.

5.1 Suggestions for future research

- When analysing energy inputs on farms, increasing the number of samples and testing more variables over a longer period of time (ideally at least five years due to seasonal variations) can help analyse energy use trends more accurately. Using the most correct method for data collection plays a critical role in this type of study, to indicate which operations and energy sources have the greatest potential for setting targets to improve energy efficiency. Investigating the potential effects of changes in farmers' behaviour on energy use on their farms would be useful for predicting future scenarios.
- Exploring further known and unknown links and correlations between parameters such as production, crop rotations and other factors should be undertaken by a team of experts.
- Matching tractor power with equipment, and selecting the right operation at the right time, can reduce fuel consumption. Using field equipment that can reduce the number of tractor passes needed can significantly reduce fuel consumption, farm expenditure and soil compaction. The effect of more powerful tractors and larger equipment on fuel and energy

consumption should be investigated. New tractors and machinery are more energy efficient; however, they may need more energy inputs to produce, service and maintain.

- New designs of irrigation systems have higher efficiencies. However, due to the shape and size of paddocks, some irrigator designs are constrained. Comparing the energy use of different irrigators based on the shape and size of paddocks and designing a practical model, would be useful.
- Standardising national energy conversion coefficients would increase the accuracy of the relevant statistics. Chemistry, physics, engineering, transport and social sciences have roles to play. After estimating the energy conversion coefficients, expert teams should monitor them, as well as energy use, for a range of agricultural products to identify energy conservation opportunities.
- Establishment of an international standard to estimate energy use in agricultural production would enable different energy analyses to be compared. This would be helpful for the adoption of more energy-efficient farming systems globally and to find the most important barriers to reducing energy use on farms.

6 Where to look for further information

In the last 20 years, several papers have been published on energy management in different countries. Unfortunately, it is difficult to verify some of the published articles. Also, due to lack of clear standards for data collection and conversion, it is not easy to compare the results of some of the papers. The researchers in this field should have a very clear idea, why they are calculating energy usage on farms and how they will use the results in future before starting the projects. They should realise the main limitations and find practical solutions based on the research area. Another important factor is finding acceptable sample size. Without enough samples, it is not possible to develop a clear picture of energy usage on farms.

The long time investigation on link between energy usage and other technical and social factors such as the farmers' age, education, farm size, etc. in different countries will be a very interesting topic to study.

7 References

Abadia, R., Rocamora, C., Ruiz, A. and Puerto, H. (2008). Energy efficiency in irrigation distribution networks I: Theory. *Biosystems Engineering* 101(1), 21–27.

Alfoldi, T., Niggli, U., Spiess, E. and Messon, J. M. G. (1994). *Input and Output of Energy for Different Crops in Bio-dynamic*. Paper presented at the 3rd ESA Congress.

Al-Ghandoor, A., Jaber, J. O., Al-Hinti, I. and Mansour, I. M. (2009). Residential past and future energy consumption: Potential savings and environmental impact. *Renewable and Sustainable Energy Reviews* 13(6–7), 1262–1274.

Alvarez, R. (2009). Predicting average regional yield and production of wheat in the Argentine Pampas by an artificial neural network approach. *European Journal of Agronomy* 30(2), 70–77.

ASAE (2006). *ASAE D496.2, Feb 2006, Agricultural Machinery Managment Data*. St. Joseph, MI: ASAE.

ASAE (2009). *ASAE D497.6, June 2009, Agricultural Machinery Managment Data*. St. Joseph, MI: ASAE.

Ashrafi Zadeh, S. R. and Kushwaha, R. L. (2006). *Development of a Tillage Energy Model Using a Simple Tool*. Paper presented at the CSBE/SCGAB 2006 Annual Conference, Edmonton, Alberta, Canada.

Bailey, J. A., Gordon, R., Burton, D. and Yiridoe, E. K. (2008). Energy conservation on Nova Scotia Farms: Baseline energy data. *Energy* 33(7), 1144–1154.

Baker, C. J., Saxton, K. E., Food and Agriculture Organization of the United Nations and C. A. B. International (2007). *No-Tillage Seeding in Conservation Agriculture* (2nd edn.). Rome, Wallingford, Oxfordshire, UK: FAO; CABI.

Barber, A. (2004). *Seven Case Study Farms: Total Energy & Carbon Indicators for New Zealand Arable & Outdoor Vegetable Production*. AgriLINK New Zealand Ltd. www.agrilink.co.nz/Portals/Agrilink/Files/Arable_Vegetable_Energy_Use_Main_Report.pdf

Baruah, D. C. and Bora, G. C. (2008). Energy demand forecast for mechanized agriculture in rural India. *Energy Policy* 36(7), 2628–2636.

Bertocco, M., Basso, B., Sartori, L. and Martin, E. C. (2008). Evaluating energy efficiency of site-specific tillage in maize in NE Italy. *Bioresource Technology* 99(15), 6957–6965.

Bowers, C. G. J. (1985). Southeastern tillage energy data and recommended reporting. *Transactions of the ASAE* 28(3), 731–737.

Boyle, G., Everett, B. and Ramage, J. (2003). *Energy Systems and Sustainability*. Oxford, UK: Oxford University Press in association with the Open University.

Carley, J. (2020). Mechancial weeders and robots reducing chemical dependence. https://www.foodandfarmingtechnology.com/features/the-mechanical-weeders-and-robots-reducing-chemical-dependence.html.

Centre for Advanced Engineering (1996). *Energy Efficiency: A Guide to Current and Emerging Technologies*. Christchurch, New Zealand: Centre for Advanced Engineering, University of Canterbury.

Christiaensen, L., Rutledge, Z. and Taylor, J. E. (2020). *The Future of Work in Agriculture, Some Reflections*. World Bank Group.

CIGR (1999). *CIGR Handbook of Agricultural Engineering. International Commission of Agricultural Engineering*. St. Joseph, MI: American Society of Agricultural Engineers.

Coley, D. A. (2008). *Energy and Climate Change: Creating a Sustainable Future*. Chichester, West Sussex; Hoboken, NJ: John Wiley.

Conforti, P. and Giampietro, M. (1997). Fossil energy use in agriculture: An international comparison. *Agriculture, Ecosystems and Environment* 65(3), 231–243.

Da Rosa, A. V. (2005). *Fundamentals of Renewable Energy Processes*. Amsterdam; Boston, MA: Elsevier Academic Press.

Dalgaard, T., Halberg, N. and Porter, J. R. (2001). A model for fossil energy use in Danish agriculture used to compare organic and conventional farming. *Agriculture, Ecosystems and Environment* 87(1), 51–65.

Devi, R., Singh, V., Dahiya, R. P. and Kumar, A. (2009). Energy consumption pattern of a decentralized community in northern Haryana. *Renewable and Sustainable Energy Reviews* 13(1), 194–200.

Fang, Q., Hanna, M. A., Haque, E. and Spillman, C. K. (2000). Neural network modeling of energy requirments, for size reduction of wheat. *Transactions of the ASAE* 43(4), 947–952.

FAO (2000a). *Current World Fertilizer Trend and Outlook to 2005/2006*. Rome, Italy: Food and Agriculture Organization, United Nations.

FAO (2000b). *World Agriculture: Towards 2015/2030, Technical Interim Report*. Rome, Italy: Food and Agriculture Organization, United Nations.

FAO (2002). *Crop and Drops: Making in Best Use of Water for Agriculture*. Rome, Italy: Food and Agriculture Organization, United Nations.

FAO (2008). FAO Stat, Available at: http://faostat.fao.org/.

Fluck, R. C. (1992). *Energy in World Agriculture* (vol. 6). Amsterdam: Elsevier.

Fluck, R. C. and Baird, C. D. (1980). *Agricultural Energetics*. Westport, CT: AVI Pub. Co.

Gajri, P. R., Arora, V. K. and Prihar, S. S. (2002). *Tillage for Sustainable Cropping*. New York: Food Products Press.

Grastina, L., Temple, S. and Diana, B. (1995). Energetic analysis of four farming system During the conversion from conventional to organic farming system in the Sacramento valley- California. *Rivista di Agronomia* 29.

Grisso, R. D., Kocher, M. F. and Vaughan, D. H. (2004). Predicting tractor fuel consumption. *Applied Engineering in Agriculture* 20(5), 553–561.

Hatirli, S. A., Ozkan, B. and Fert, C. (2006). Energy inputs and crop yield relationship in greenhouse tomato production. *Renewable Energy* 31(4), 427–438.

Hornacek, M. (1979). Energetic analysis of 14 agricultural practices. *Etude du Centre National d'Etude et d'Expérimentation de Machinisme Agricole, France* 457.

Hülsbergen, K. J. and Kalk, W. D. (2001). *Energy Balances in Different Agricultural Systems – Can They Be Improved?* Paper presented at the The International Fertiliser Society.

Ilyas, H. M. A., Safa, M., Bailey, A., Rauf, S. and Khan, A. (2020). Energy efficiency outlook of New Zealand dairy farming systems: An application of data envelopment analysis (DEA) approach. *Energies* 13(1), 251.

Jebaraj, S. and Iniyan, S. (2006). A review of energy models. *Renewable and Sustainable Energy Reviews* 10(4), 281–311. doi: 10.1016/j.rser.2004.09.004.

Joseph, S., Peters, I. and Friedrich, H. (2019). Can regional organic agriculture feed the regional community? A case study for Hamburg and North Germany. *Ecological Economics* 164, 106342. doi: 10.1016/j.ecolecon.2019.05.022.

Karkacier, O. and Gokalp Goktolga, Z. (2005). Input-output analysis of energy use in agriculture. *Energy Conversion and Management* 46(9–10), 1513–1521.

Karkacier, O., Gokalp Goktolga, Z. and Cicek, A. (2006). A regression analysis of the effect of energy use in agriculture. *Energy Policy* 34(18), 3796–3800.

Kitani, O. (1999). *CIGR Handbook of Agricultural Engineering*. St. Joseph, MI: American Society of Agricultural Engineers.

Kizilaslan, H. (2008). Input-output energy analysis of cherries production in Tokat Province of Turkey. *Applied Energy* 86(7–8), 1354–1358.

Kuesters, J. and Lammel, J. (1999). Investigations of the energy efficiency of the production of winter wheat and sugar beet in Europe. *European Journal of Agronomy* 11(1), 35–43. doi: 10.1016/S1161-0301(99)00015-5.

Lin, X., Sun, X., Manogaran, G. and Rawal, B. S. (2021). Advanced energy consumption system for smart farm based on reactive energy utilization technologies.

Environmental Impact Assessment Review 86, 106496. doi: 10.1016/j.eiar.2020.106496.

Liu, J. (2009). A GIS-based tool for modelling large-scale crop-water relations. *Environmental Modelling and Software* 24(3), 411–422.

Manaloor, V. and Sen, C. (2009). *Energy Input Use and CO_2 Emissions in the Major Wheat Growing Regions of India*. Paper presented at the International Association of Agricultural Economists Conference.

Mani, I., Kumar, P., Panwara, J. S. and Kanta, K. (2007). Variation in energy consumption in production of wheat-maize with varying altitudes in hilly regions of Himachal Pradesh, India. *Energy* 32(12), 2336–2339.

McChesney, I. G., Pearson, R. G., Bubb, J. W. and Joint Centre for Environmental Sciences (N.Z.) (1978). *Energy Use on Canterbury Mixed Cropping Farms: A Pilot Survey*. Lincoln, UK: Joint Centre for Environmental Sciences, Lincoln College.

McLaughlin, N. B., Drury, C. F., Reynolds, W. D., Yang, X. M., Li, Y. X., Welacky, T. W. and Stewart, G. (2008). *Energy Inputs for Conservation and Conventional Primary Tillage Implements in a Clay Loam Soil*. ASABE, American Society of Agricultural and Biological Engineering.

McLaughlin, N. B., Gregorich, E. G., Dwyer, L. M. and Ma, B. L. (2002). Effect of organic and inorganic soil nitrogen amendments on mouldboard plow draft. *Soil and Tillage Research* 64(3–4), 211–219.

Meisterling, K., Samaras, C. and Schweizer, V. (2009). Decisions to reduce greenhouse gases from agriculture and product transport: LCA case study of organic and conventional wheat. *Journal of Cleaner Production* 17(2), 222–230. doi: 10.1016/j.jclepro.2008.04.009.

Meul, M., Nevens, F., Reheul, D. and Hofman, G. (2007). Energy use efficiency of specialised dairy, arable and pig farms in Flanders. *Agriculture, Ecosystems and Environment* 119(1–2), 135–144. doi: 10.1016/j.agee.2006.07.002.

Mohtasebi, S. S., Behroozi Lar, M., Safa, M. and Chaichi, M. R. (2008). Comparison of direct and indirect energy coefficients for seeding and fertilizing in irrigated wheat production. *World Applied Sciences* 3(3), 353–358.

Mukherji, A. (2007). The energy-irrigation nexus and its impact on groundwater markets in eastern Indo-Gangetic basin: Evidence from West Bengal, India. *Energy Policy* 35(12), 6413–6430.

Murphy, J. D. and Power, N. M. (2008). How can we improve the energy balance of ethanol production from wheat? *Fuel* 87(10–11), 1799–1806.

Nemecek, T., von Richthofen, J.-S., Dubois, G., Casta, P., Charles, R. and Pahl, H. (2008). Environmental impacts of introducing grain legumes into European crop rotations. *European Journal of Agronomy* 28(3), 380–393.

Nguyen, M. L. and Hignett, T. P. (1995). Energy and labour efficiency for three pairs of conventional and alternative mixed cropping (pasture-arable) farms in Canterbury, New Zealand. *Agriculture, Ecosystems and Environment* 52(2–3), 163–172.

Nielsen, R. L. (2020). Historical corn grain yields in the U.S. https://protect-au.mimecast.com/s/D0W2CjZ1ZnuYz84mcR8MQA?domain=agry.purdue.edu; https://www.agry.purdue.edu/ext/corn/news/timeless/YieldTrends.html.

Odum, H. T. (1994). *Ecological and General Systems: An Introduction to Systems Ecology* (rev. edn.). Niwot, CO: University Press of Colorado.

Odum, H. T. and Odum, E. C. (1976). *Energy Basis for Man and Nature*. New York: McGraw-Hill.

Outlaw, J. L., Collins, K. J. and Duffield, J. A. (2005). *Agriculture as a Producer and Consumer of Energy*. Wallingford, Oxfordshire, UK; Cambridge, MA: CABI Publishing.

Ozkan, B., Akcaoz, H. and Fert, C. (2004). Energy input-output analysis in Turkish agriculture. *Renewable Energy* 29(1), 39–51.

Pellizzi, G. (1992). Use of energy and labour in Italian agriculture. *Journal of Agricultural Engineering Research* 52, 111–119.

Pellizzi, G., Guidobono Cavalchini, A., Lazzari, M. and Commission of the European Communities (1988). *Energy Savings in Agricultural Machinery and Mechanization*. London; New York: Elsevier Applied Science.

Pimentel, D. (2009). Reducing energy inputs in the agricultural production system. *Monthly Review* 61(3), 99–101.

Pimentel, D., Gardner, J., Bonnifield, A., Garcia, X., Grufferman, J., Horan, C., Schlenker, J. and Walling, E. (2007). Energy efficiency and conservation for individual Americans. *Environment, Development and Sustainability* 11, 523–546.

Pimentel, D., Hepperly, P., Hanson, J., Douds, D. and Seidel, R. (2005). Environmental, energetic, and economic comparisons of organic and conventional farming systems. *BioScience* 55(7), 573.

Pimentel, D., Marklein, A., Toth, M. A., Karpoff, M. N., Paul, G. S., McCormack, R., Kyriazis, J. and Krueger, T. (2009). Food Versus biofuels: Environmental and economic costs. *Human Ecology* 37(1), 1–12.

Pimentel, D. and Pimentel, M. (2008). *Food, Energy, and Society* (3rd edn.). Boca Raton, FL: CRC Press.

Pimentel, D. and Pimentel, R. (2014). *Integrated Pest Management: Pesticide Problems* (vol. 3). Springer Science & Business Media.

Randolph, J. and Masters, G. M. (2008). *Energy for Sustainability: Technology, Planning, Policy*. Washington, DC: Island Press.

Riethmuller, G. (1989). *Draft Requirements of Tillage Equipment in the Western Australian Wheatbelt*. Paper presented at the Conference on Agricultural Engineering 1988: An Australasian Conference to Celebrate the Australian Bicentennial.

Roller, W. L., Keener, H. M., Kline, R. D., Mederski, H. J. and Curry, R. B. (1975). *Grown Organic Matter as a Fuel Raw Material Resource*. Wooster, OH: Ohio Agricultural Research and Development Center.

Safa, M. and Samarasinghe, S. (2011). Determination and modelling of energy consumption in wheat production using neural networks: "A case study in Canterbury Province, New Zealand". *Energy* 36(8), 5140–5147. doi: 10.1016/j.energy.2011.06.016.

Safa, M. and Samarasinghe, S. (2013). Modelling fuel consumption in wheat production using artificial neural networks. *Energy* 49, 337–343.

Safa, M., Samarasinghe, S. and Mohssen, M. (2010). Determination of fuel consumption and indirect factors affecting it in wheat production in Canterbury, New Zealand. *Energy* 35(12), 5400–5405. doi: 10.1016/j.energy.2010.07.015.

Safa, M., Samarasinghe, S. and Mohssen, M. (2011). A field study of energy consumption in wheat production in Canterbury, New Zealand. *Energy Conversion and Management* 52(7), 2526–2532. doi: 10.1016/j.enconman.2011.01.004.

Safa, M., Samarasinghe, S. and Nejat, M. (2015). Prediction of wheat production using artificial neural networks and investigating indirect factors affecting it: Case study in Canterbury Province, New Zealand. *Journal of Agricultural Science and Technology* (17), 791–803.

Safa, M. and Tabatabaeefar, A. (2002). *Energy Consumption in Wheat Production*. Paper presented at the IACE, Wuxi, China.

Sartori, L., Basso, B., Bertocco, M. and Oliviero, G. (2005). Energy use and economic evaluation of a three year crop rotation for conservation and organic farming in NE Italy. *Biosystems Engineering* 91(2), 245–256. doi: 10.1016/j.biosystemseng.2005.03.010.

Sauerbeck, D. R. (2001). CO_2 emissions and C sequestration by agriculture - perspectives and limitations. *Nutrient Cycling in Agroecosystems* 60(1/3), 253–266.

Saunders, C., Barber, A. and Taylor, G. (2006). *Food Miles: Comparative Energy/Emissions Performance of New Zealand's Agriculture Industry*. Lincoln, New Zealand: Agribusiness & Economics Research Unit, Lincoln University.

Serrano, J. M., Peça, J. O., Marques da Silva, J., Pinheiro, A. and Carvalho, M. (2007). Tractor energy requirements in disc harrow systems. *Biosystems Engineering* 98(3), 286–296.

Siemens, J. and Bowers, W. (1999). *Machinery Management: How to Select Machinery to Fit the Real Needs of Farm Managers*. East Moline, IL: John Deere Publishing.

Singh, H., Mishra, D. and Nahar, N. M. (2004). Energy use pattern in production agriculture of a typical village in arid zone--Part III. *Energy Conversion and Management* 45(15-16), 2453-2472.

Singh, S. and Mittal, J. (1992). Energy in Production Agriculture. Mittal Publishing Company, Mittal Publication, India.

Smil, V. (1991). *General Energetics: Energy in the Biosphere and Civilization*. New York: Wiley.

Smil, V. (1994). *Global Ecology: Environmental Change and Social Flexibility*. London; New York: Routledge.

Smil, V. (2008). *Energy in Nature and Society: General Energetics of Complex Systems*. Cambridge, MA: The MIT Press.

Sözen, A. (2009). Future projection of the energy dependency of Turkey using artificial neural network. *Energy Policy* 37(11), 4827–4833. doi: 10.1016/j.enpol.2009.06.040.

SSSA (1997). *Glossary of Soil Science Terms*. Medison, WI: Soil Science Society of America.

Stout, B. A. (1990). *Handbook of Energy for World Agriculture*. London; New York: Elsevier Science Publishing Co.

Tester, J. W. (2005). *Sustainable Energy: Choosing among Options*. Cambridge, MA: MIT Press.

Tzilivakis, J., Warner, D. J., May, M., Lewis, K. A. and Jaggard, K. (2005). An assessment of the energy inputs and greenhouse gas emissions in sugar beet (*Beta vulgaris*) production in the UK. *Agricultural Systems* 85(2), 101–119. doi: 10.1016/j.agsy.2004.07.015.

UNESCO (2001). *Securing the Food Supply*. World Water Assessment Programme. Paris: United Nations Education Scientific and Cultural Organization.

USDA (2008). *U.S. Agriculture and Forestry Greenhouse Gas Inventory: 1990–2005*. Washington, DC: United States Department of Agriculture (USDA).

Vlek, P. L. G., Rodríguez-Kuhl, G. and Sommer, R. (2004). Energy use and CO_2 production in tropical agriculture and means and strategies for reduction or mitigation. *Environment, Development and Sustainability* 6(1/2), 213–233.

Wallgren, C. and Höjer, M. (2009). Eating energy--Identifying possibilities for reduced energy use in the future food supply system. *Energy Policy* 37(12), 5803–5813. doi: 10.1016/j.enpol.2009.08.046.

WCED (1987). *Our Common Future*. World Commission on Environment and Development, United Nations through the Oxford University Press.

Wells, C. (2001). *Total Energy Indicators of Agricultural Sustainability: Dairy Farming Case Study* (No. 0478079680 (pbk.)). Wellington, New Zealand: Ministry of Agriculture and Forestry.

Witney, B. (1988). *Choosing and Using Farm Machines*. Harlow, Essex, UK; New York: Longman Scientific & Technical; Wiley.

Worldbank (2021). *Employment in Agriculture (% of Total Employment) (Modeled ILO Estimate)*. Available at: https://data.worldbank.org/indicator/SL.AGR.EMPL.ZS.

Ziesemer, J. (2007). *Energy Use in Organic Food Systems*. Rome: Natural Resources Management and Environment Department, Food and Agriculture Organization of the United Nations.

Chapter 2

Advances in energy-efficient lighting and ventilation for food production systems

Tom Tabler, Mississippi State University, USA

1 Introduction

Fifteen per cent of agricultural production costs in industrialized agriculture is energy related, according to the U.S. Department of Agriculture (Sands et al., 2011). Even though lighting is an essential part of most farms, only recently has lighting been seen as an opportunity to reduce energy costs. Energy-efficient lighting options offer farmers and growers new opportunities to reduce electricity costs and manage operations more sustainably. In the past decade, animal and greenhouse agriculture has migrated from incandescent to compact fluorescent lights (CFLs) to cold cathode lamps and finally has appeared to settle on light-emitting diode (LED) technology as the best option from an energy efficiency standpoint.

Advancements in ventilation may not be as great as in the lighting sector but opportunities also exist to become energy smart and more efficient. Much of commercial swine and poultry production today occurs inside confinement buildings that require ventilation to maintain the proper environment. The greenhouse industry also relies on ventilation to maintain proper conditions

http://dx.doi.org/10.19103/AS.2022.0100.02

for controlled environment crop production. Efficient use of fans, heaters etc. is critical to managing energy costs and maintaining sustainability. Opportunities include everything from elbow grease to direct-drive motors to virtual reality. This chapter addresses advances in lighting and ventilation and provides a case study on sustainability and food security.

2 Light sources: benefits and shortcomings

While LEDs appear to be the lighting technology now and for the future, various other options exist and are regularly used. Hy-Line International (2017) offered the following information on various light sources.

2.1 Sunlight

2.1.1 Benefits

- In equatorial regions, light from the sun is consistent from season to season.
- Full-spectrum light ranges from ultraviolet (UV) to infrared (IR).
- Responses to sunlight change from day to day and season to season and are naturally inherent in both domestic and wild fowls.
- Poultry houses designed to utilize natural daylight may require little to no artificial light, thereby saving on energy costs.

2.1.2 Shortcomings

- Spectral composition and intensity of sunlight change from dawn to noon to dusk, from season to season, sunrise to sunset, with location and with cloud cover.
- Light intensity will change throughout the day as light comes in from different areas of the house.
- Light intensity is much higher from the sun than from an artificial lamp and overcoming seasonal changes in day length can be difficult. A bright sunny day can be 60 000–100 000 lux (6000–10 000 foot-candle (fc)).
- High light intensity may cause aberrant behaviours in poultry such as nervousness, feather pulling, pecking and cannibalism.

2.2 Incandescent light

2.2.1 Benefits

- Inexpensive.
- Good red spectrum output.
- Excellent light distribution.

- Quick to turn on.
- No difference in performance when used in cold weather.

2.2.2 Shortcomings

- Short lifespan and lamps must be frequently changed.
- Usually constructed of metal and glass and are prone to breakage.
- More than 90% of the energy used by the lamp goes to producing heat rather than light.
- Many types of incandescent lamps do not comply with new energy efficiency standards.

2.3 Compact fluorescent light (CFL)

2.3.1 Benefits

- Energy efficient.
- Relatively inexpensive.
- Similar colour spectra as incandescent bulbs.
- Available in both warm and cool spectra (Kelvin rating).
- Proven success in layer, broiler and breeder poultry industries.

2.3.2 Shortcomings

- Contain mercury.
- Uncovered spiral tubes may be difficult to clean.
- Made of metal and glass and are prone to breakage.
- Lamps do not dim well, with a potential to burn out more quickly when dimmed.
- While appearing to be white light, CFLs are composed of light spectrum peaks depending on the colour spectra of the phosphors utilized in the lamp.
- Lamps require several minutes to reach maximum light intensity when turned on.
- Poor performance in cold weather.
- Not ideal in situations where light must be turned on and off multiple times per day.
- Requires an electronic ballast to regulate current and voltage supplied to the lamp.

2.4 Light-emitting diode (LED)

2.4.1 Benefits

- Provides a full spectrum of light.
- Typically, the most efficient light lamp as measured in lumens per watt.

- Because LEDs do not emit IR radiation (heat), they can be constructed out of non-glass materials that are water resistant and shatterproof.
- Typically manufactured from non-toxic materials.
- Can be designed to focus light onto desired areas.
- Colour spectrum of the light can be adjusted depending on the phosphors used.
- Easier to dim than CFL lamps.
- Dimming can extend a lamp's lifespan.
- Very long life – up to 10 years at 16 h per day (50 000–60 000 h).
- Rapidly reaches peak light intensity after being turned on.
- Ideal for areas where lights are frequently turned on and off.
- Efficient in cold weather with no change in performance.

2.4.2 Shortcomings

- Expensive (although prices continue to decrease).
- Must use the proper dimmer; otherwise lamps may flicker and burn out more quickly.
- LED light is directional and requires an appropriate lens to focus light or appropriate diffusers to cover a broader area.
- May need to change sockets and wiring to fit the ideal LED electrical specifications.
- Efficiency of heat fins is reduced with dust build-up, poor ventilation around the lamp or putting the lamp in a 'jelly jar' for waterproofing.
- Lights may not burn out after the expected lifespan but may be dimmed to greater than 70% of the original lumen output. As a result, baseline lux testing in the animal or greenhouse may be required to determine when lamps should be changed.
- Cheaper LEDs may not have an appropriate heat sink, spectrum, hardware or warranty for agricultural environments.

LED-based lighting, the primary type of solid-state lighting today, is revolutionizing the lighting market (Stober et al., 2017). LED lighting has surged to the top above all conventional lighting technologies in terms of energy efficiency, life expectancy, versatility and colour quality. In addition, because of their increasing cost competitiveness, LED products are successfully competing in a variety of lighting applications. A U.S. Department of Energy (DOE) study forecasted that LED lighting will represent 86% of all lighting sales by 2035, resulting in an annual primary energy savings of 390 PJ (3.7 quadrillion British thermal units) (U.S. DOE, 2016).

Light quality affects production efficiency, animal welfare and growth rate. The swine and dairy industries are increasingly recognizing the benefits of LED lighting. Also, the poultry industry has rapidly adopted LED lighting systems across all segments of the industry including layers, pullets, breeders and broilers. The four main features associated with light are intensity, photoperiod, source and spectrum (Manser, 1996). The colour of light is determined by the various outputs from different wavelengths that make up the visible spectrum. White light contains all wavelengths of the visible spectrum, but it differs in colour temperature (Kelvin rating) depending on the power emitted from different individual wavelengths within the spectrum.

It's not only animal agriculture that has seen the light where LED technology is concerned. Since plants first emerged on Earth, they have relied on light from the sun to power the process of photosynthesis. Today, however, thanks to numerous technological advances, not the least of which are innovations in LED technology, plants can be grown on a commercial basis without direct sunlight. UV light can be scientifically manipulated on both seeds and seedlings to beneficially trigger biological mechanisms that increase plant growth, vigour and yields (https://www.biolumic.com/ 2020). In addition, LEDs benefit both animal and plant agriculture by using less energy and giving off less heat compared to other forms of electrical illumination such as incandescent, CFLs, metal halide and high-pressure sodium lamps.

Because LED technology appears to be the future in energy-efficient lighting, this chapter will focus on how best to manage and take advantage of the technology. Despite remarkable energy savings offered by LEDs, there are things we must better understand such as dimming capabilities, compatibility with the light dimmer, wiring and socket quality, matching Kelvin rating to the need and choosing the right LED. It's not as simple as unscrewing an incandescent lamp and replacing it with an LED lamp. Benefits and challenges are discussed throughout the lighting section.

3 Energy savings

As early as 2008, reports from the poultry industry appeared related to cost and energy savings associated with switching to more energy-efficient lamps (Tabler et al., 2008). Dramatic reductions in energy use for lighting were observed at a University of Arkansas commercial four-house broiler farm (18 000 birds/house; 72 000/flock) when 60-watt (W) incandescent lamps were replaced with 23-W dimmable CFLs and 8-W dimmable cold cathode lamps (Table 1). A lighting cost savings of US$520 per flock was also reported on the same four-house farm (Table 1).

However, over time, the dimmable cold cathode and CFL lamps proved to be less durable under demanding poultry house conditions (Watkins et al., 2012). As a result, growers that switched to these technologies were soon reporting frequent lamp failures and loss of light intensity because of dirt accumulation on the spiral CFL designs. Challenges identified with using these

Table 1 Use and cost of electricity for lighting per flock for two periods comparing incandescent lamps with more efficient CFL and cold cathode lighting

	April–December 2006		April–August 2008	
House #	kWh used 60 W Inc.[2]	Power costs[1] 60 W Inc.[2]	kWh used CFL and CC[3]	Power costs[1] CFL and CC[3]
1	2114	$148	510	$36
2	2131	$149	531	$37
3	2604	$182	479	$34
4	2584	$181	476	$33

[1] Power costs are in USD and were the same for both periods at US$0.07/kWh.
[2] All houses had 60-W incandescent lamps. Houses 3 and 4 had higher ceilings with additional lamps to provide similar light intensity at floor level.
[3] 23-W CFL lamps in houses 1 and 2 and 8-W cold cathode lamps in houses 3 and 4.

lamps included lamp failures when lamps were dimmed below 10% and also due to incompatibility between lamps and the metal used in light sockets. Lamp failures were also reported in association with power surges related to the operation of standby generators.

As an alternative, LEDs soon began to emerge as a potential lighting source for the poultry industry. LED lamps are more efficient, durable, versatile and longer lasting than incandescent or fluorescent lamps. Light is produced by the movement of electrons in a semiconductor material, resulting in the illumination of tiny LEDs, whereas incandescent lamps operate by passing an electric current through a metal filament until it becomes so hot that it glows, producing both light and heat. LED lamps release only a small amount of heat into a heat sink above the diodes.

Savings associated with LED lamps are even more impressive than with other energy-efficient alternatives. Energy savings associated with LED lamps to give the same amount of light are as much as 80–85% compared to incandescent lamps (Tabler et al., 2015), as proven multiple times in field trials across the U.S. poultry industry from Arkansas and Texas to the Delmarva Peninsula. A recent report from the U.S. Department of Energy (Lee et al., 2020) indicated major developments in indoor animal lighting that included the following:

- If all animal lighting today was converted to LED technology, annual lighting energy consumption for animal production would be reduced to 2.8 terawatt hours (TWh) of on-site electricity, which represents lighting energy savings of around 25% equating to US$96 million.
- The poultry industry has the largest total illuminated floor area, with 0.2 billion m^2 for broilers and 0.1 billion m^2 for layers. The hog industry

has an estimated 77 million m^2 and the dairy industry has an estimated 66 million m^2.

- As of 2019, broiler farms had seen the highest adoption of LED lighting, at 58%. LED products made up 47% of lights in hog production, 36% in dairy farms and only 26% for layers.
- As of 2019, lighting installations for animal production in the United States consumed 3.7 TWh of on-site electricity per year (Brumm, 2019), of which, 60% comes from lighting in dairy farms, 24% from broilers and 7% each from lighting in poultry layers and hog facilities.

Energy savings associated with LEDs have never been in question. The question for the poultry industry has always been whether birds will perform as well under LEDs as they do under incandescent lamps. The answer is yes - *if* - light levels are adequate during brood and grow periods, light dimmers do their job correctly and birds are cared for properly. Birds perform just as well under LEDs as under other typical light sources found in chicken houses. In the early days of LEDs in the poultry industry, most of the comparative work was done with broiler chickens. More recent work with pullets, layers and broiler breeder flocks has since proven just as successful (Tabler et al., 2015).

A commercial broiler breeder grower in Mississippi, who switched a four-house farm from high-pressure sodium lamps to 10-W non-dimmable 2700 Kelvin (K) A19 LED lamps, was rewarded by a consistent US$1500 per month savings on the farm's electric bill (Tabler et al., 2015). LED lamps were distributed in two rows to the length of the house above the nests at 3 m centres (Fig. 1) with an additional two rows on 1.6 m centres above the slats between the nests and the cool cells. These run the length of the cool cells to prevent darker spots appearing behind the cool cells (Fig. 2).

Picking the correct Kelvin rating on the lamp is critical to flock performance. The layer, pullet and broiler breeder segments of the poultry industry prefer a 2700–3000 K lamp because the orange/red in the colour spectrum helps with achieving sexual maturity in the flock. The broiler segment prefers a 5000 K lamp because the light blue colour in the 5000 K spectrum helps to keep broilers calm, increases weight gain and reduces fear and stress level (Archer, 2017). Broilers are harvested before sexual maturity and, therefore, the orange/red part of the light spectrum is not as critical.

Regardless of Kelvin rating, it's important that a poultry farmer looks closely at the price, warranty and designated use of the lighting system. All LEDs, regardless of price, can save significant amounts of energy that can translate into big monetary savings. However, it's important to choose a lamp that has a track record in an agricultural environment such as a chicken house (Tabler et al., 2015). Omnidirectional residential LEDs are not the best choice for such

Figure 1 LED lamps at 3 m centres in two rows above the nesting boxes in a poultry shed.

Figure 2 LED lamps located above the floor slats and behind the cool cells prevent any darker spots.

a setting because they direct too much light to the walls and ceiling; this light is wasted in a poultry house because the birds are on the floor.

A good choice of the lamp should be competitively priced, can withstand the harsh poultry house environment, is compatible with the light dimmer and will deliver the light levels required. This is probably not the least expensive lamp on the market, but it need not be the most expensive lamp either. Help is available to select the correct lamp for a particular operation from poultry intgerators, growers that have already switched to LEDs and university/ extension personnel with LED lighting expertise.

4 Shifting from incandescent to LED lamps

In less than a decade, the world has made a dramatic shift from incandescent lamps to dimmable LED lamps. Despite some growing pains, energy savings and the long-life potential of LED technology made the frustration worth it. However, within the poultry industry, at least in the United States, widespread problems with rapid lumen depreciation, erratic dimming performance and premature total lamp failure have recently been reported on LED lamps that were only 2–3 years old; in many cases, this was several years before the warranty had expired (Tabler et al., 2020a).

Most poultry houses in the United States are equipped with some version of a triode alternating current (TRIAC) light dimmer to adjust lighting levels as needed across the flock. A TRIAC is a small semiconductor device, similar to a diode or transistor. Resistors rapidly turn the light circuit off and on to reduce the total amount of energy flowing through the circuit. The lamp circuit is switched off many times per second. The switching cycle is built around a typical alternating current (AC) that has varying voltage polarity in an undulating sine wave fluctuating from a positive to a negative voltage.

In the United States, this alternating cycle happens 60 times per second, a frequency referred to as 60 Hertz (Hz), meaning the current changes direction (or 'alternates') 120 times per second (equivalent to 60 cycles). A light dimmer interrupts the sine wave depending upon the dimmer setting. It automatically shuts the light circuit off every time the current changes direction. This happens twice per cycle (120 times per second). It will turn the light circuit back on when the voltage reaches a certain point. This 'turn-on point' is based on where the dimmer switch is set (Tabler et al., 2019).

If the dimmer switch is turned to a brighter setting, the cut-on point is relatively soon after the cut-off point. Therefore, the circuit is on for most of the cycle and the overall light level remains bright. However, if the dimmer is set for a low light level, the cut-on point is later in the cycle and the circuit is off for much of the cycle, keeping the light level relatively dim. This is ideal for incandescent lighting because the tungsten filament in incandescent lamps is slow to heat up and cool down and the human eye sees light output as a constant level of decreased brightness.

There is research indicating that it is possible to raise broilers under brighter light levels when using LEDs without lowering growth or efficiency and still attaining the positive benefit savings of LED lighting on fear and stress. Archer (2016) found that LED lighting at 20 lux did not have detrimental effects on fear or stress as compared to incandescent lighting, when also at 20 lux, that resulted in more stress in the birds. It was concluded that light intensity and its effect on broiler chickens does not apply equally when using LEDs (Archer, 2016).

Mendes et al. (2013) found that birds raised under LEDs performed better overall than birds reared under CFLs. LED lighting has also been shown to improve weight gain and feed conversion over CFL and incandescent lighting (Archer, 2015; Huth and Archer, 2015). Low-intensity light is used to increase the feed efficiency and keep the birds calm (Appleby et al., 1992; Prescott et al., 2003) compared to high-intensity light that has been observed to lower body weight and produce poor feed conversion (Charles et al., 1992; Deaton et al., 1976; Lien et al., 2008). However, these intensity studies utilized lighting types other than LEDs (Archer, 2016), perhaps making them less relevant.

5 Issues with light dimmers

Lighting programmes are common practice on modern poultry farms to improve bird performance and meet welfare criteria (Tabler et al., 2019). An important aspect to consider today is the difference between omnidirectional and directional LED lamps (Fig. 3). Omnidirectional lamps are not ideal in a chicken house because they send roughly 35–40% of their light either out or up. However, since the chickens are on the floor, most of the light needs to go down to the floor, not to the walls or ceiling, regardless of whether lights are dimmed or not.

To have a better understanding of dimming issues, it's important to become familiar with a few lighting terms. When considering LEDs, it is important to understand information on the label (Fig. 4) and on the lamp (Fig. 5). This information may include:

- Lumens – the amount of light output from a lamp (brightness).
- Watts – the amount of energy required to create the light output (power demand).
- Dimmable or non-dimmable.
- Average rated life – that for a non-LED lamp is the typical time it takes for 50% of lamps to fail. For LEDs, usable life (L70; see Fig. 4) is the point in time when LEDs produce only 70% of their initial light output.

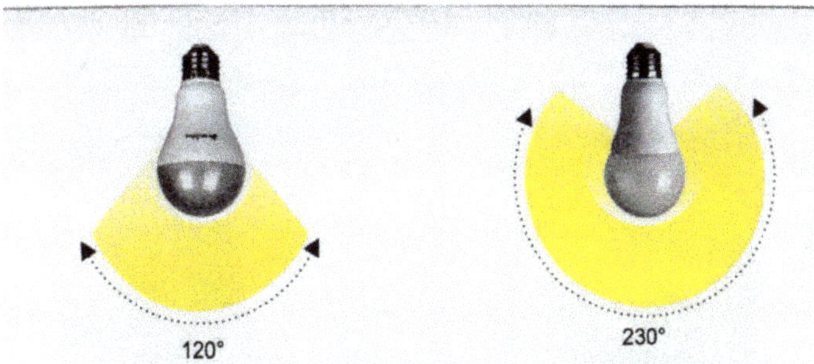

6W A19 Overdrive	Equal to =	10W Household OMNI
6W	Lamp Wattage	9.5W
545	Lamp Rated Lumen	800
100%	%age Light on floor est.	65%
545	Effective Lumen on Floor	520
$16	Energy Cost over 25,000 hr	$26
$10	Saving in Energy Cost over Life	-
5 Year	**Warranty**	**3 Year**

Figure 3 Differences between directional and omnidirectional LEDs (courtesy of Forrest Tabor and Tabor Group Inc.).

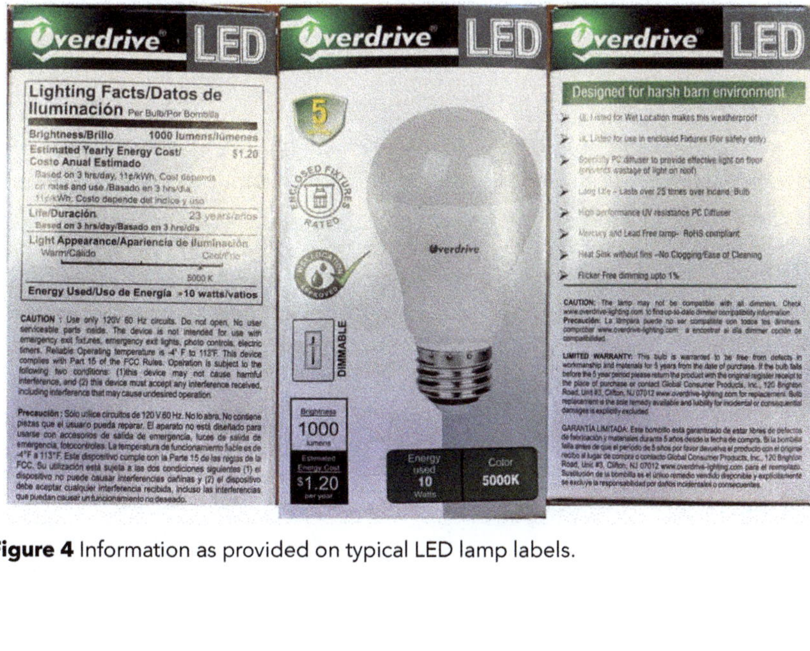

Figure 4 Information as provided on typical LED lamp labels.

Figure 5 Information provided on an LED lamp.

- Colour Rendering Index (CRI) is a measure of the accuracy of the colour perceived under the light of the lamp as compared to the colour perceived under natural daylight. A lower CRI value indicates that some colours may appear unnatural when illuminated by the lamp, with a CRI of 100 representing the maximum value. For poultry applications, LEDs should have a CRI of 80 or greater.
- Kelvin (K) – the colour or appearance of the light that the lamp emits. Warm/soft white light (similar to incandescent light) has a range of 2700–3000 K. Natural white light has a range of 3000–4500 K and cool white light has a range of 4500–6500 K. Broilers today are often grown under 5000 K lamps while layers, pullets and broiler breeders are more likely grown under 2700–3000 K lamps.

There are two main dimming methods used to interrupt or 'chop' the AC sine wave that controls the light level (Tabler et al., 2020a).

1. Leading edge dimming – utilizes a current that is turned off as the AC sine wave begins, just after it crosses zero into positive territory. It cuts the front edge of each wave's half cycle. Leading edge dimming creates

a rush of voltage every half cycle, resulting in a rush of current to the light source. Leading edge dimmers can produce spikes in current that can cause increased stress to electronic drivers (Liao, 2014). Most LED chip manufacturers have indicated that exposure to current greater than 300 mA (milliamperes) will cause irreversible damage to the chips.

Leading edge dimmer switches are simpler, less expensive and more common than trailing edge dimmers. They typically use a TRIAC switch to control power. TRIAC dimmers, used since around 1960, were originally designed to dim incandescent and halogen lamps and wire-wound magnetic transformers. Many leading edge dimmer switches have a high minimum load, which limits their use with modest-load LED circuits. This explains why it is sometimes necessary to put an incandescent lamp at the end of the line to help the dimmer find enough load to properly do its job.

2 Trailing edge dimming – trailing edge dimming utilizes a current that turns off as the AC sine wave ends, with dimmers designed specifically for use with low-wattage LED lamps. Trailing edge dimmers often use a metal oxide semiconductor field effect transistor (MOSFET) or insulated gate bipolar transistor (IGBT) switch (ERP Power, 2016). Trailing edge dimmers have a lower minimum load and are more suitable for powering modest-wattage LED lamps. They avoid current spikes by switching the light phase circuit on just as the current changes direction, allowing the voltage to rise gradually before turning it off later in the half cycle.

6 Novel ultraviolet (UV) technologies

UV light is a non-ionizing form of invisible light situated in the portion of the electromagnetic spectrum between the visible light and X-rays (NASA, 2019). UV light comprises the wavelength range between 100 nm and 400 nm (Delorme et al., 2020) divided into three bands of long wave UV-A (315–400 nm), UV-B (280–315 nm) and short-wave UV-C (100–280 nm).

While the classical use of longer wave UV light has been in biological safety cabinets in laboratories throughout the world, its use has recently been extended to the inactivation of microorganisms in the food processing industry, potable water and wastewater (Gómez et al., 2011). UV light inactivates microorganisms by forming pyrimidine dimmers in their DNA and RNA, interfering with transcription and replication (Cutler and Zimmerman, 2011; Goosen and Moolenaar, 2008). The germicidal effect of UV light treatment is dependent on microbial exposure, but when used on opaque foods with irregular surfaces, UV light may cause less microbial destruction (Pereira et al., 2014).

UV irradiation is widely used as a disinfection method for drinking water treatment (Pullerits et al., 2020). Disinfection resulting from exposure to UV light is mainly by causing damage to the nucleic acids by irradiation (Jungfer et al., 2007). Nucleotides absorb UV light with wavelengths of between 200 and 300 nm, with a peak absorption between 260 and 265 nm (USEPA, 2006). Both pyrimidines and purines can absorb UV light, although pyrimidines are considered more photoreactive (Rastogi et al., 2010; Ravanat et al., 2001). Some microorganisms can repair UV damage either by photoreactivation or dark repair (Harm, 1968; Sinha and Häder, 2002).

Much is known about long-wave UV-A light, and there is abundant literature about the biological and physiological effects of UV-B light and the signalling and metabolic pathways it triggers and influences (Barnes et al., 2015; Urban et al., 2016; Wargent and Jordan, 2013). Much less is known about short-wave UV-C light that exerts direct and indirect inhibitory and damaging effects on living cells (Urban et al., 2016). Recent observations suggest that UV-C light can be exploited to stimulate the production of health-promoting phytochemicals and to extend the shelf life of fruits and vegetables, perhaps related to its stimulating effect on the production of reactive oxygen species (Urban et al., 2016).

Foodborne pathogens can lead to illness and death of the consumer. In the United States, for example, each year there are 48 million cases of illnesses, 301 000 hospitalizations and 4300 deaths (Scallan et al., 2011). *Salmonella* spp., *Campylobacter*, *Listeria monocytogenes* and *Escherichia coli* O157:H7 are among the most common pathogens based on the number of cases caused by these agents and/or the severity of the ensuing diseases (Fan et al., 2017). UV-C stands out as one of the most promising innovative food processing technologies to control pathogens with great commercialization potential (Jermann et al., 2015; Koutchma, 2009).

It is considered to be a non-thermal and non-chemical intervention technology that uses only physical energy and so is safe and non-toxic. Lamps require relatively low investment and maintenance costs when compared to other food preservation methods and do not cause environmental impacts (Keyser et al., 2008).

UV-C rays (with short wavelengths in the range of 200–280 nm) are described as germicides because they have the most effective lethal effect on different microorganisms such as bacteria, viruses, protozoa, fungi and algae (Shin et al., 2016). Application is capable of disrupting the DNA of microorganisms, thereby altering its metabolism and reproduction, leading to cell death (Guerrero-Beltráin and Barbosa-Cánovas, 2004). This short-wave portion of the UV spectrum lies outside what is visible to the human eye; however, some portions of UV light are perceivable to insects and birds (Stanford University, n.d.). The uses and benefits of the three main classes of UV spectrum are as follows:

- UV-A (320–400 nm), the least harmful UV light with the longest wavelength, is more commonly known as the 'black light'. It has applications in UV curing, counterfeit detection and forensics, as well as agriculture, given its ability to trigger desired plant responses. Many insects and birds can perceive this type of UV light visually.
- UV-B (280–320 nm) is mostly (95%) absorbed by the Earth's atmosphere. It is associated with sunburn leading to an increased risk of skin cancer in humans. However, it has been shown to have antimicrobial effects related to agricultural infections and pests, such as powdery mildew and spider mites.
- UV-C (100–280 nm) is considered extremely harmful and is almost completely absorbed by the Earth's atmosphere. Because of its germicidal effects, it is commonly used as a disinfectant in food, air and water as it can kill microorganisms, including pathogenic bacteria and viruses, by destroying their cells' nucleic acids, DNA and RNA (Escalona et al., 2010; Harm, 1980; Hijnen et al., 2006). The advantages to food processors associated with UV-C light are many, as it does not leave behind any chemical residue, has no legal restrictions and does not require extensive worker protection equipment (Yousef and Marth, 1988). This enhances the technology's relevancy because of the association of fresh produce with foodborne illnesses, the suitability of UV-C treatment and the difficulty of decontaminating fresh produce (Ramos et al., 2013).

UV light disinfection is today a well-established technology for the treatment of air, water and surfaces. Particularly in the last decade, UV technology has found more applications in the food production chain because it is an economic, effective and versatile dry processing method that can potentially improve both food safety and the preservation of quality parameters (Koutchma et al., 2018). For example, food safety is at the forefront of agricultural issues that address potential public health risks associated with an increase in foodborne disease outbreaks (Breitenmoser et al., 2011; CDC, 2015; Fan et al., 2008; Mazari-Hiriart et al., 2008).

Agricultural water is an important carrier for human pathogens (Cooley et al., 2007; Ijabadeniyi et al., 2011; Park et al., 2012). If contaminated when used for irrigation of agricultural crops, this water poses a high risk of foodborne disease outbreaks (Olaimat and Holley, 2012). The Food Safety Modernization Act (FSMA) Produce Safety Rule (PSR) has identified agricultural water as an important source of microbial contamination to fresh produce (Fan et al., 2008). The rule requires that water used for irrigation and production/processing of fresh produce must be safe and of adequate sanitary quality for its intended use.

Adhikari et al. (2019) evaluated the potential of using UV-C treatment in reducing the microbial population in agricultural water and found that UV-C treatment could be a promising strategy in reducing produce safety risks associated with irrigation water. Koutchma et al. (2018) indicated that UV-C LEDs are the next wave in the LED revolution that can have numerous advantages for the safety of food products and food plant facilities. Potentially, UV-C LEDs could be used for the treatment of beverages, disinfection of produce surfaces, packaging and other food contact and noncontact surfaces.

Despite the great potential offered by the novel UV technology, particularly to the food processing industry, there are challenges that must be dealt with in order to enhance its value. UV-C can easily penetrate through clear or transparent liquids. For example, UV-C light would lose 30% of its intensity when passing through 40 cm of distilled water, 10 cm of sea water or 5 cm of 10% sucrose solution (Bintsis et al., 2000). However, its penetration capacity into solid foods is extremely limited since UV-C light is easily absorbed in opaque solid materials (Fan et al., 2017).

The greatest challenge to the commercial application of UV-C technology is this penetration power. Its effectiveness is restricted to inactivating only those microorganisms that the light can reach. It will be important to determine dose uniformity as well as minimum and maximum UV-C doses that each food item receives because complicated shapes of many food items make it difficult to apply uniform UV-C doses (Trivittayasil et al., 2015). The minimum UV-C dose must achieve the targeted benefits (such as specific reductions of human pathogens), while the maximum dose is what the food can tolerate without changing product quality (Fan et al., 2017).

Although UV light radiation can inactivate a wide range of microorganisms and is proving to be an economic alternative to thermal pasteurization, there are several aspects that require further investigation. To further elucidate the effects of UV light, factors such as microbial survival, bacteria recovery and the impact on post-treatment stability should be investigated (Delorme et al., 2020). Furthermore, the effects of this alternative technology should be evaluated against conventional treatments.

7 Greenhouse and controlled environment lighting applications

The use of electronic systems for controlled environment agriculture (CEA) in greenhouses has steadily increased in recent years as smart food production continues to advance. These systems use sensors to automate the environmental management of greenhouses and can lead to significantly larger yields per square metre. However, mismanagement of a CEA system can lead to the loss of entire crops. Due to the sensitivity of crops in CEA and potential losses, only fully trained personnel are usually allowed to work directly with CEA systems.

To address this issue, and to increase opportunities for students and less qualified staff to interact with CEA systems, Carruth et al. (2020) created an immersive virtual greenhouse simulation. Implementing realistic interfaces for physical tasks and for the CEA system provides 'hands on' training in the virtual greenhouse. Despite challenges that limit the implementation of physical actions and realistic interfaces in virtual reality, the next generation of energy-smart farmers can now hone their skills at managing such factors as light, temperature, humidity and ventilation in the virtual world before taking the helm in the real world.

Light condition and air temperature are the two most important environmental factors for plant growth. They are intrinsically related, and one cannot be optimized without considering the other (Shamshiri et al., 2018b). The optimal combination of air temperature, relative humidity and light will result in maximum crop yield, assuming that other factors such as CO_2, soil pH and nutrients are not limiting growth potential.

The most common artificial light sources used in modern greenhouses and CEA production systems are incandescent/halogen lamps, discharge lamps (fluorescent tubes, metal halide and high-pressure sodium lamps) and LEDs. LEDs have gained much favour in recent years because of their cost efficiency, compact design, durability, light quality and low thermal energy generation (Shamshiri et al., 2018b). Although research on LEDs as a light substitute for plant growth began in the 1980s, it was only after the mid-2000s that they became practical and economically feasible for large-scale production operations.

LED technology represents a totally different and energy-efficient approach for the greenhouse industry that has proficient advantages over high-pressure sodium lamps currently installed in many greenhouses (Mitchell et al., 2012; Morrow, 2008). A properly designed LED light system can provide highly efficient performance and longevity well beyond any traditional lighting source (Bourget, 2008).

LED lighting on various vegetables has shown good results in terms of maximal productivity and optimal nutritional quality, paving the way for wider acceptance of LED technology in greenhouses in the future (Singh et al., 2015). LEDs allow the control of spectral composition and the adaptation of light intensity to be matched to plant photoreceptors in order to furnish better growth and to influence plant morphology as well as the different physiological processes such as flowering and photosynthetic efficiency (Yeh and Chung, 2009).

A downside to incandescent and fluorescent lamps is their relatively short lifespan (approximately 1000 h for incandescent and 20 000 h for fluorescent) and the fact that they generate heat and consume relatively large amounts of electrical power. LEDs have lifetimes of 50 000 h or more and, because of their low radiant heat production, can be placed close to plants and configured to emit high light fluxes even at high light intensities (Barta et al., 1992; Tennessen et al., 1994).

A major advantage of LEDs over all other lamp types for greenhouse lighting is that the technology is evolving in electrical-use efficiency at a rapid pace (Olle and Viršilė, 2013). For example, blue LEDs that were 11% efficient in 2006 were reported to be 49% efficient, converting electrical energy to photon energy in 2011 (Mitchell et al., 2012). LEDs do not typically burn out but do have lumen depreciation over time. Therefore, their lifetime is measured by the time it takes for the LED to dim to 70% of its original intensity (Olle and Viršilė, 2013).

Energy inputs contribute 20–30% of the total production cost in the heated greenhouse industry (Brumfield, 2007; Langton et al., 2006). With lighting costs a significant share, it is critical to provide appropriate crop lighting, particularly in regions where the seasonal photoperiod fluctuates and there is not sufficient natural light for optimal plant growth (Singh et al., 2015). High-pressure sodium lamps have been the most commonly used light, with LEDs representing energy-efficient cost savings together with other technical advantages (Bourget, 2008).

LEDs can be manufactured to emit broad-band (white) light or narrow-spectrum (coloured) wavelengths specific to the desired applications, such as targeted plant responses (Nanya et al., 2012). As a result, LEDs offer a simple replacement of current greenhouse lighting sources with better control on spectral composition (Singh et al., 2015). A controlled spectrum composition is more advantageous for plants since it targets performance in flowering time, low heat stress and high photosynthetic efficiency (Johkan et al., 2010; Kim et al., 2004; Singh et al., 2015). Although plants cultivated under blue and red light in a closed environment look purplish-grey to the human eye, making it difficult to visually evaluate plant health. Massa et al. (2008) indicated a small flux of green light (up to 20%) helps alleviate this problem.

LEDs offer the following advantages to the greenhouse industry (Massa et al., 2008; Vänninen et al., 2010):

- reduction in energy consumption up to 70% compared to traditional light sources;
- fast switching and steady-state operation;
- simple electronic dimming function;
- reduction of electricity cable gauge (and hence lower cost and weight);
- control of the spectral composition of blue, green, red, and far-red wavelengths;
- high relative quantum efficiency (RQE) with red light having the highest, meaning it is the most efficient at photosynthesis, whereas blue light is about 70–75% as efficient;
- stable temperature inside the greenhouse or growth chamber;

- reduction of heat stress on plants;
- reduction in watering and ventilation maintenance; and
- lifetime, reliability and compact lamp size as the major technical advantages.

Olle and Viršilė (2013) also highlighted the safety features of LEDs. There is no fragile glass envelope to break, no high touch temperatures and no hazardous materials, such as mercury. Their relatively small size allows versatile design allowing lighting systems to be configured to emit high light fluxes.

The greenhouse industry in the United States is divided into three categories: (1) supplemented greenhouses, (2) non-stacked indoor farms and (3) vertical farms. Supplemented greenhouses employ electric lighting to extend the hours of light provided to plants, to supplement low levels of sunlight on days with inclement weather and/or to disrupt periods of darkness for purposes of altering plant growth. Non-stacked indoor farms are used for simple growing operations and/or tall plants and represent an application where plants are grown in a single layer on the floor under ceiling-mounted electric lighting.

Vertical farms are more sophisticated farming operations where small plants are stacked along vertical shelving from floor to ceiling such that the grow area can be increased in the same building floorspace. As a result of the unique architecture associated with vertical farms, lighting is typically mounted within the vertical shelving units and much closer to the plants themselves than in either indoor non-stacked farms or supplemented greenhouses. A 2017 U.S. Department of Energy analysis (Stober et al., 2017) and a similar analysis by Lee et al. (2020) reported a summary of horticultural lighting analyses (Table 2). They also reported a summary of U.S. animal lighting analysis (Table 3).

The United States is not the only major player in the greenhouse industry that has expanded rapidly in many regions throughout the world in recent decades (Fernández et al., 2018). Canada has the largest concentration of greenhouses in North America, and the Netherlands has more than 90 sq. km of greenhouse land surface area, roughly 3.5 times the size of the Canadian greenhouse industry (Sondag, 2020). Greenhouse farming has demonstrated its efficiency in intensifying food production and is a feasible alternative for ensuring a sufficient food supply, which constitutes one of the greatest challenges faced by humankind in the twenty-first century (Aznar-Sánchez et al., 2020).

The industry continues developing new strategies and technologies to resolve specific crop limitations, of which LED lighting has proven to be one of the most beneficial. In addition, soil-less crops (Magwaza et al., 2020); control of the factors that make up the micro-climate inside the greenhouse (Ghoulem et al., 2019; Shamshiri et al., 2018a); creation of vertical agroecosystems that

Table 2 U.S. horticultural lighting analyses for indoor plant production systems in 2017 and 2019[1]

Analysis output		Vertical farming		Supplemented greenhouse		Non-stacked indoor		Total[2]	
Year	Units	2017	2019	2017	2019	2017	2019	2017	2019
Estimated total light growing area	km^2	0.05	0.07	2.49	5.26	1.74	2.96	4.27	8.30
Annual operating hours	hours/year	6570	6570	2000	2000	5200	5200	-	-
Technology mix									
LED	%	66	100	2	2	4	11	-	-
HPS/MH[4]	%	<1	0	98	98	89	86	-	-
Fluorescent	%	34	0	0	0	7	3	-	-
Annual energy consumption									
Current	GWh/year	60	81	588	1202	5300	8307	5940	9591
Theoretical (if change to all LEDs)		55	81	416	832	3100	5395	3570	6307
Theoretical light energy savings[3]	%	10	0	29	31	41	35	40	34

[1] Adapted from Stober et al. (2017) and Lee et al. (2020).
[2] All values may not add due to rounding.
[3] Theoretical percentage of light energy savings were converted to LEDs, which is the percentage of difference in electricity consumption between the current and theoretical scenarios.
[4] High-pressure sodium/metal halide.

Table 3 Intensive animal production house lighting demands in the United States for 2019[1]

Analysis outputs	Units	Poultry broiler	Poultry layer	Dairy	Pig	Total[2]
Estimated total animal count	Million	8890	368	10	235	9,503
Estimated total illuminated area	km^2	221.76	102.66	66.43	77.57	468.42
Technology mix:						
LED	%	58	26	36	47	-
HPS/MH[4]	%	0	25	22	19	-
Fluorescent	%	25	23	42	34	-
Incandescent	%	17	25	0	0	-
Energy consumption:						
Current	GWh/year	884	263	2263	258	3667
Theoretical (if all changed to LED)	GWh/year	503	122	1918	224	2766
Theoretical % light energy savings[3]	%	43	54	15	13	25

[1] Adapted from Lee et al. (2020).
[2] All values may not add due to rounding.
[3] Theoretical percentage of light energy savings were converted to LEDs, which is the percentage of difference in energy consumption of the current and theoretical scenarios.
[4] High-pressure sodium/metal halide.

can be located in urban environments (Al-Kodmany, 2018); development of specific environmentally sustainable solutions for supplying greenhouse production with renewable energy (Chel and Kaushik, 2011; Dolgikh et al., 2019); and the development of new materials and structures capable of optimizing production (Petropoulos et al., 2019; Sahdev et al., 2019) are now possible.

Sondag (2020) reported the greenhouse industry in the Netherlands has become the most energy efficient in the world, driven in part by relatively high energy prices. Dutch greenhouses are efficiently controlled environments in which water, light, energy, nutrients and temperature can be precisely managed by innovative software, thus saving space, money and energy while maximizing yields. For example, 1 ha of a tomato greenhouse yields ten times the amount of tomatoes as 1 ha of tomatoes in a field.

After the United States, the Netherlands is the largest exporter of agricultural produce in the world (Ministry of Agriculture, Nature and Food Quality, 2020). The Dutch do this by precision farming on a landmass that is approximately 240 times smaller than the United States. The agricultural and horticultural sectors account for 10% of the Dutch economy and employment. More than half of the total Dutch land area is used for agriculture and horticulture, and there are extraordinary greenhouse complexes, with some units covering 0.7 km^2. With 80% of cultivated land being under greenhouse glass, crops can grow around the clock in any kind of weather (Civil Engineer, 2020).

The Dutch government has focused efforts on drastically reducing energy and gas consumption in the greenhouse sector. From 2020, all new greenhouses must be climate-neutral (Ministry of Agriculture, Nature and Food Quality, 2020). The Dutch government plans to reach this goal through innovative technologies such as:

- increased use of solar energy;
- light-saving measures (use of natural light and more energy-efficient lamps);
- energy-efficient growth strategies;
- geothermal applications;
- use of biofuels (biogas and wood residues); and
- generation and use of more sustainable electricity.

8 Ventilation advances in livestock housing

While providing on average 17% of food calories and more than a third of protein to human diets (Herrero et al., 2009), livestock is consuming almost 60% of the total global biomass harvest (Krausmann et al., 2008), uses around 30% of agricultural water withdrawals (Mekonnen and Hoekstra, 2010; Peden et al., 2007) and dominates the agricultural nitrogen cycle (Bodirsky et al., 2012, 2014; Bouwman et al., 2013). However, livestock farming reduces vulnerability to hunger from environmental risks for 600 million poor smallholder farmers and their families (Steinfeld et al., 2006; Thornton and Herrero, 2010) and provides other services beyond food production (Herrero et al., 2009; Steinfeld et al., 2006).

Food security is critical to achieving sustainable growth, poverty reduction and political and economic stability, especially in regions of the world undergoing rapid population growth, such as sub-Saharan Africa, where the population is projected to double by 2050 (UN, 2020). In 2018, the total population of sub-Saharan Africa was approximately 1.08 billion; by 2060, the population could have grown to 2.7 billion (UN, 2020). Smallholder farmers provide up to 80% of the food supply in sub-Saharan Africa through cropping and livestock production (Tabler et al., 2020b) and play a major role in food security in developing countries (Dixon et al., 2004).

Although it impacts many critical environmental impacts, including greenhouse gas emissions, livestock production is crucial to the human protein supply, particularly for many poor and undernourished people in the developing world (Weindl et al., 2015). As a result, livestock is intertwined with many aspects of the challenge to sustainably feed a growing population and achieve a balance between livelihoods, food security and the environment (Herrero and Thornton, 2013). Therefore, the need exists to fill large gaps in knowledge of how climate change could affect livestock production and how a transformation of livestock production systems could contribute to climate-smart agriculture and energy-smart farming.

In Organisation for Economic Cooperation and Development countries, swine and poultry production occurs predominantly inside livestock confinement buildings (Gerber et al., 2013), characterized by a mechanical ventilation system, high stocking density and insulation (Schauberger et al., 2019). The increase in heat stress for farm animals due to anthropogenic global warming in recent decades is well known, especially in livestock confinement buildings. These effects are relevant not only from an animal welfare perspective (Huynh et al., 2005) but also from an economic point of view (St-Pierre et al., 2003).

Heat stress directly impairs the productivity of meat, milk, egg yields and quality, as well as reproductive performance, animal health and welfare (Gaughan, 2012; Lara and Rostagno, 2013; Nardone et al., 2010; Thornton et al., 2009). Rötter and van de Geijn (1999) assumed that the vulnerability of confined livestock systems to a warming climate is lower than that of alternative outdoor systems due to the opportunity for the application of adaptation measures. However, Mikovits et al. (2019) indicated that, for Central Europe, the indoor climate of livestock confinement systems is more sensitive to heat stress compared to the outdoor system.

This sensitivity is caused primarily by the higher indoor temperatures compared with outdoor temperatures due to the highly sensitive heat load of the animals. Therefore, future global warming will affect requirements for the control of thermal conditions of livestock buildings, such as through mechanical ventilation systems (Olesen and Bindi, 2002). Such adaptation measures can help reduce heat stress and its impact on welfare, health and productivity. While these measures will increase the direct input costs associated with intensive livestock production, they should be partly offset by the beneficial effects on livestock productivity and animal welfare.

Adaptive measures may influence the immediate thermal vicinity of the animals. They include floor cooling (Bull et al., 1997; Cabezon et al., 2017; Silva et al., 2009), higher air velocity at the animal level to increase convective heat release (known as the wind chill effect), tunnel ventilation, booster fans, hybrid ventilation systems (Zhang and Bjerg, 2017), radiative cooling by a cooled cover of the laying zone (Pang et al., 2010), cooled drinking water (Jeon et al., 2006; Renaudeau et al., 2012) and the provision of wallows (Bracke, 2011; de Mello et al., 2017) or water baths (Huynh et al., 2006).

Of all these adaptive measures, tunnel ventilation is likely the most widely used, particularly in the swine and poultry industries, but also by the greenhouse industry. Fans are essential components of mechanically ventilated greenhouses, poultry houses, swine barns and dairy sheds. Fan performance characteristics, especially power consumption, are critical for the optimization of environmental control and energy conservation. Attention should be focused on improving the energy efficiency of tunnel ventilation fans with shutter type and location key to an energy-efficient system and to optimize the overall fan performance. Traditional shutters may be located in the interior (intake side; Fig. 6) or exterior (exhaust side) of the building. Exterior butterfly shutters (Fig. 7) are significantly better than traditional interior or exterior shutters.

They offer little resistance to the flow of air in to or out of a fan. In addition, the fan does not have to work against gravity to open the shutter and performance can be further improved because dust and dirt build-up doesn't increase the force required to open a butterfly shutter as it does with traditional shutters. As a result, fans with butterfly shutters are 5–10% more efficient than clean traditional shutters and 10–20% more efficient than dirty traditional shutters. However, butterfly shutters are not without their issues.

They tend to scare animals when they 'slam' close when the fan shuts off, and they allow more light into the house, which may be detrimental when lights are dimmed. However, the most significant issue is that they are prone

Figure 6 Traditional shutters on the intake side of tunnel fans installed in the wall of a livestock house.

Figure 7 Tunnel fans with butterfly shutters on the exterior wall of a typical livestock building.

to condensation problems that can cause moisture to pool at the bottom of the fan and run down interior walls, possibly damaging the fan and the wall over time. Therefore, because of the condensation issues, butterfly shutters are better suited for warmer climates.

Figure 8 Discharge cones on tunnel fans installed on a livestock building.

Fan performance has been greatly improved by installing discharge cones on tunnel fans (Fig. 8) that reduce the exit pressure on fans and tend to increase the air flow approximately 5% without increasing power demand. Yet another way to improve fan performance is to increase the size. All other things being equal, larger fans are more efficient than smaller fans, although it is useful to understand some basic fan laws. For example, the volume of air flow (cubic feet per minute; cfm) is proportional to fan speed; increasing fan speed by 10% increases fan output by 10%.

Installing a larger motor pulley to speed up the fan blades would increase the fan air volume movement capacity, but since power demand is exponentially related to fan speed, increasing it by 10% will increase power demand by around 30%. Therefore, installing a bigger pulley to move more air will eventually burn out the electric motor driving the fan unless a higher capacity motor is installed.

However, fan laws work both ways. If fan speed is decreased by 10%, the power usage is decreased by 30%. It may be more energy efficient to install more fans with a smaller pulley moving less air than installing fewer fans with a larger pulley moving more air. An alternative to changing pulleys would be to install variable-speed motors and controllers. Variable-speed, direct-drive fans offer several advantages including power savings of 30–50%. They also clean easily, come with a motorized butterfly shutter, are well insulated and are much quieter when running at full speed.

However, there are disadvantages to direct-drive, variable-speed fans as well. The initial cost is at least twice that of a comparable belt-driven fan. In addition, some controllers are not capable of controlling multiple variable-speed motors. There are more electronic components which may mean more risk of lightening damage. Variable-speed fans may also have a problem operating at higher static pressures when running at low speeds. Most variable-speed fans move 50% of their rated airflow volume at approximately 65% of their full speed rotations per minute (Harmon et al., 2012).

Fans should be selected based on air flow and efficiency ratings. Choose fans that are in the upper 25% of rated fan efficiencies. The Bioenvironmental and Structural Systems (BESS) Laboratory at the University of Illinois tests all major agricultural fans with results presented on their website (http://bess.illinois.edu/index2.htm). All manufactured fans should be tested by such an independent laboratory to indicate air flow and efficiency as a function of static pressure. A fan should not be selected on its diameter alone and it cannot be assumed that two fans of equal size will have equal performance. Different motors, blade curvature and other attributes greatly affect performance.

Fans must be clean to be most efficient, whether for the swine, poultry or greenhouse industries. Regular cleaning and maintenance of fans and fan accessories is essential to maintaining fan performance (Zhang et al., 2015).

Guard screens, shutters and other accessories, as well as fan wear and tear, affect airflow rate and efficiency (Person et al., 1979; Simmons et al., 1997).

Zhang et al. (2015) reported significant differences in power consumption between fans before and after cleaning, as well as considerable variation in both airflow and power consumption among otherwise identical fans. In addition, maintenance of proper tension on drive belts and replacement of worn belts in a timely manner were found to be important in maintaining fan performance. A small reduction in fan speed from a worn or slipping belt has a large effect on airflow, which increases power consumption and costs. Therefore, while having the right equipment is important, it requires good maintenance to make energy-smart farming efficient and sustainable.

9 Conclusion and future trends

Rapid advances have been made recently in energy-efficient lighting systems and, to a lesser extent, ventilation systems, as used in intensive farming and crop production. After using several different types of lamps that are all the more energy efficient than incandescent lamps, LEDs have become the popular choice for both livestock and CEA. Improvements in the technology will continue since there are issues regarding dimmers, wiring and sockets that affect performance, longevity and lumen depreciation over time.

However, energy savings associated with LEDs can't be ignored in as much as 80–85% compared to incandescent lamps as demonstrated in numerous trials undertaken by the poultry industry across the broiler growing regions of the United States. If light levels are adequate during brooding and growing periods, the light dimmer does its job and the birds are cared for properly, then poultry flocks perform just as well under LEDs as under other typical light sources that tend to be more costly.

Food safety is currently at the forefront of agricultural issues that address potential public health risks associated with an increase in foodborne disease outbreaks. Novel UV lighting technologies offer many possibilities to both water treatment and food processing industries. UV light disinfection is a well-established technology for the treatment of air, water and surfaces. Such technology has found numerous applications in the food production chain over the past decade because it is an economical, effective and versatile dry processing method that can potentially improve both food safety and the preservation of quality parameters.

Use of UV-C light in reducing the microbial population in agricultural water is a promising strategy in reducing produce safety risks associated with irrigation water. UV-C LEDs may be the next wave in the LED revolution that can have numerous advantages for the safety of food products and food plant facilities. Potentially, UV-C LEDs could be used for the treatment of beverages and disinfection of produce surfaces, packaging and other food contact and noncontact surfaces.

Fans are critical components of mechanically ventilated greenhouses, poultry houses, swine barns, dairy sheds etc. Guard screens, shutters and other accessories, as well as fan aging, affect airflow rate and efficiency. Proper fan design, operation and maintenance are keys to successful intensive livestock housing and controlled environment crop production. Regular cleaning and maintenance of fans and accessories is essential to maintaining performance characteristics, especially power consumption. These are critical for the optimization of environment control and energy conservation.

10 Where to look for further information

- Bioenvironmental and Structural Systems Laboratory – University of Illinois (http://bess.illinois.edu/index2.htm).
- Sustainable Poultry Center – Mississippi State University Extension Service (http://sustainablepoultrycenter.extension.msstate.edu/topics/general-information).
- Mississippi State University Poultry Science Extension (https://www.poultry.msstate.edu/extension/index.php).
- Centers for Disease Control and Prevention (https://www.cdc.gov/).
- Key journals/conferences
  - Applied Engineering in Agriculture (https://www.asabe.org/Applied).
  - Applied Animal Behaviour Science (https://www.journals.elsevier.com/applied-animal-behaviour-science).
  - Food Science and Nutrition (https://onlinelibrary.wiley.com/journal/20487177).
  - Livestock Science (https://www.journals.elsevier.com/livestock-science).
  - Horticulture Science (https://www.agriculturejournals.cz/web/hortsci/).
  - Journal of Food Science (https://onlinelibrary.wiley.com/journal/17503841).
  - Poultry Science (https://www.journals.elsevier.com/poultry-science).
  - Journal of Applied Poultry Research (https://www.sciencedirect.com/journal/journal-of-applied-poultry-research).
  - International Poultry Scientific Forum (https://www.ippexpo.org/ipsf/).
  - Poultry Science Association Annual Meeting (https://poultryscience.org/PSA-Annual-Meeting).
  - Journal of Animal Science (https://academic.oup.com/jas).
  - American Society of Animal Science (https://www.asas.org/#_).
- Poultry science departments at major universities
  - Auburn University (https://poul.auburn.edu/).
  - Mississippi State University (https://www.poultry.msstate.edu/).

- North Carolina State University (https://cals.ncsu.edu/prestage-department-of-poultry-science/).
- Texas A&M University (https://posc.tamu.edu/).
- University of Arkansas (https://poultry-science.uark.edu/).
- University of Georgia (https://poultry.caes.uga.edu/).

11 References

Adhikari, A., Parraga-Estrada, K. J., Chhetri, V. S., Janes, M., Fontenot, K. and Beaulieu, J. C. (2019). Evaluation of ultraviolet (UV-C) light treatment for microbial inactivation in agricultural waters with different levels of turbidity, *Food Sci. Nutr.* 8, 1234-1243.

Al-Kodmany, K. (2018). The vertical farm: a review of develops and implications for the vertical city, *Buildings* 8(24). 36 pages.

Appleby, M. C., Hughes, B. O. and Elson, A. H. (1992). *Poultry Production Systems: Behaviour, Management and Welfare*, Wallingford, Oxon: CAB International.

Archer, G. S. (2015). Comparison of incandescent, CFL, LED and bird level LED lighting: growth, fear and stress, *Int. J. Poult. Sci.* 14(8), 449–455.

Archer, G. S. (2016). Comparison of raising broilers under light emitting diode or incandescent light at differing intensities on growth, fear and stress, *Int. J. Poult. Sci.* 15, 425–431.

Archer, G. S. (2017). Color temperature of light-emitting diode lighting matters for optimum growth and welfare of broiler chickens, *Animal* 12(5), 1015–1021.

Aznar-Sánchez, J. A., Velasco-Munoz, J. F., López-Felices, B. and Roman-Sanchez, I. (2020). An analysis of global research trends on greenhouse technology: towards a sustainable agriculture, *Int. J. Environ. Res. Public Health* 17(644). 22 pages.

Barnes, P. W., Flint, S. D., Ryel, R. J., Tobler, M. A., Barkley, A. E. and Wargent, J. J. (2015). Rediscovering leaf optical properties: new insights into plant acclimation to solar UV radiation, *Plant Physiol. Biochem.* 93, 94–100.

Barta, D. J., Tibbits, T. W., Bula, R. J. and Morrow, R. C. (1992). Evaluation of light emitting diode characteristics for a space-based plant irradiation source, *Adv. Space Res.* 12(5), 141–149.

Bintsis, T., Litopoulou-Tzanetaki, E. and Robinson, R. K. (2000). Existing and potential applications of ultraviolet light in the food industry: a critical review, *J. Sci. Food Agric.* 80(6), 637–645.

Bodirsky, B. L., Popp, A., Weindl, I., Dietrich, J. P., Rolinski, S., Scheiffele, L., Schmitz, C. and Lotze-Campen, H. (2012). N2O emissions from the global agricultural nitrogen cycle: current state and future scenarios, *Biogeosciences* 9(10), 4169–4197.

Bodirsky, B. L., Popp, A., Lotze-Campen, H., Dietrich, J. P., Rolinski, S., Weindl, I., Schmitz, C., Müller, C., Bonsch, M., Humpenöder, F., Biewald, A. and Stevanovic, M. (2014). Reactive nitrogen requirements to feed the world in 2050 and potential to mitigate nitrogen pollution, *Nat. Commun.* 5, 4858. Available at: https://www.nature.com/articles/ncomms4858.pdf. (Accessed: 15 September 2020).

Bourget, C. M. (2008). An introduction to light emitting diodes, *Hortscience* 43(7), 1944–1946.

Bouwman, L., Goldewijk, K. K., Van Der Hoek, K. W., Beusen, A. H., Van Vuuren, D. P., Willems, J., Rufino, M. C. and Stehfest, E. (2013). Exploring global changes in nitrogen and phosphorus cycles in agriculture induced by livestock production over the 1900–2050 period, *Proc. Natl. Acad. Sci. U. S. A.* 110(52), 20882–20887.

Bracke, M. B. M. (2011). Review of wallowing in pigs: description of the behaviour and its motivational basis, *Appl. Anim. Behav. Sci.* 132(1–2), 1–13.

Breitenmoser, A., Fretz, R., Schmid, J., Besl, A. and Etter, R. (2011). Outbreak of acute gastroenteritis due to a washwater-contaminated water supply, Switzerland, 2008, *J. Water Health* 9(3), 569–576.

Brumfield, R. (2007). Dealing with rising energy costs, *Greenhouse Prod. News* 17, 24–31.

Brumm, M. C. (2019). Space allocation decisions for nursery and grow-finish facilities, *eXtension*. Available at: https://swine.extension.org/space-allocation-decisions-for-nursery-and-grow-finish-facilities/. (Accessed: 24 October 2020).

Bull, R. P., Harrison, P. C., Riskowski, G. L. and Gonyou, H. W. (1997). Preference among cooling systems by gilts under heat stress, *J. Anim. Sci.* 75(8), 2078–2083.

Cabezon, F. A., Schinckel, A. P. and Stwalley III, R. M. (2017). Thermal capacity of hog-cooling pad, *Appl. Eng. Agric.* 33(6), 891–899.

Carruth, D. W., Hudson, C., Fox, A. A. A. and Deb, S. (2020). User interface for an immersive virtual reality greenhouse for training precision agriculture, in: *Virtual, Augmented and Mixed Reality. Industrial and Everyday Life Applications, Lecturer Notes in Computer Science, vol 12191*, Cham: Springer, 35–46.

Centers for Disease Control and Prevention (CDC) (2015). List of selected multistate foodborne outbreak investigations. Available at: http://www.cdc.gov/foodsafety/outbreaks/multistate-outbreaks/outbreaks-list.html. (Accessed: 8 September 2020), Centers for Disease Control and Prevention.

Charles, R. G., Robinson, F. E., Hardin, R. T., Yu, M. W., Feddes, J. and Classen, H. L. (1992). Growth, body composition and plasma androgen concentration of male broiler chickens subjected to different regimens of photoperiod and light intensity, *Poult. Sci.* 71(10), 1595–1605.

Chel, A. and Kaushik, G. (2011). Renewable energy for sustainable agriculture, *Agron. Sustain. Dev.* 31(1), 91–118.

Civil Engineer. (2020). Dutch greenhouses have revolutionized modern farming, *TheCivilEngineer.org*. Available at: https://www.thecivilengineer.org/news-center/latest-news/item/1544-dutch-greenhouses-have-revolutionized-modern-farming. (Accessed: 2 December 2020).

Cooley, M., Carychao, D., Crawford-Miksza, L., Jay, M. T., Myers, C., Rose, C., Keys, C., Farrar, J. and Mandrell, R. E. (2007). Incidence and tracking of *Escherichia coli* O157:H7 in a major produce production region in California, *PLoS ONE* 2(11), e1159.

Cutler, T. D. and Zimmerman, J. J. (2011). Ultraviolet irradiation and the mechanisms underlying its inactivation of infectious agents, *Anim. Health Res. Rev.* 12(1), 15–23.

Deaton, J. W., Reece, F. N., Kubena, L. F. and May, J. D. (1976). Effect of varying light intensity on broiler performance, *Poult. Sci.* 55(2), 515–519.

Delorme, M. M., Guimarães, J. T., Coutinho, N. M., Balthazar, C. F., Rocha, R. S., Silva, R., Margalho, L. P., Pimentel, T. C., Silva, M. C., Freitas, M. Q., Granato, D., Sant'Ana, A. S., Duart, M. C. K. H. and Cruz, A. G. (2020). Ultraviolet radiation: an interesting technology to preserve quality and safety of milk and dairy foods, *Trends Food Sci. Technol.* 102, 146–154.

de Mello, J. L. M., Berton, M. P., de Cassia Dourado, R., Giampietro-Ganeco, A., de Souza, R. A., Ferrari, F. B., de Souza, P. A. and Borba, H. (2017). Physical and chemical characteristics of the longissimus dorsi from swine reared in climate controlled and uncontrolled environments, *Int. J. Biometeorol.* 61(10), 1723–1731.

Dixon, J., Tanyeri-Abur, A. and Wattenbach, H. (2004). Framework for analysing impacts of globalization on smallholders, *Smallholders, Globalization, and Policy Analysis*. Available at: http://www.fao.org/3/y5784e/y5784e02.htm#bm02. (Accessed: 16 September 2020).

Dolgikh, P. P., Parshukov, D. V. and Shaporova, Z. E. (2019). Technology for managing thermal energy flows in industrial greenhouses, *IOP Conf. Ser. Mater. Sci. Eng.* 537, 062041.

ERP Power. (2016). The trailing edge dimmer for LED lighting control, *Energy Recovery Production Power*, Moorpark, CA. Available at: https://www.erp-power.com/the-trailing-edge-dimmer-for-led-lighting-control/. (Accessed: 5 August 2020).

Escalona, V. H., Aguayo, E., Martínez-Hernández, G. B. and Artés, F. (2010). UV-C doses to reduce pathogen and spoilage bacterial growth in vitro and in baby spinach, *Postharvest Biol. Technol.* 56(3), 223–231.

Fan, X., Annous, B. A., Beaulieu, J. C. and Sites, J. E. (2008). Effect of hot-water surface pasteurization of whole fruit on shelf life and quality of fresh-cut cantaloupe, *J. Food Sci.* 73(3), M91–M98.

Fan, X., Huang, R. and Chen, H. (2017). Application of ultraviolet C technology for surface decontamination of fresh produce, *Tends Food Sci. Technol.* 70, 9–19.

Fernández, J. A., Orsini, F., Baeza, E., Oztekin, G. B., Muñoz, P., Contreras, J. and Montero, J. I. (2018). Current trends in protected cultivation in Mediterranean climates, *Eur. J. Hortic. Sci.* 83(5), 294–305.

Gaughan, J. B. (2012). Basic principles involved in adaption of livestock to climate change, in: Sajian, V., Naqvi, S. M. K., Ezeji, T., Lakritz, J. and Lal, R. (Eds), *Environmental Stress and Amelioration in Livestock Production*, Berlin: Springer, pp 245–261.

Gerber, P., Steinfeld, H., Henerson, B., Mottet, A., Dijkman, J., Falcucci, A. and Tempio, G. (2013). *Tackling Climate Change through Livestock. A Global Assessment of Emissions and Mitigation Opportunities*, Rome: Food and Agriculture Organization of the United Nations (FAO).

Ghoulem, M., El Moueddeb, K., Nehdi, E., Boukhanouf, R. and Kaiser Calautit, J. (2019). Greenhouse design and cooling technologies for sustainable food cultivation in hot climates: review of current practice and future status, *Biosys. Eng.* 183, 121–150.

Gómez, P. L., Welti-Chanes, J. and Alzamora, S. M. (2011). Hurdle technology in fruit processing, *Annu. Rev. Food Sci. Technol.* 2, 447–465.

Goosen, N. and Moolenaar, G. F. (2008). Repair of UV damage in bacteria, *DNA Repair* 7(3), 353–379.

Guerrero-Beltrán, J. A. and Barbaso-Cánovas, G. V. (2004). Advantages and limitations on processing foods by UV light, *Food Sci. Technol. Int.* 18, 464–473.

Harm, W. (1968). Effects of dose fractionation on ultraviolet survival of *Escherichia coli*, *Photochem. Photobiol.* 7(1), 73–86.

Harm, W. (1980). *Biological Effects of Ultraviolet Radiation*, Cambridge: Cambridge University Press.

Harmon, J., Brumm, M. and Petersen, D. (2012). Managing swine ventilation controller settings to save energy, in: *Iowa State University Extension and Outreach*, PM 2089T. March.

Herrero, M. and Thornton, P. K. (2013). Livestock and global change: emerging issues for sustainable food systems, *Proc. Natl. Acad. Sci. U. S. A.* 110(52), 20878–20881.

Herrero, M., Thornton, P. K., Gerber, P. and Reid, R. S. (2009). Livestock, livelihoods and the environment: understanding the tradeoffs, *Curr. Opin. Environ. Sustain.* 1(2), 111–120.

Hijnen, W. A. M., Beerendonk, E. F. and Medema, G. J. (2006). Inactivation credit of UV radiation for viruses, bacteria and protozoan (oo)cysts in water: a review, *Water Res.* 40(1), 3–22.

Huynh, T. T. T., Aarnink, A. J. A., Gerrits, W. J. J., Heetkamp, M. J. H., Canh, T. T., Spoolder, H. A. M., Kemp, B. and Verstegen, M. W. A. (2005). Thermal behaviour of growing pigs in response to high temperature and humidity, *Appl. Anim. Behav. Sci.* 91(1–2), 1–16.

Huynh, T. T. T., Aarnink, A. J. A., Truong, C. T., Kemp, B. and Verstegen, M. W. A. (2006). Effects of tropical climate and water cooling methods on growing pigs' responses, *Livest. Sci.* 104(3), 278–291.

Hy-Line International. (2017). Understanding poultry lighting: a guide to LED bulbs and other sources of light for egg producers, *Hy-Line International*, Technical Update.

Ijabadeniyi, O. A., Debusho, L. K., Vanderlinde, M. and Buys, E. M. (2011). Irrigation water as a potential preharvest source of bacterial contamination of vegetables, *J. Food Saf.* 31(4), 452–461.

Jeon, J. H., Yeon, S. C., Choi, Y. H., Min, W., Kim, S., Kim, P. J. and Chang, H. H. (2006). Effects of chilled drinking water on the performance of lactating sows and their litters during high ambient temperatures under farm conditions, *Livest. Sci.* 105(1-3), 86–93.

Jermann, C., Koutchma, T., Margas, E., Leadley, C. and Ros-Polski, V. (2015). Mapping trends in novel and emerging food processing technologies around the world, *Innov. Food Sci. Emerg. Technol.* 31, 14–27.

Johkan, M., Shoji, K., Goto, F., Hashida, S. and Yoshihara, T. (2010). Blue light-emitting diode light irradiation of seedlings improves seedling quality and growth after transplanting in red leaf lettuce, *HortScience* 45(12), 1809–1814.

Jungfer, C., Schwartz, T. and Obst, U. (2007). UV-induced dark repair mechanisms in bacteria associated with drinking water, *Water Res.* 41(1), 188–196.

Keyser, M., Miller, I. A., Cilliers, F. P., Nel, W. and Gouws, P. A. (2008). Ultraviolet radiation as a non-thermal treatment for the inactivation of microorganisms in fruit juice, *Innov. Food Sci. Emerg. Technol.* 9(3), 348–354.

Kim, H. H., Goins, G. D., Wheeler, R. M. and Sager, J. C. (2004). Green-light supplementation for enhanced lettuce growth under red-light and blue-light emitting diodes, *HortScience* 39(7), 1617–1622.

Koutchma, T. (2009). Advances in ultraviolet technology for non-thermal processing of liquid foods, *Food Bioprocess Technol.* 2(2), 138–155.

Koutchma, T., Green, A. and Popovic, V. (2018). UV light-emitting diodes: the upcoming solution for enhanced food safety, *Food Safety*, August/September. Available at: https://www.food-safety.com/articles/5904-uv-light-emitting-diodes-the-upcoming-solution-for-enhanced-food-safety#:~:text=UV%20Light%2DEmitting%20Diodes%3A%20The%20Upcoming%20Solution%20for%20Enhanced%20Food%20Safety,-August%2013%2C%202018&text=Ultraviolet%20(UV)%20light%20disinfection%20is,air%2C%20water%2C%20and%20surfaces.&text=In%20the%20UV%2DC%20or,cysts%2C%20mold%2C%20and%20yeasts.

Krausmann, F., Erb, K., Gingrich, S., Lauk, C. and Haberl, H. (2008). Global patterns of socioeconomic biomass flows in the year 2000: a comprehensive assessment of supply, consumption, and constraints, *Ecol. Econ.* 65(3), 471–487.

Langton, A., Plackett, C. and Kitchener, H. (2006). Energy saving in poinsettia production, *Hortic. Dev. Counc. Fact Sheet* 7, 1–12.

Lara, L. J. and Rostagno, M. H. (2013). Impact of heat stress on poultry production, *Animals* 3(2), 356–369.

Lee, K., Elliott, C. and Pattison, M. (2020). Energy savings potential of SSL in agricultural applications, *United States Department of Energy, Office of Energy Efficiency & Renewable Energy*. June. Available at: https://www.energy.gov/sites/prod/files/2020/07/f76/ssl-agriculture-jun2020.pdf. (Accessed: 26 October 2020).

Liao, A. (2014). LEDs: a deep dive in dimming, *Archit. Lighting*. Available at: https://www.archlighting.com/technology/leds-a-deep-dive-in-dimming_o. (Accessed: 20 August 2020).

Lien, R. J., Hess, J. B., McKee, S. R. and Bilgili, S. F. (2008). Effect of light intensity on live performance and processing characteristics of broilers, *Poult. Sci.* 87(5), 853–857.

Magwaza, S. T., Magwaza, L. S., Odindo, A. O. and Mditshwa, A. (2020). Hydroponic technology as decentrialsed system for domestic wastewater treatment and vegetable production in urban agriculture: a review, *Sci. Total Environ.* 698, 134154.

Manser, C. (1996). Effects of lighting on the welfare of domestic poultry: a review, *Anim. Welf.* 5, 341–360.

Massa, G. D., Kim, H. H., Wheeler, R. M. and Mitchell, C. A. (2008). Plant productivity in response to LED lighting, *Hortscience* 43(7), 1951–1956.

Mazari-Hiriart, M., Ponce-de-León, S., López-Vidal, Y., Islas-Macías, P., Amieva-Fernández, R. I. and Quiñones-Falconi, F. (2008). Microbiological implications of periurban agriculture and water reuse in Mexico City, *PLoS ONE* 3(5), e2305.

Mekonnen, M. M. and Hoekstra, A. Y. (2010). *The Green, Blue, and Grey Water Footprint of Farm Animals and Farm Products*, Delft, the Netherlands: Unesco-IHE.

Mendes, A. S., Paixao, S. J., Restelatto, R., Morello, G. M., Jorge de Moura, D. and Possenti, J. C. (2013). Performance and preference of broiler chickens exposed to different lighting sources, *J. Appl. Poult. Res.* 22(1), 62–70.

Mikovits, C., Zollitsch, W., Hörtenhuber, S. J., Baumgartner, J., Niebuhr, K., Piringer, M., Anders, I., Andre, K., Hennig-Pauka, I., Schönhart, M. and Schauberger, G. (2019). Impacts of global warming on confined livestock systems for growing-fattening pigs: simulation of heat stress for 1981 to 2017 in Central Europe, *Int. J. Biometeorol.* 63(2), 221–230.

Ministry of Agriculture, Nature and Food Quality. (2020). Agriculture and horticulture, *Government of the Netherlands*. Available at: https://www.government.nl/topics/agriculture/agriculture-and-horticulture. (Accessed: 2 December 2020).

Mitchell, C. A., Both, A., Bourget, C. M. and Burr, J. (2012). LEDs: the future of greenhouse lighting, *Chron. Hortic.* 55, 6–12.

Morrow, R. C. (2008). LED lighting in horticulture, *HortScience* 43(7), 1947–1950.

Nanya, K., Ishigami, Y., Hikosaka, S. and Goto, E. (2012). Effects of blue and red light on stem elongation and flowering of tomato seedlings, *Acta Hortic.* 956, 261–266.

Nardone, A., Ronchi, B., Lacetera, N., Ranieri, M. S. and Bernabucci, U. (2010). Effects of climate changes on animal production and sustainability of livestock systems, *Livest. Sci.* 130(1–3), 57–69.

NASA (National Aeronautics and Space Administration) (2019). The electromagnetic spectrum. Available at: https://imagine.gsfc.nasa.gov/science/toolbox/emspectrum1.html. (Accessed: 9 September 2020).

Olaimat, A. N. and Holley, R. A. (2012). Factors influencing the microbial safety of fresh produce: a review, *Food Microbiol.* 32(1), 1-19.

Olesen, J. E. and Bindi, M. (2002). Consequences of climate change for European agricultural productivity, land use and policy, *Eur. J. Agron.* 16(4), 239-262.

Olle, M. and Viršilė, A. (2013). The effects of light-emitting diode lighting on greenhouse plant growth and quality, *Agric. Food Sci.* 22(2), 223-234.

Pang, Z., Li, B., Xin, H., Yuan, X. and Wang, C. (2010). Characterisation of an experimental water-cooled cover for sows, *Biosyst. Eng.* 105(4), 439-447.

Park, S., Szonyi, B., Gautam, R., Nightingale, K., Anciso, J. and Ivanek, R. (2012). Risk factors for microbial contamination in fruits and vegetables at the preharvest level: a systematic review, *J. Food Prot.* 75(11), 2055-2081.

Peden, D., Tadesse, G., Misra, A. K., Awad Amed, F., Astatke, A., Ayalneh, W., Herrero, M., Kiwuwa, G., Kumsa, T., Mati, B., Mpairwe, D., Wassenaar, T. and Yimegnuhal, A. (2007). Water and livestock for human development. In: Molden, D. (Ed), *Water for Food, Water for Life: A Comprehensive Assessment of Water Management in Agriculture*, Oxford: Oxford University Press, pp 485-514. Available at: http://www.iwmi.cgiar.org/assessment/Water%20for%20Food%20Water%20for%20Life/Chapters/Chapter%2013%20Livestock.pdf. (Accessed: 15 September 2020).

Pereira, R. V., Bicalho, M. L., Machado, V. S., Lima, S., Teixeira, A. G., Warnick, L. D. and Bicalho, R. C. (2014). Evaluation of the effects of ultraviolet light on bacterial contaminants inoculated into whole milk and colostrum, and on colostrum immunoglobulin G, *J. Dairy Sci.* 97(5), 2866-2875.

Person, H. L., Jacobson, L. D. and Jordan, K. A. (1979). Effect of dirt, louvers and other Attachments on fan performance, *Trans. ASAE* 22(3), 612-616.

Petropoulos, S. A., Fernandes, Â., Katsoulas, N., Barros, L. and Ferreira, I. C. (2019). The effect of covering material on the yield, quality and chemical composition of greenhouse-grown tomato fruit, *J. Sci. Food Agric.* 99(6), 3057-3068.

Prescott, N. B., Wathes, C. M. and Jarvis, J. R. (2003). Light, vision and the welfare of poultry, *Anim. Welf.* 12, 269-288.

Pullerits, K., Ahlinder, J., Holmer, L., Salomonsson, E., Öhrman, C., Jacobsson, K., Dryselius, R., Forsman, M., Paul, C. J. and Rådström, P. (2020). Impact of UV irradiation at full scale on bacterial communities in drinking water, *npj Clean Water* 3(1). Available at: https://www.nature.com/articles/s41545-020-0057-7.pdf. (Accessed: 17 September 2020).

Ramos, B., Miller, F. A., Brandão, T. R. S., Teixeira, P. and Silva, C. L. M. (2013). Fresh fruits and vegetables–an overview on applied methodologies to improve its quality and safety, *Innov. Food Sci. Emerg. Technol.* 20, 1-15.

Rastogi, R. P., Richa, Kumar, A., Tyagi, M. B. and Sinha, R. P. (2010). Molecular mechanisms of ultraviolet radiation-induced DNA damage and repair, *J. Nucelic Acids* 592980.

Ravanat, J. L., Douki, T. and Cadet, J. (2001). Direct and indirect effects of UV radiation on DNA and its components, *J. Photochem. Photobiol. B* 63(1-3), 88-102.

Renaudeau, D., Collin, A., Yahav, S., de Basilio, V., Gourdine, J. L. and Collier, R. J. (2012). Adaptation to hot climate and strategies to alleviate heat stress in livestock production, *Animal* 6(5), 707-728.

Rötter, R. and van de Geijn, S. C. (1999). Climate change effects on plant growth, crop yield and livestock, *Clim. Change* 43(4), 651-681.

Sahdev, R. K., Kumar, M. and Dhingra, A. K. (2019). A comprehensive review of greenhouse shapes and its applications, *Front. Energy* 13(3), 427-438.

Sands, R., Westcott, P., Price, J. M., Beckman, J., Leibtag, E., Lucier, G., McBride, W., McGranahan, D., Morehart, M., Roeger, E., Schaible, G. and Wojan, T. R. (2011). Impacts of higher energy prices on agriculture and rural economies, *ERR-123, U. S. Department of Agriculture, Economic Research Service*, August. Available at: https://www.ers.usda.gov/webdocs/publications/44894/6814_err123_1_.pdf?v=41432. (Accessed: 1 October 2020).

Scallan, E., Hoekstra, R. M., Angulo, F. J., Tauxe, R. V., Widdowson, M. A., Roy, S. L., Jones, J. L. and Griffin, P. M. (2011). Foodborne illness acquired in the United States: major pathogens, *Emerg. Infect. Dis.* 17(1), 7–15.

Schauberger, G., Mikovits, C., Zollitsch, W., Hörtenhuber, S. J., Baumgartner, J., Niebuhr, K., Piringer, M., Knauder, W., Anders, I., Andre, K., Hennig-Pauka, I. and Schönhart, M. (2019). Global warming impact on confined livestock in buildings: efficacy of adaptation measures to reduce heat stress for growing-fattening pigs, *Clim. Change* 156(4), 567–587.

Shamshiri, R. R., Jones, J. W., Thorp, K. R., Ahmad, D., Man, H. C. and Taheri, S. (2018a). Review of optimum temperature, humidity, and vapour pressure deficiet for microclimate evaluation and control in greenhouse cultivation of tomato: a review, *Int. Agrophys.* 32(2), 287–302.

Shamshiri, R. R., Kalantari, F., C. Ting, K., R. Thorp, K., A. Hameed, I., Weltzien, C., Ahmad, D. and Mojgan Shad, Z. (2018b). Advances in greenhouse automation and controlled environment agriculture: a transition to plant factories and urban agriculture, *Int. J. Agric. Biol. Eng.* 11(1), 1–22.

Shin, J. Y., Kim, S. J., Kim, D. K. and Kang, D. H. (2016). Fundamental characteristics of deep-UV light emitting diodes and their application to control foodborne pathogens, *Appl. Environ. Microbiol.* 82(1), 2–10.

Silva, B. A. N., Oliveira, R. F. M., Donzele, J. L., Fernandes, H. C., Lima, A. L., Renaudeau, D. and Noblet, J. (2009). Effect of floor cooling and dietary amino acids content on performance and behaviour of lactating primiparous sows during summer, *Livest. Sci.* 120(1–2), 25–34.

Simmons, J. D., Hannigan, T. E. and Lott, B. D. (1997). Reduction in poultry ventilation fan output due to shutters, *Appl. Eng. Agric.* 13(5), 671–673.

Singh, D., Basu, C., Meinhardt-Wollweber, M. and Roth, B. (2015). LEDs for energy efficient greenhouse lighting, *Renew. Sustain. Energy Rev.* 49, 139–147.

Sinha, R. P. and Häder, D. P. (2002). UV-induced DNA damage and repair: a review, *Photochem. Photobiol. Sci.* 1(4), 225–236.

Sondag, M. (2020). 'The Netherlands, a significant partner in Canada's greenhouse industry', *Hortdaily*, Feb. 7. Available at: https://www.hortidaily.com/article/9186769/the-netherlands-a-significant-partner-in-canada-s-greenhouse-industry/. (Accessed: 2 December 2020).

Stanford University (n.d.). UV light, *Stanford Solar Center*. Available at : http://solar-center.stanford.edu/about/uvlight.html. (Accessed: 8 September 2020).

Steinfeld, H., Gerber, P., Wassenaar, T., Castel, V., Rosales, M. and de Haan, C. (2006). *Livestock's Long Shadow: Environmental Issues and Options*, Rome: Food and Agriculture Organization of the United Nations (FAO).

Stober, K., Lee, K., Yamada, M.and Pattison, M. (2017). Energy savings potential of SSL in horticultural applications, *U. S. Department of Energy, Office of Energy Efficiency & Renewable Energy*, December. Available at: https://www.energy.gov/sites/prod/files/2017/12/f46/ssl_horticulture_dec2017.pdf. (Accessed: 20 August 2020).

St-Pierre, N. R., Cobanov, B. and Schnitkey, G. (2003). Economic losses from heat stress by US livestock industries, *J. Dairy Sci.* 86, E52–E77.

Tabler, G. T., Watkins, S. E. and Watkins, P. A. (2008). Energy efficiency associated with poultry house lighting, *Avian Advice* 10(3), 1–3.

Tabler, T., Farnell, M. and Wells, J. (2015). LED Bulbs: much to offer the poultry industry, in: *Mississippi State University Extension Service,* Publ No. 2894(July).

Tabler, T., Khaitsa, M. L. and Wells, J. (2019). Smallholder farmer constraints to village chicken production, in: *Mississippi State University Extension Service*, Publ No. 3315, January.

Tabler, T., Khaitsa, M. L., Mbaga, S. H., Jeckoniah, J. N., Moon, J. and Wells, J. (2020a). Gender issues in ag Extension programming in sub-Saharan Africa, in: *Mississippi State University Extension Service,* Publ No. 3501, August.

Tabler, T., Khaitsa, M. L., Mbaga, S. H., Jeckoniah, J. N., Moon, J. and Wells, J. (2020b). Poultry Extension personnel needed across East Africa, in: *Mississippi State University Extension Service,* Publ No. 3494, July.

Tennessen, D. J., Singsaas, E. L. and Sharkey, T. D. (1994). Light-emitting diodes as a light source for photosynthesis research, *Photosynth. Res.* 39(1), 85–92.

Thornton, P. K. and Herrero, M. (2010). Potential for reduced methane and carbon dioxide emissions from livestock and pasture management in the tropics, *Proc. Natl. Acad. Sci. U.S.A.* 107(46), 19667–19672.

Thornton, P. K., van de Steeg, J., Notenbaert, A. and Herrero, M. (2009). The impacts of climate change on livestock and livestock systems in developing countries: a review of what we know and what we need to know, *Agric. Syst.* 101(3), 113–127.

Trivittayasil, V., Nashiro, K., Tanaka, F., Hamanaka, D. and Uchino, T. (2015). Inactivation characteristics and modeling of mold spores by UV-C radiation based on irradiation dose, *Food Sci. Technol. Res.* 21(3), 365–370.

United Nations (U.N.). (2020). United Nations: population. Available at: https://www.un.org/en/sections/issues-depth/population/index.html#:~:text=Africa%3A%20fastest%20growing%20continent,projected%20to%20double%20by%202050. (Accessed: 16 September 2020).

Urban, L., Charles, F., Alcântara de Miranda, M. R. A. and Aarrouf, J. (2016). Understanding the physiological effects of UV-C light and exploiting its agronomic potential before and after harvest, *Plant Physiol. Biochem.* 105, 1–11.

U.S. Department of Energy Solid State Lighting Program (2016), Energy savings forecast of solid-state lighting in general illumination applications, *U.S. Department of Energy Prepared by Nagivant Consulting*, August. Available at: https://energy.gov/sites/prod/files/2016/10/f33/energysavingsforecast16_0.pdf. (Accessed: 20 August 2020).

USEPA (2006). *Ultraviolet Disinfection Guidance Manual for the Final Long Term 2 Enhanced Surface Water Treatment Rule*, United States Environmental Protection Agency, 815-R-06-007.

Vänninen, I., Pinto, D. M., Nissinen, A. I., Johansen, N. S. and Shipp, L. (2010). In the light of new greenhouse technologies: plant-mediates effects of artificial lighting on arthropods and tritrophic interactions, *Ann. Appl. Biol.* 157(3), 393–414.

Wargent, J. J. and Jordan, B. R. (2013). From ozone depletion to agriculture: understanding the role of UV radiation in sustainable crop production, *New Phytol.* 197(4), 1058–1076.

Watkins, S., Sullivan, S., Tabler, T., McCreery, D. and Clark, T. (2012). LED lighting evaluation with commercial poultry production, *University of Arkansas Cooperative Extension*

Service. Available at: https://www.uaex.edu/farm-ranch/animals-forages/docs/Evaluation%20of%20LED%20Lighting.pdf. (Accessed: 21 August 2020).

Weindl, I., Lotze-Campen, H., Popp, A., Müller, C., Havlík, P., Herrero, M., Schmitz, C. and Rolinski, S. (2015). Livestock in a changing climate: production system transitions as an adaptation strategy for agriculture, *Environ. Res. Lett.* 10(9), 094021.

Yeh, N. and Chung, J. P. (2009). High-brightness LEDs: energy efficient lighting sources and their potential in indoor plant cultivation, *Renew. Sustain. Energy Rev.* 13(8), 2175–2180.

Yousef, A. H. E. and Marth, E. L. H. (1988). Inactivation of *Listeria monocytogenes* by ultraviolet energy, *J. Food Sci.* 53(2), 571–573.

Zhang, G. and Bjerg, B. S. (2017). Developments of thermal environment techniques of animal housing in hot climate: a review, in: Animal Environment and Welfare: Proceedings of the International Symposium, Beijing: China Agriculture Press, pp 375–384.

Zhang, Z., Gates, R. S., Zhirong, Z. and Xiaohui, H. (2015). Evaluation of ventilation performance and energy efficiency of greenhouse fans, *Int. J. Agric. Biol. Eng.* 8(1), 103–110.

Chapter 3

Improving farm machinery operation and maintenance to optimise fuel use efficiency

T. A. Jensen and J. N. Tullberg, University of Southern Queensland, Australia; and D. L. Antille, University of Southern Queensland and CSIRO Agriculture and Food, Australia

1 Introduction

It has been estimated that 15% of agricultural production costs on-farm are energy related (Sims et al., 2015). This chapter reviews recent research on ways of reducing these costs as well as the environmental impacts of on-farm energy use. Using existing energy sources more efficiently and greater adoption of renewable forms of energy at the farm scale can both save costs and reduce greenhouse gas (GHG) emissions. Efforts to measure and reduce energy use can be traced back to the two 'oil shocks' of the 1970s and early 1980s which saw significant increases in prices as well as supply shortages in some regions, resulting partly from actions by members of the Organisation of Petroleum Exporting Countries (OPEC) (Kesicki, 2010). In more recent times, concerns about GHG emissions and their contribution to global warming and climate change, as exemplified by the Paris Climate Agreement (2015) and the Glasgow Climate Pact (2021), are causing many national and local

http://dx.doi.org/10.19103/AS.2022.0100.03

governments, as well as agricultural organisations, to examine the means of reducing such emissions. Global energy demand is the principal contributor to the anthropogenic release of GHGs, in particular CO_2, to the atmosphere. Global warming and climate change caused by GHG emissions are strongly linked to fossil energy production and utilization (Höök and Tang, 2013; IPCC, 2021). The global food supply chain is responsible for around one-third of end-use energy consumption and a quarter of total GHG emissions (FAO, 2011).

This chapter discusses energy use and emissions from direct fossil fuel combustion on farms to power farm machinery. It briefly covers indirect energy use (e.g. use of fossil fuels in manufacturing inputs such as animal feed or fertiliser), but does not discuss other on-farm energy uses (such as for heating or ventilation of buildings) or other sources of GHG emissions such as biogenic methane from ruminant livestock or nitrous oxide emissions arising from animal manure and nitrogenous fertiliser use. Most farms depend on diesel fuel and, to a lesser extent, on gasoline to run tractors, farm vehicles, harvesters, and other farm equipment. Such fossil fuels are the main energy costs for arable farm operations, in particular, representing up to 75% of the total energy spent on modern industrial-scale farms, depending on the cropping system (Warwick HRI, 2007; Woods et al., 2010)[1]. Cutting fuel costs can be relatively simple, starting with operating machinery in a smarter manner, e.g., by correctly setting up and maintaining vehicles and equipment, and developing a fuel management plan. As an example, US research suggests that the basic practice of maintaining clean fuel and air filters can save around 100 gallons of diesel fuel annually (Hoy et al., 2014). Buying more fuel-efficient vehicles can also result in significant fuel savings.

2 Techniques for optimising tractor fuel consumption

2.1 Adaptive driving: gear-up and throttle back (GUTB)

GUTB is a fuel-saving practice that has been popular around the world since the fuel crisis of the mid-1970s (Hoy et al., 2014). This technique is implemented to save fuel when drawbar loads are light (<75% of rated power) and power take-off (PTO) speed can be reduced. According to Grogan et al. (1987), this method improves engine efficiency by maintaining high engine load and an engine speed at 60–80% of rated speed. Pang et al. (1985) found significant improvements in fuel and tractive efficiency from this method with two- and four-wheel drive tractors in Canada.

For maximum operating efficiency, a tractor engine should be operated near its rated capacity. Many field operations such as light tillage, planting,

1 Small-scale and subsistence farms in developing countries use more animal and human labour, including for transport of products to markets, and hence can be less dependent on fossil fuel inputs.

cultivating, spraying and hay-raking, however, do not require full tractor power. This is particularly true when older implements that were sized for smaller tractors are used with today's high-horsepower tractors or existing tractors are used in controlled traffic farming (CTF)/zero-tillage operations where power requirements are much lower. Many operations require fixed ground speeds. For these lighter operations, substantial fuel savings are possible by operating the tractor on a higher gear and lower engine speed whilst maintaining the desired field speed. The GUTB method will not suit all tractor operations. Certain PTO implements require high engine speeds for the correct operation of the equipment and, as such, their use is incompatible with the GUTB method. When evaluating options for a new tractor purchase, it is recommended to clarify how energy efficiency can be optimised for PTO applications and how the PTO affects general engine efficiency.

The most basic efficiency measure is to avoid overloading the engine by using a gear that is too high for the task. Adaptive driving depends on the operator receiving accurate feedback on both engine performance and driver performance. Depending on the age and capabilities of the tractor, the key sources of data are:

- in-cab, real-time monitoring of engine performance;
- listening to and feeling the engine; and
- learning the specific signals that indicate the limits of engine speed reduction possible for a given gearing and load. (Engine sound can be misleading, so black smoke coming from the exhaust is also a good indicator, as is lack of engine speed response to an increase in throttle setting).

2.2 Optimising tyre inflation pressure

Tyres should be always inflated to the correct pressure given load, speed, and surface conditions (ground carrying capacity and slope). When carrying a load over a firm or hard surface the tyre deflects, causing an increase of the contact patch area, until the inflation pressure acting on the contact patch area can support the load. A loaded tyre is deflected near the contact patch and is being continuously flexed as it rolls along. For most agricultural tyres, the maximum deflection is limited to about 18–20% of the section height, in order to prevent the damage of casing (Inns and Kilgour, 1978). With an increasing load on the tyre, the inflation pressure must be raised to maintain acceptable deflections, avoid damage to the tyre and also for safety reasons.

In soft (off-road) conditions, the soil deforms to allow for increased contact patch area, thus reducing the deflection of the tyre for a given load and inflation pressure (Bygden et al., 2003; Antille et al., 2013). Hence, in softer soil conditions, the tyre inflation pressure may be reduced compared with the recommended pressure for the same load on a firmer surface. This may help to reduce rolling resistance (that is, the pull required to move a wheel across a horizontal surface) and improve traction under such ground conditions, but the recommended tyre inflation pressure must be restored for firmer surface use (Hunt, 1986).

Rolling resistance increases almost proportionally to an increase in load which the tyre is supporting (Inns and Kilgour, 1978). The rolling resistance has two main components (Komandi, 1999):

- the internal rolling resistance (which is caused by the loss of energy resulting from the continuous flexing of the tyre's carcass as the wheel rotates in contact with the ground); and
- the external rolling resistance (which arises from the energy that the wheel has to expend in deforming the soil surface).

In off-road conditions, the rolling resistance caused by soil deformation is much greater (about 5 times or more) than the internal resistance, depending upon the tyre construction. A rut formed by the pass of the wheel over the soil is evidence of the expenditure of energy in deforming that soil, and in general, the larger the cross-section of the tyre rut, the greater the rolling resistance (Raper et al., 1995). Rolling resistance, slip and compaction may be minimised (and thrust maximised) by using power by travelling at higher speeds, making use of the cohesive component of shear force in heavier soils. This is achieved by maximising contact area and reducing weight, but if weight cannot be reduced, by increasing contact area to minimise contact pressure. Increasing contact pressure can be achieved by increasing tyre diameter, section width or both (although the former is preferable), operating the tyre at the minimum allowable (safe) inflation pressure or fitting dual tyres at low inflation pressure, and where possible, using repeated wheeling (Wong, 2009). An increase in contact area through increased tyre diameter is preferable to section width because it will minimise rut width and reduce rolling resistance (McAllister, 1983; Crossley et al., 2001; Kurjenluoma et al., 2009). Increased contact area will also reduce slip and therefore rolling resistance (Inns and Kilgour, 1978; Komandi, 1999). The correct inflation is determined by the tyre size, tyre construction type, load to be carried and speed, and it can be obtained from the tyre manufacturer's website or the tractor's operating manual.

2.3 Wheel slip and tractor ballasting

Most tractor operators know proper ballasting is important to transfer as much engine power as possible to the drawbar. Exactly how to accomplish this ballasting, however, frequently remains a mystery. Too little weight or ballast results in excessive slippage of the drive wheels and hence an obvious waste of fuel. Conversely, carrying too much ballast on a tractor dramatically lowers wheel slip but results in greater rolling resistance as the tractor sinks too far into the soil, causing wheels to be constantly climbing out of a deep rut (Hanna et al., 2010).

Many larger modern tractors have an option to display wheel slippage on the instrument panel for the operator. To maximize the transfer of power from drive axles to the drawbar, optimum amounts of wheel slippage depend on the soil surface. On firm, untilled soil, wheel slip should be in a range of about 6–13%. Higher slippage of 8–16%, is acceptable on a tilled surface with slightly more on a non-cohesive sandy soil. Conversely, on concrete or tar seal surfaces, optimal wheel slip is about 4–8%.

Checking wheel slippage on tractors equipped to display this information makes it easy to check to determine if the tractor is optimally applying power to the drawbar and therefore using fuel efficiently. If tractor wheel slip is outside suggested ranges for operation with drawbar loads, check the operator's manual for ballasting suggestions. Table 1 also provides guidance on gross tractor weights. The total gross tractor weight required for optimal ballasting is a function of tractor type (two-wheel drive (2WD), mechanical front-wheel assist (FWA) drive, four-wheel drive (4WD)), and travel speed in the field.

Since only wheels on powered axles supply traction, it is important to distribute ballast properly between the front and rear axles. Optimal weight on each axle is affected by tractor style and whether the attached implement is pulled or mounted (Table 2). Equipment such as manure tank wagons and grain carts have significant tongue weight on the drawbar and can be considered 'fully mounted' loads when calculating the proper weight split between front

Table 1 Gross tractor weight, kg/kW (after Hanna et al., 2010)

Tractor type	Gross tractor weight, kg kW^{-1}		
Speed, km h^{-1}	<7.0	8	>9.0
2WD or FWA	80	73	67
4WD	67	61	55

Table 2 Front-to-rear axle weight ratios as a percentage of total weight (% front/% rear) (after Hanna et.al., 2010)

Tractor type	Towed/drawbar	Semi-mounted	Fully mounted
2WD	25/75	30/70	35/65
FWA	35/65	35/65	40/60
4WD	55/45	55/45	60/40

and rear axles because they add weight to the tractor's rear axle similar to fully mounted implements.

An important exception to this ballasting procedure occurs when lighter drawbar loads are used that require less than half of the available tractor power. Examples include a pull-behind sprayer, a small planter, or a field cultivator that does not require much power. Ballast previously added for primary tillage or heavy drawbar loads simply adds to tractor weight, increases rolling resistance, and can increase fuel use. Although adding and removing cast iron ballast as wheel weights or front weights can be daunting, it is necessary for optimum performance. Ballast needs to be removed for lighter drawbar work, such as planting and spraying and added to the tractor prior to heavy tillage operations which require more of the tractor's engine power to be transferred to the drawbar.

Optimum total tractor weights and percentage weight-splits between the front and rear axles are shown in Tables 1 and 2 as a guide. Because tractor fuel and power efficiency are optimized over a range of wheel slippages, fuel use is not likely to increase substantially with a 5% deviation from these values. However, increased fuel use may become evident if weights differ by 10% or more. Incorrect ballasting can have a substantial negative effect on fuel efficiency and on tractor performance overall. When the ballast is set around the right general range, little fuel-efficiency benefit can be achieved from constantly tweaking ballast to compensate for different soil conditions or to allow for variations in the weight of mounted equipment (Zoz and Wiley, 1995). If wheel slip cannot be easily checked and tractor axle weights are not known, they should be measured to gain confidence that fuel is not being wasted. Total tractor weight as well as the weight being carried on each axle (Table 2) can be determined by a weigh bridge on the farm, use of a public weigh bridge, or using portable scales under each wheel.

There are two ways to add ballast to a tractor: attaching cast-iron weights or filling the tyres with fluid. Cast weights can be more expensive but make it easier to quickly adjust ballast to suit changing conditions. Adding fluid to the front tyres on a front-wheel drive (FWD) tractor is an efficient method for adding ballast to the front axle when needed and controlling power hop without exceeding the optimum inflation pressure. When fluid is added to tyres on the rear axle, exceeding 40% fill should be avoided. All tyres on the same axle should be filled to the same level. The added weight from adding dual wheels to gain traction under soft soil conditions should be considered as ballast on the rear axle. Dual tyres are typically used when a second wheel is required to support axle weight or to improve floatation or stability. Dual tyres can also decrease rolling resistance or improve traction if soil conditions are wet or marginal. Unpowered front wheels on a 2WD tractor are necessary for steering control but do not help tractive propulsion, although some 2WD models offer optional FWD so that they help pull the load rather than simply creating rolling resistance when being passively pushed through the soil. To create traction, the peripheral speed of lugs on the front wheels are slightly faster than those on rear tyres. To prevent extra wear on the drive transmission, manufacturers sometimes recommend disengaging FWD during road travel when added traction may not be beneficial.

2.4 Estimating tractor power needs

Fuel consumption for most tillage operations is directly related to tillage depth. The goal of the tillage operation should be to set the correct depth rather than simply pulling the implement as deeply as the tractor's power will allow. Travel speed affects the time required to do the job and impacts productivity. In most cases, farmers choose to accomplish work as quickly as possible, so reducing field speed is not an appealing option. Fortunately, although a tractor's speed of operation impacts on energy use, in some cases fuel consumption may only be marginally impacted, such as when reduced engine speed and a higher gear is used for faster travel speed. Fuel consumption may occasionally decrease with faster tillage speed if small changes in drawbar load are balanced by operating the tractor engine at a more fuel-efficient combination of greater torque and lower engine speed.

2.5 Purchasing a fuel-efficient tractor

When upgrading to a new tractor, the manufacturers' claims of fuel consumption should be compared with the aim to minimise overall fuel consumption over the life of the tractor. However, the real measure of fuel efficiency is litres per hectare, not litres per hour. Fuel consumption is a combination of engine and tractor design and operator performance. Using less fuel per hour but taking more time to complete a task may mean that optimum fuel savings are not being achieved.

3 Developments in fuel-efficient tractor design

3.1 Power/load matching

The major tractor manufacturers (such as John Deere, Kubota, Massey Ferguson etc.) have all adopted power/load matching methods that are indicated by a range of acronyms (e.g. IVT CVT) but offer the same functionality. For example, Howard et al. (2013) compared a John Deere 8295R infinitely variable transmission (IVT) tractor with a John Deere 8295R PowerShift (PST) with continuously variable transmission (CVT). When shifting up two gears and throttling back, they found fuel consumption was linearly related to drawbar power. This relationship alternated between the PST being more economical at lower loads (<52% maximum load) with the CVT being more efficient above that level.

The John Deere IVT provides a seamless range of speeds and no clutching is required to start or stop the tractor. The IVT has a fully integrated electronic management system (EMS) that allows the engine to communicate 100 times a second with the transmission. Electronic communication and interaction between the engine and the transmission enable optimum productivity and efficiency at any engine speed. This is achieved because the operating conditions are being constantly monitored and relayed to the EMS which then automatically determines whether the tractor is in a near fully loaded condition or a light- to no-load condition. It then makes the appropriate adjustments accordingly.

The EMS works with the IVT transmission and engine to maintain the selected travel speed at reduced engine speed (rpm) when the IVT selector activates the system. This results in increased efficiency and reduced fuel consumption. There are generally several 'mode settings' as outlined below:

- Off-operator selects the engine speed and ground speed.
- Primarily for PTO-powered applications, the engine speed can only be reduced minimally in order to maintain the correct standard PTO operating speed (of either 540 rpm or 1000 rpm). When a load from an implement is placed on the tractor, the engine will first increase power to maintain PTO rpm. Once maximum engine power is reached and additional load is placed on the tractor, the IVT transmission will reduce forward speed enough to keep the tractor's engine at around maximum power output level until the tougher conditions have passed. Then the transmission will increase the forward speed back up to the required speed as selected. Once the load is reduced further, engine speed will again drop automatically to reduce fuel consumption. All of this is done through the EMS ensuring communication between the engine and the transmission.
- For heavy draught and tillage applications, or when hydraulic oil flow is needed, the tractor will try to maintain the selected ground speed at reduced engine speed for improved efficiency and reduced fuel consumption. The minimum engine speed is adjusted by the CommandCenter™ to customize the tractor to the specific application. Since the PTO is no longer involved, the speed is allowed to go lower than when powering an implement through the PTO. When the load increases on the tractor, the engine increases power until maximum power is reached, then, if needed, the IVT transmission will ratio down (to slow the tractor forward speed) enough to keep engine power at maximum. Once the tractor has passed

through this tougher area of the paddock, the IVT will ratio up (and hence increase forward speed) back to the selected speed and the engine will reduce rpm speed. All of these operations are performed automatically by the EMS.

- For transport and light tillage applications, the engine rpm speed will reduce compared to either the PTO or heavy draught operations to save additional fuel. A tractor is typically capable of a full transport road speed of around 40 km/h at very low engine speed (sometimes as low as 1200 rpm). As with the John Deere 8295R example, the minimum engine rpm speed is adjustable by the CommandCenter which allows the operator to customize the tractor performance to suit the application.

In this example, the EMS operates similarly to the traditional 'gear up and throttle back' operation that can be performed manually by good operators. However, because the EMS continuously communicates with the engine and transmission, making appropriate adjustments based on inputs received from the tractor, it responds quickly and precisely to changing conditions. This allows maximum efficiency and productivity to be obtained from the tractor. Given that the changes are automatic, the system reduces the reliance on operator experience to make the appropriate manual adjustments at the correct time.

3.2 Fuel injection and exhaust after-treatment systems

Traditional diesel engines used in tractors have a mechanical injection system. Recent developments such as high-pressure and electronically controlled fuel systems are being increasingly installed by manufacturers. High-pressure injection systems operate at a pump pressure of around 1500 bar compared with 240 bar for traditional designs. The injection pump transfers the fuel to a high-pressure line known as the "common rail" where each fuel injection unit is opened by pressure (Schlosser et al., 2020). The use of high-pressure injection systems allows changes in the injection time, the opening pressure of the injection unit and the fuel injection point (Horrocks, 2010). This improves use efficiency and emission control (Basavarajappa et al., 2015). In addition, it enables the use of alternative biofuels, with engine performance being maintained using different fuel compositions and blends.

Electronically controlled injection systems have enabled the development of injection strategies based on engine sensors. Electronically controlled sensors regulate the angular speed of the crankshaft and the position of the accelerator. These sensors transmit information on these parameters to the electronic control unit (ECU) which varies the amount of fuel injected to meet the load. In addition, sensors at air and fuel inlets monitor possible obstructions and engine temperature. This enables current emission standards, such as the US

Environmental Protection Agency (EPA) Tier 2/3/4/5 standards (North America) and EURO standards (Europe), to be met. Through advances in design, modern diesel engines have been transformed into one of the cleanest prime moving systems available today. These engine improvements include high-pressure common rail fuel injection systems, electronically controlled injector solenoids and advanced turbocharging. Manufacturers can meet emissions standards by adjusting the electronically controlled high-pressure injection system (Özkan, 2015; Jungler and Diotto, 2018).

While these improvements have enabled engine manufacturers to meet Tier 2 and Tier 3 emissions standards, in order to meet the more stringent Tier 4 and Tier 5 standards, exhaust after-treatment is also required. The main technologies used for exhaust after-treatment are as follows:

- Selective catalytic reduction (SCR). This works by combining the exhaust gases with ammonia (urea or diesel emissions fluid (DEF)) and passing this mixture over a catalyst. Roughly 1 L of DEF is required to treat emissions from 20 L of diesel fuel.
- Diesel particulate filter (DPF). This works by using a mechanical filter to trap soot particles after they have been partially oxidized by a catalyst. At certain intervals during operation, the trapped particles are incinerated.
- Exhaust gas recirculation (EGR). A small amount of cooled exhaust gas is recirculated back into the combustion chamber. This reduces the combustion temperature and effectively reduces the production of NOx. However, EGR increases particulate emissions, so a DPF will also be needed to meet Tier 4 final regulations.

Tier 4 emissions legislation is leading to widespread adoption of exhaust after-treatment in off-highway applications for engines with power greater than 18 kW (24 HP). Manufacturers of large engines are faced with the significant challenge of packaging a multitude of catalyst technologies into existing designs, contending with the fuel consumption consequences of the increased back pressure/higher operating temperature/greater cooling requirement, as well as dealing with the incremental cost and weight associated with after-treatment equipment (Subramaniam et al., 2011).

4 Measuring tractor performance

Interest in data to monitor the in-field performance of an agricultural tractor is not a new topic. In the late 1990s transducers were mounted on tractors to measure operational parameters such as engine performance, wheel and ground speeds with additional devices used to measure fuel consumption (Yule et al., 1999; Baillie et al., 2018). Modern tractors now have a performance-monitoring system that provides real-time fuel consumption, engine and ground speed, and wheel-slip with telemetry enabling data to be transmitted and stored for future reference. Real-time data can be analysed to give immediate feedback to assess different machine settings and configurations as well as operator performance (Al-Aani et al., 2016). Recent advances in data processing and exchange (e.g., CAN standard ISO 11898) as well as developments such as Real-Time Kinematic (RTK) global positioning systems which allow for more precise tracking, also allows for comparisons and benchmarking of machines undertaking similar tasks.

An example of the use of such data is tractor tests completed as part of detailed on-farm energy assessments conducted by Foley et al. (2015). The tests were conducted on the black-cracking clay soils of a large, irrigated cotton farm in the Darling Downs region of Queensland in Australia, using a 4.5 m wide fixed-tyne ripper behind a 2010 John Deere 8220 tractor. The tractor's 8.1- L engine was electronically controlled and turbocharged to give a rated power of 225 kW (168 hp). For the trial, the tractor was put through a series of tillage runs at varying depths and across a range of engine speeds and gear selections to compare ground speed and fuel consumption. Hourly fuel consumption rates varied at two ripping depths as a result of changing engine speed and gear selection (Fig. 1). Reducing the engine speed of the tractor reduced fuel use by between 6 and 9 L/h across three selected pairs of equal ground speed and ripping depths (shown by the red circles and bars in Fig. 1) with an average reduction of 7 L/h. This confirmed that throttling back and gearing up reduces fuel consumption.

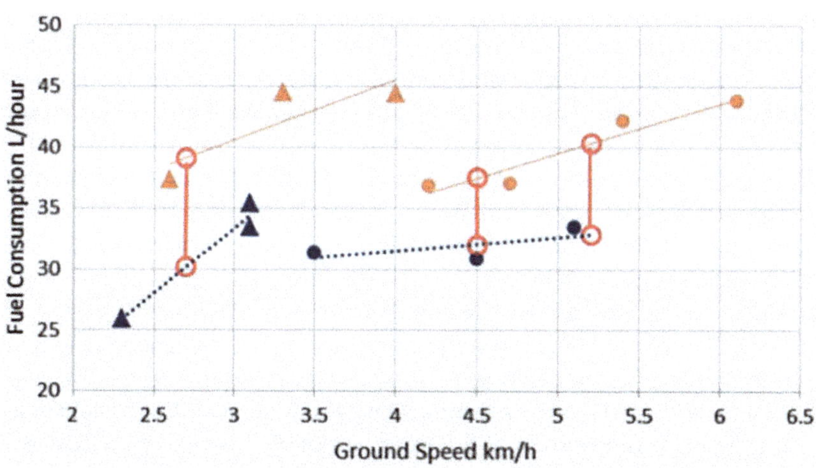

Figure 1 Fuel consumption for engine speeds of 2300 rpm (lighter coloured solid lines) and 1830 rpm (darker coloured dotted lines) at ripping depths of 350 mm (triangles) and 250 mm (solid circles) (after Foley et al., 2015).

For a heavy tillage operation, at a work rate of around 2 ha/h and a depth of 250 mm, shifting gears from 5th up to 7th, and throttling back the engine speed from 2200 to 1830 rpm, reduced fuel consumption by 6.7 L/h while maintaining the same ground speed and work rate. On a per hectare basis, this equated to a reduction in fuel use of 3.6 L/ha, or about 20%, which represented a reduction of 0.14 GJ of energy per hectare and 10 kg CO_2 per hectare emissions. Tests also showed that, for every 25 mm increase in the depth of heavy tillage, tractor fuel consumption increased by 2.0 L/ha. A 10% reduction in fuel costs and emissions per hectare was possible where tillage depth could be reduced by 25 mm (Foley et al., 2015).

5 Energy effects of farming systems: no-tillage and controlled traffic farming

Tillage has been the basis of most farming systems for many centuries. It is the traditional way to kill weeds, bury residues, relieve compaction and prepare a seedbed. It is also energy intensive, expensive, accelerates moisture loss, and leaves the soil vulnerable to erosion by wind or water. When effective herbicides appeared in the 1960s, no-till cropping research and development followed rapidly. On-farm adoption was much slower, as farmers and (often small) machinery makers worked to develop and adapt equipment and systems for no-till cropping. No-tillage is a generic term, which precludes any pre-seeding tillage, but includes systems using tine seeders (e.g. wide-tine or knife types), and minimal-disturbance disc seeders (the latter sometimes referred to as zero-tillage).

The economic and environmental costs of tillage ensured that its role was questioned in almost all cropping systems. In more humid environments (e.g., northern Europe) issues of managing high residue loads have slowed adoption, but there is ongoing farmer interest in no-till, and a widespread move towards reduced tillage systems (Stroud, 2020). The advantages of no-till were clearest in more arid and erosion-prone environments, such as Australia, where almost 85% of grain production is no-till (Umbers, 2017). Issues facing long-term no-till farming include the development of herbicide tolerance by some weeds, nutrient stratification and soil compaction in some environments. On-farm energy costs are clearly reduced by no-till (Pimental, 2009), but overall energy benefits are less obvious when the embodied energy of herbicides is included (Tullberg, 2014). Traffic-induced surface compaction is a challenge for effective seeding in many no-till systems, and regular deep tillage is used in some environments to ameliorate the compaction caused by uncontrolled field traffic. Deep tillage is energy-intensive, but most compaction issues can be avoided, and other benefits are achieved by the use of CTF, together with no-tillage.

Controlled traffic systems (sometimes referred to as zero-traffic farming systems) restrict all field traffic to narrow permanent lanes (sometimes termed 'tramlines'), leaving the rest of the field area unaffected by wheels or tracks. The concept can be traced as far back as the 1850s and was originally associated with gantry or wide-span implement carriers that were designed to reduce the soil problems associated with the heavy steam traction engines. Gantry systems allowed a single vehicle to cover a larger area in a single pass. Most of the detailed research occurred in the USA and the UK in the 1960s and 70s with one company, Dowler, commercialising a 12 m-wide unit. In practice, these early units were sold largely to research organisations, facilitating the accumulation of a significant body of literature on zero-traffic agriculture. A number of gantry units have subsequently been developed, and some have been used for specialised commercial applications such as vegetable harvesting. Chamen et al. (1992) investigated the effects of a partial 12 m-wide gantry system on energy consumption and concluded that it could potentially reduce fuel consumption by up to 44%. However, take-up has been limited, partly because of the difficulty of adapting them for harvesting grain, forage, or most root crops. Some standard tractors are designed to allow modification of their wheel track 'gauge' width to match that of most grain and other types of harvesters - usually ~ 3 m. The concept was first demonstrated at field scale in Texas in the USA (Morrison 1985), in Scotland in the UK (Douglas and Crawford, 1989) and in Queensland in Australia (Tullberg and Hilton, 1985). This system provided a harvester-compatible tractor track gauge by removing the inner pair of the dual tyres from 2WD row-crop tractors and extending the non-driven front steering axle width. Other minor modifications such as outrigger support bearings were sometimes used to avoid overloading rear axle bearings.

Anecdotal reports of improved soil condition and crop yields came from a number of Australian farmers (largely on the Darling Downs region of Queensland, Australia), who adopted 3-m controlled traffic systems in the 1980s. Similar systems were adopted in many countries, and on a large scale in the central highlands of Queensland as a result of an on-farm R&D program in the 1990s (Yule, 1995). At about the same time, work in the high rainfall zones of Victoria and Western Australia successfully addressed the issue of waterlogging using a similar approach by forming 1.5–2.0 m 'raised beds' that a tractor and harvester could then straddle.

Permanent traffic lanes could be achieved in these systems by using a seeder and harvester of the same width, with a sprayer and fertiliser spreader a multiple of that width. In grain production, this allowed permanent traffic lanes to occupy 10–15% of the field area. This system was adopted by many farmers, but the non-trafficked 'bed' area was easily compromised by poor steering accuracy. In the mid-1990s, the development of RTK GPS autosteer by a small Australian company, Beeline Technologies, completed what is now recognised as a basic CTF system. The essentials of the system, as defined by the Australian Controlled Traffic Farming Association (ACTFA), include (after Antille et al., 2015):

- all machinery having the same, or modular, working and track gauge widths, to allow the establishment of permanent traffic lanes;
- all machinery being capable of precise guidance along these permanent traffic lanes; and
- farm, paddock, and permanent traffic lane layouts arranged to optimise surface drainage and logistics.

In isolation, CTF provides significant advantages. It is also the basis for integrating a number of other practices and technologies such as:

- minimal soil disturbance, preferably zero-till or, at most, strip-tillage;
- more intensive cropping frequency and cover crops to maximise biomass production and provide a greater return of crop residues to the system;
- precise management such as inter-row seeding and accurate application of chemicals and fertiliser;
- spatial monitoring, mapping and management (e.g. yield mapping and subsequent zonal management, if required) at a progressively finer scale within a defined spatial framework, resulting in permanent traffic and crop zones; and
- accurate and repeatable on-farm research trials based on a defined spatial framework and spatial technologies.

Several major farm machinery manufacturing companies now provide equipment that is more readily compatible with CTF. Smaller companies have made a valuable contribution by providing properly engineered extensions for tractors with steerable, front-driven wheels. The scale of CTF operations has increased so that the largest commercial systems are now based on an 18 m module (that is, 18 m seeder and cereal harvester, 54 m sprayer, and an extendable 'catcher' on the grain chaser bin allowing it to stay on the neighbouring set of traffic lanes when the harvester is unloading into the bin). The broad productive and environmental benefits of CTF have been set out by Tullberg et al. (2007) and put in a broader environmental context by Antille et al. (2015, 2019). Large-scale surveys indicated CTF adoption levels are >50% of the total grain crop area for the northern region, and about 30% of the total Australian grain area (Umbers, 2017). CTF adoption in the grain industry has increased considerably since then. There has also been substantial adoption of CTF in cotton production, but effectiveness has been limited by the dual tyre requirement of modern cotton-picking machines. Adoption is also increasing in sugarcane. The preponderance of different, contractor-operated equipment has restricted CTF adoption in horticulture (McPhee et al., 2015), except where growers accept that their crop options are limited to those where a modular equipment track gauge and operating width is possible. CTF systems are now also in use in areas of grain production in Alberta (Canada) and Iowa (USA). CTF is also used in several countries in Europe; for example, in the United Kingdom (Godwin et al., 2015) and Slovakia (Galambošová et al., 2017).

6 Energy savings with controlled traffic farming systems

CTF was originally looked at as a method of overcoming the problems of soil compaction, but both research and farmer experience have demonstrated a number of other substantial advantages some of which are a direct consequence of reduced compaction. These include (after Bluett et al., 2019):

- reduced energy requirements for field traffic;
- improved soil structure and rainfall infiltration;
- reduced run-off and erosion, and therefore nutrients and sediment transport to water courses;
- enhanced soil macro-biota, particularly earthworms;
- reduced nitrous oxide emissions and improved nitrogen uptake efficiency by plants; and
- improved field access, that extends the possible operating period for an operation such as spraying.

As discussed earlier, tractive power, and its loss in travel reduction 'wheel slip' and motion 'rolling' resistance, is central to fuel efficiency and overall tractor performance. The magnitude of the energy lost in rolling resistance can be approximated, knowing that the best tractive efficiency of 4WD tractors on firm soil surfaces rarely exceeds 80%. (Zoz and Grisso, 2003). If 80% of axle power is converted to tractive power, and only 5–7% can normally be attributed to travel loss due to wheel-slip, the balance of 13–15% power loss can be attributed to rolling resistance (Inns and Kilgour, 1978). Tyre flexure is largely elastic so less than 1.5% of rolling resistance power loss is usually attributed to the tyres. The balance of between 10% and 14% of power output can be accounted for by soil deformation (that is, soil compaction). This is largely plastic deformation of the moist soil. Detailed consideration of compaction mechanics is beyond the scope of this chapter. Suffice to say that in general, compaction damage increases with heavier axle loads, greater contact pressures, and, for most practical purposes, greater soil moisture content. The first pass of any given traffic event produces the greatest damage, with repeated passes resulting in much smaller effects, if other factors remain unchanged.

Tractive power, and its efficient transmission, was an important concern when tillage was the major task but, even then, surveys indicated that most tractor tillage operations were carried out at wheel slip levels substantially less than the optimum. This implies that tractive efficiency was less than optimal. However, the economic value of a small degree of tractive 'inefficiency' was seen as less significant than the inefficiency and inconvenience resulting from difficulties working through wet or heavy soil areas of a paddock. Tractive efficiency loss from wheel slip is only relevant to draught field operations like tillage where the tractor's main function is the provision of tractive power via the drawbar or hydraulic linkage. Rolling resistance, on the other hand, is relevant to all field operations where machines traverse the soil. This resistance to forward motion is related to machine weight, soil conditions, and machinery running gear (tyres, tracks or belts). It is the largest single power requirement for most spraying and logistics operations, and a significant aspect in harvesting (Luhaib et al., 2017). Optimising wheel slip and reducing rolling resistance result in lower soil damage and reduced fuel consumption.

CTF might be expected to provide a significant reduction in rolling resistance if all field traffic is confined to hard permanent lanes, but data is relatively scarce, perhaps because it is so variable. It has been observed, for instance, that when one wheel follows another, the motion resistance of the second, similar wheel can be halved (ASABE, 2020). Under practical farm conditions, however, there will be substantial differences between wheel parameters of the various wheelsets of the various machines used (e.g. seeding tractor/air cart, sprayer, and harvester/chaser bin). There will also be a substantial time lapse between the passage of various machines, during which wetting and/or drying will change surface soil conditions.

Recent tests in the low rainfall zone of the state of Victoria in Australia demonstrated that the motion resistance of equipment operating on permanent traffic lanes of these relatively rigid soils was approximately 25% less than that of wheels on non-trafficked soil. This result also corresponded with the mean of the grower's assessments of the fuel impact of CTF (Mitchell et al., 2020). Rather greater CTF effects have been measured on the more plastic soils of Southern Queensland. CTF was also shown to provide yield benefits and substantial fuel savings on the North China plain (Chen and Yang, 2015). These observations suggest that CTF has the potential to significantly reduce the energy requirements of all soil-engaging operations. Tullberg (2000) showed that CTF could reduce energy requirements of tillage operations by up to 50%. Modelling work conducted by Gasso et al. (2013) showed that the lack of soil compaction in the crop zones in CTF systems led to a significant reduction of draft force required in tillage operations with a resulting 23% reduction in fuel consumption compared to random traffic operations. When comparing conventional no-till with no-till operations using CTF, Chen and Yang (2015) also found a reduced fuel consumption of just over 23%, representing a saving of over 19 L/ha. Luhaib et al. (2017) reported that the benefits of CTF in minimising traffic-induced compaction and thus tillage draft, as well as improved traction on compacted permanent traffic lanes, resulted in 40% energy savings compared to the energy required for random traffic operations. A broader assessment of overall CTF effects on the energy requirements of crop production can be found in Tullberg (2014). Farmers often note the timing benefits provided by improved trafficability of permanent traffic lanes and more rapid field access after rain. This is particularly important in herbicide weed control but can apply to most farming operations. In one of the few studies on this topic, McPhee et al. (1995) observed that CTF allowed seeding an average of 8 days earlier than non-CTF cropping in the Burdekin area of Queensland, and its major impact on the reliability of double cropping in that area.

7 Other benefits of controlled traffic farming systems

Soil compaction from using heavy field traffic can be defined as an increase in bulk density. This implies a loss of voids between and within soil particles and structural units of the soil profile. The larger voids are generally the first to be lost, with the constriction of progressively smaller voids with increased degrees of compaction. Loss of voids might be expected to influence parameters such as infiltration rate as confirmed by Li et al. (2001) on a Vertosol in Southeast Queensland. This was a four-year study to assess the impact of tillage and traffic on run-off from replicated, annually cropped plots. The results demonstrated that controlled traffic reduced mean annual run-off by 31% compared with annually wheeled treatments, and that no-till reduced run-off by 19%. Controlled traffic with no-till reduced annual run-off by 46% and increased grain yield by 16% compared with wheeled, tilled treatments. Results by Li et al. (2001) agreed with those reported by Ngo-Cong et al. (2021) who showed that a 10–20% increase in soil bulk density, due to compaction, reduced cumulative infiltration (steady-state) by 55–82%, and the available water storage capacity by 3–49%, depending upon the soil type. Wheel impact from heavy vehicles on infiltration was further investigated by Li et al. (2009). Rainfall simulator tests were used on a range of wheeling treatments and residue cover levels on previously non-tilled, non-wheeled soil. An inverse relationship between steady-state infiltration and the energy dissipated in the soil by wheeling treatments was demonstrated, providing the soil surface was adequately protected by crop residues. Surface degradation was shown to be related to rainfall energy, but subsoil degradation was related to imposing heavy wheel energy. McHugh et al. (2009), observed the impact of field traffic on a neighbouring degraded vertosol using a number of parameters of soil porosity and structure. These assessments considered soil structure in 100 mm increments to 300 mm depth.

This cropping soil had been ameliorated through multiple natural wetting and drying events over the course of two crop cycles. Wheel damage to soil structure remained unchanged through this process, while amelioration moved down through the profile of non-wheeled soils beyond 300 mm. Available water for growing plants in the top 300 mm of non-wheeled soil was 50% greater than that of wheel compacted soil. These research plots were untouched for 2 years after completion of this study, during which time they grew a sequence of weed populations. This provided the opportunity for a new set of soil structural assessments which showed little change in the ameliorated condition of non-wheeled soil. The wheel compacted soil treatment had been ameliorated to some extent. However, a single pass by a tractor wheel returned the structural condition and plant available water capacity of the non-wheeled treatment to its original, degraded state. These same replicated run-off plots were also used to investigate tillage and traffic effect on soil biota. Pangnakorn et al. (2003) sampled them at approximately six-week intervals over a two-year period that included three cropping cycles; times when the soil profile was close to the wilting point; and times when it was near field capacity. Soil samples were taken to a depth of 150 mm, then earthworms and other macrofauna were assessed by manual sieving, with mesofauna and microfauna extracted by standard funnel methodology. This study confirmed anecdotal reports of greater earthworm numbers in CTF. Compared with tilled and wheel compacted soil, earthworm numbers were roughly doubled in wheeled no-till treatments. They have doubled again in non-wheeled, no-till treatments that replicated CTF. Pangnakorn et al. (2003) also noted that soil arthropods (such as springtails, mites etc.) were slightly more common under controlled traffic conditions, an effect consistent with results from Tasmania (Rodgers et al., 2018). Burrows formed by these earthworms and arthropods are likely to have contributed to the improved infiltration capacity of no-till and non-wheeled treatments.

Wheeling and tillage effects on soil microbiota (nematodes, bacteria and fungi) appeared to be relatively small, and there were no significant treatment effects on total nematode numbers. Interestingly these results demonstrated that no-till CTF soils hosted a much greater percentage of free-living (non-parasitic) nematodes and a much smaller percentage of parasitic nematodes. Independent studies of permanent bed vegetable production (Stirling, 2008) have noted similar effects and beneficial outcomes. Given the well-known association between waterlogging and loss of soil nitrogen, CTF might be expected to reduce the loss of soil N. This should also reduce the accompanying emissions of nitrous oxide (N_2O) – an important 'greenhouse' gas with almost 300 times the global warming impact of carbon dioxide (CO_2) over a 100-year period. This was demonstrated by Tullberg et al. (2018) who assessed soil emissions from CTF beds, freshly wheeled CTF beds and CTF permanent traffic lanes between seeding and harvesting, taking an average of 14 sets of replicated chamber measurements per crop from

15 different grain crops in various regions over four years. Results showed that N_2O emissions from wheeled soils were normally at least twice those of non-wheeled CTF beds. These beds also absorbed small amounts of methane, which was emitted in small quantities by wheeled areas. In terms of CO_2 equivalent per winter cereal crop, average emissions from wheeled and non-wheeled CTF beds in temperate southern regions were 562 kg/ha and 265 kg/ha respectively, indicating that CTF should reduce emissions by between 150 kg/ha and 200 kg/ha and suggesting a reduction in denitrification of>10 kg N/ha. Equivalent data for sub-tropical northern region crops (4 winter cereals, 2 summer sorghum) indicated that reductions in both emissions and nitrogen loss were approximately 50% of southern region values. Nitrogen saved in N_2O emissions translates into improved fertiliser nitrogen use efficiency (Hussein et al., 2021a,b).

8 Controlled traffic farming operations in complex topographies

On undulating land, contour operations by tractors and machinery, usually between contour banks, has long been recommended practice, such as in the northern cropping regions of Australia. The objective is to reduce water erosion following high-intensity rainfall. Run-off water will initially run close to the contour, between the ridges formed by the tillage or planting equipment. It does this until it reaches a low point, or the flow increases enough to break out from the ridges and then flow directly down the slope to the contour bank. This will often form rills, but the soil will settle in the contour channel, and so is not lost. In severe rainfall events, these rills can be substantial and soil movement to the contour bank can block the channel so that the water then overflows and breaks out of the contour bank. This concentrated run-off can cascade from one contour bank to the next progressively down a hillside paddock, producing significant gullies and making it unfarmable. Yule (1995), aware that contour banks were usually designed for the '10-year recurrence interval' storm, appreciated that major erosion events were unavoidable under such a system. He also observed that severe damage occurred only when the run-off was concentrated and proposed a solution described as 'downslope' CTF operation with widely spaced, broad-based drive-over contour banks. Orienting field operations at 90° to the contour will keep most run-off moving within the crop row, or at least within a CTF machine width. With residues anchored to protect the soil surface from droplet impact and better soil structure, infiltration is optimised, and run-off minimised. While controversial when first proposed, downslope CTF is now common in some regions where occasional high-intensity rainfall events occur. It is generally regarded as effective in avoiding major water erosion, but at the cost of limited traffic lane erosion, particularly at the slope change that occurs immediately above the contour channel.

The limited data on soil and nutrient loss from these systems suggest that – compared with conventional operations on the contour, downslope systems are effective in limiting damage from major rainfall events, but results are inconclusive with respect to soil and nutrient loss in minor events. Downslope operations also ensure that traffic lanes drain and largely overcome the problems posed by wet spots delaying field operations. Growers frequently comment on the value of firm permanent tracks in extending the 'operating window', allowing CTF operations to proceed when non-CTF operations cannot. This can produce substantial economic benefits, such as those found by a Queensland CTF farmer who harvested one crop and planted the next crop just a few days after a minor rainfall event. Delay in the field access to a neighbouring non-CTF operation resulted in substantial losses when harvesting the crop, and a lost planting opportunity. These aspects can be remarkably important when farm operations must be completed in periods of relatively frequent rainfall. Undisturbed and unseeded traffic lanes are common in Australia's northern region, where water is the major erosion hazard, but in many southern areas wind erosion is the major threat. In this situation the dry surface soil of bare traffic lanes can be the initiation point of 'blowouts'[2]. Traffic lanes are normally seeded in these conditions, but orientation parallel to the prevailing wind is avoided where possible to reduce this possibility. Downslope seeded permanent traffic lanes are common elsewhere, but the topography is obviously a major consideration.

Field layout for CTF can be quite straightforward on broad, uniform, shallow slopes, but much more challenging in complex topographies. Regardless of traffic lane orientation, efficient logistics must always be a consideration when planning a CTF system. This might sometimes require changes to paddock boundaries, and careful siting of access pathways across long runs in large paddocks. CTF is currently being adapted to different environments with the aim to minimise the area of permanent traffic lanes, minimise soil disturbance, and manage run-off or wind erosion effects.

9 Embodied energy

Embodied energy has been defined as the sum of the energy requirements associated directly or indirectly with the delivery of a good or service, that is the energy incorporated or 'embodied' in a product from raw material production through to manufacture, distribution, use and potentially disposal.

2 These are strips of significant soil loss sometimes found near the crest of ridges after severe winds blowing parallel to strips of bare soil. These typically occur on traffic lanes used repeatedly by sprayers and spreaders, so the problem can usually be dealt with by 'rotating' between different sets of traffic lanes available because the working width of these units is typically a factor of 2 or 3 greater than the seeder and harvester.

9.1 Machinery

Energy embodied in machinery can be determined from the mass of the different materials of construction and the energy requirements for their fabrication and transport to the user. This can be simplified by using published values of the energy intensity (i.e. embodied energy/mass ratios) across equipment categories. For most farm equipment, this usually ranges between 50 MJ/kg and 80 MJ/kg (Zentner et al., 2004). Energy intensity can also be expressed per unit purchase price (MJ/$), or as a percentage of operating energy when depreciated over the expected machine life. Large economic and embodied energy investment is a direct consequence of the low levels of equipment utilization. Equipment contract and sharing arrangements are often suggested to improve utilization, but local arrangements are problematic when most farmers within a district all need to seed (or harvest) at the same time. Substantial improvements can be achieved only when cooperative arrangements or contractors can extend equipment operating seasons by working across a range of climate zones.

9.2 Fertilisers

Fertiliser manufacture, particularly nitrogen fertilisers, is often identified as a major energy cost of modern agriculture. Commonly used values for energy embodied in fertilisers (Table 3) are generally comparable to those quoted by Pimental (2009) and other sources. Fertiliser manufacture has been estimated to consume >1% of total global energy production, which is largely a consequence of the energy intensity of ammonia production. In New Zealand, Balance Fertiliser Company has invested in wind turbines to produce 'green hydrogen' through electrolysis in order to displace the hydrogen from the steam reforming of natural gas, with the consequent release of CO_2, in their urea manufacturing plant. Poor efficiency of nitrogen and phosphate fertiliser application is also a matter of concern, and global estimates by Smil (1999) suggested that 50% (or less) of nitrogen applied as fertiliser is recovered in harvested crops. The balance may be lost via volatilisation, denitrification, leaching and runoff (Dalal et al., 2003; Masters et al., 2008; Melland et al.,

Table 3 Energy embodied in various common fertilisers (after Zentner et al., 2004)

Fertilisers	MJ kg^{-1}
Nitrogen	75.63
Phosphate (P_2O_5)	9.53
Potassium (K_2O)	9.85
Sulphur (SO_3)	1.12

2017; Antille and Moody, 2021). Regardless of such loss mechanisms, pollution by nitrogen and other fertiliser components represents serious threats to the environment and to human health in many regions of the world.

Fertiliser nitrogen and phosphorus have both been identified as the major cause of algal blooms and eutrophication of waterways and coastal waters. Nitrate levels at or exceeding World Health Organization recommendations are a threat to healthy water supplies in both the developed and the developing world. Improved fertiliser use efficiency thus has many benefits in addition to those of greater energy efficiency in food production (Tullberg et al., 2018). Denitrification is an additional environmental concern because nitrous oxide (a powerful GHG) is produced under near-waterlogged soil conditions when nitrate and carbon are available in the soil. Methane, another significant GHG, is absorbed in small quantities by healthy aerated soils, but produced in rather larger quantities when soil is totally waterlogged or flooded such as in rice-paddy fields (Dalal et al., 2003). Warmer environments exacerbate this problem.

9.3 Herbicides

Reduced on-farm fuel use in no-till systems is offset to some extent by the energy embodied in additional herbicide inputs. Values for a number of common herbicides are quoted in Table 4. The original data source is from a 1999 Canadian publication but more recent, well-supported data on this topic are rare, perhaps because manufacturing technology is proprietary. Values quoted by Helsel (2019) are similar but slightly lower overall, perhaps reflecting improvements in herbicide manufacturing technology. The embodied energy of a number of herbicides expressed as MJ/ha, is large and of similar magnitude to the direct energy requirements of light tillage. No-till advocacy rarely acknowledges the embodied energy of herbicides in tillage system comparisons but it should be noted that the values in Table 4 are for overall herbicide application. They take no account of the scope to reduce active

Table 4 Energy embodied in herbicides (after Zentner et al., 2004)

	Embodied energy	
Product	Active ingredient, MJ kg^{-1}	At usual application rates, MJ ha^{-1}
Glyphosate	511	225
Paraquat	538	292
Diquat	75	141
2,4-D dicamba	336	37
Trifluralin	167	233
Metolachlor+atrazine	313	1045

application rates by improved technology and better-targeted application, or the opportunities to avoid herbicide use by improved crop system management or organic farming methods (Friedrich and Kassam, 2012). The development of herbicide tolerance in weeds represents a major threat to no-till systems as currently practised in most high-resource regions. In an attempt to overcome this, physical weed control systems such as targeted tillage, flame, steam, microwave energy or high-pressure water jets are being proposed that might become economic for highly resistant weed control. However, these appear to be even more energy-intensive if used to replace overall herbicide spraying. The efficacy of current herbicides can be extended by preventing the reproduction of any herbicide-resistant 'escapees'. These are the weeds that have developed some tolerance to a herbicide, and so survive normal application. This can now be achieved in a second operation using sprayer-mounted optical weed detection technology to deliver spot applications of different herbicides to those escapees. In the longer term, the destruction of herbicide escapees might be seen as the first major opportunity for large-scale use of robotics in cropping, using swarms of small autonomous vehicles to search for, identify and deliver herbicide or physical weed treatments.

10 Future trends

One potential solution to more sustainable energy use in farm machinery is to use biofuels from renewable resources**.** The reduction of GHG emissions through the substitution of diesel with biodiesel depends on the feedstock used to provide the triglycerides in the biodiesel, (such as from canola, soy oil, animal tallow); the inter-esterification process utilised in its production; the storage period; and ambient conditions. A full life-cycle analysis is needed for any useful comparison to be meaningful.

When using fuel with 40% biodiesel blended with diesel (BD40), the maximum power dropped by over 6% after 1000 h of operation. Specific fuel consumption and CO emission was increased by around 7% and 12.5% respectively and CO_2 emissions were reduced by around 6%. The poorer engine performance was probably a consequence of carbon deposits on the tip of injectors which was formed due to thermal oxidation of unsaturated fatty acids that amounted to over 90% in the used raw material for biodiesel production (Tomić et al., 2021). As countries around the world mandate minimum renewable content in diesel fuel, issues around fuel consumption will only increase. Reduction of both pollutant emissions and fossil fuel dependency is an objective of energy policies worldwide. In many countries, governments promote the use of more efficient and cleaner vehicles. A hybrid electric vehicle incorporating electric drives in a tractor presents advantages in terms of increased energy efficiency and expanded functionalities. This higher efficiency leads to a reduction in diesel fuel consumption and hence a comparable decrease in CO_2 emission.

Tractor electrification takes advantage of decoupling loads and drives from the engine which allows operating the latter at its highest efficiency point. Major advantages of machinery electrification are torque and speed control, noise reduction, and a more flexible design. A review of the state-of-the-art of high voltage electrification of agricultural machinery was undertaken by Moreda et al. (2016). The future for autonomous tractors is promising, though not new with the first commercial application for use in orchards arising from research undertaken at Reading University (UK) in the 1960s. Electric-powered tractors are near to commercialisation or are already commercially available. Where the local electricity supply has a low-carbon emission factor, this can also result in a significant reduction in emissions. A final development is the increasing potential of small lightweight robotic equipment to undertake functions currently undertaken by tractor-drawn equipment and other heavy equipment with high fuel consumption (Thomasson et al., 2018, 2019). McPhee et al., (2020) have pointed out the size and weight limitations inherent in current harvesting and transport technology noting that soil compaction will still be a problem with robotic units. They suggest that the robotic operation of medium-scale equipment within a precision CTF environment should offer more feasible and energy-efficient alternatives. A recent example is the integration of weed sensing sprayers with robotic field equipment, CTF and innovative agronomy. Each of these technologies is independently worthwhile, but it is their integration that might provide the best prospects for more sustainable, low-energy cropping systems.

Sensor-actuated spray systems such as 'Weedit' for instance, can ensure herbicide is applied only where green vegetation is detected, reducing the quantity of active ingredients required for effective weed control. When combined with robotic field vehicles (e.g. Swarmfarm units) they allow weed control operations, particularly in fallow or terminated cover crops, at frequencies that would otherwise be totally uneconomic. This ensures all weeds are killed while very small, facilitating more profitable use of cover crops and more sustainable cropping systems. This robotic-based system is already being practised on several Australian farms, but similar developments are occurring overseas, so we can expect further development of enabling technologies and innovative cropping systems. One example is the sensors and relay-controlled spray jet units. Individual units are still relatively expensive, so each unit must currently detect and treat swaths of significant width. As these become cheaper over time, narrower swath widths and greater reductions in herbicide requirements will be possible; and while current units can detect only colour change, development continues on sensors able to distinguish between (e.g. broadleaves and grasses). Lower unit costs and narrower swaths will also allow alternative methods of weed control, potentially including mechanical, steam or microwave treatments, each providing different advantages and limitations compared with herbicide control or overall tillage. A current example is the 'Weed Chipper' developed at the University of Western Australia. This uses a sweep-equipped chisel plough tine, so its initial application would be in dealing with large, hard-to-kill weeds, but there is clear potential for this technology to operate on a finer scale.

CTF is not inherently necessary for any of these applications, but it can provide more reliable field access and improved uniformity of soil and crop conditions. It reduces energy requirements and facilitates more precise relationships between crop rows and machine components or sensors. At the most basic level, better tractor/implement guidance allows interrow 'shield' spraying, on-row crop chemical application or interrow seeding. Increasing precision with RTK tractor and/or vision-based implement guidance allows precise seed placement in relation to crop residue or growing crops to improve the microenvironment for companion or succession crops.

11 References

Al-Aani, F. S., Darr, M. J., Covington, B. J. and Powell, L. J. (2016). *The Performance of Farm Tractors as Reported by Can-Bus Messages*. ASABE Paper Number: 162461746. St. Joseph, MIC: American Society of Agricultural and Biological Engineers. https://doi.org/10.13031/aim.20162461746.

Angus, J., Bell, M., McBeath, T. and Scanlan, C. (2019). Chapter 14: Nutrient-management challenges and opportunities in conservation agriculture. In: Pratley, J. and Kirkegaard, J. (Eds) *Australian Agriculture in 2020: From Conservation to Automation*. Wagga Wagga, NSW, Australia: Agronomy Australia and Charles Sturt University, pp. 221–236.

Antille, D. L., Ansorge, D., Dresser, M. L. and Godwin, R. J. (2013). Soil displacement and soil bulk density changes as affected by tire size. *Transactions of the ASABE* 56(5): 1683–1693. https://doi.org/10.13031/trans.56.9886.

Antille, D. L., Chamen, W. C. T., Tullberg, J. N. and Lal, R. (2015). The potential of controlled traffic farming to mitigate greenhouse gas emissions and enhance carbon sequestration in arable land: A critical review. *Transactions of the ASABE* 58(3): 707–731. https://doi.org/10.13031/trans.58.11049.

Antille, D. L. and Moody, P. W. (2021). Nitrogen use efficiency indicators for the Australian grains, cotton, sugar, dairy, and horticulture industries. *Environmental and Sustainability Indicators* 10: Article number: 100099. https://doi.org/10.1016/j.indic.2020.100099.

Antille, D. L., Peets, S., Galambošová, J., Botta, G. F., Rataj, V., Macák, M., Tullberg, J. N., Chamen, W. C. T., White, D. R., Misiewicz, P. A., Hargreaves, P. R., Bienvenido, J. F. and Godwin, R. J. (2019). Review: Soil compaction and controlled traffic farming in arable and grass cropping systems. *Agronomy Research* 17(3): 653–682. https://doi.org/10.15159/AR.19.133.

ASABE. (2020). Standard ASAE-D497.7-MAR2011 (R2020). *Agricultural Machinery Management Data*. St. Joseph, MIC: American Society of Agricultural and Biological Engineers.

Baillie, C. P., Lobsey, C. R., Antille, D. L., McCarthy, C. L. and Thomasson, J. A. (2018). *A Review of the State of the Art in Agricultural Automation. Part III: Agricultural Machinery Navigation Systems*. ASABE Paper No. 1801591. St. Joseph, MI: American Society of Agricultural and Biological Engineers. https://doi.org/10.13031/aim.201801801591.

Basavarajappa, D. N., Banapurmath, N. R., Khandal, S. V. and Manavendra, G. (2015). Performance evaluation of common rail direct injection (CRDI) engine fuelled with uppage oil methyl ester (UOME). *International Journal of Renewable Energy Development* 4(1): 1–10. https://doi.org/10.14710/ijred.4.1.1-10.

Bluett, C., Tullberg, J. N., McPhee, J. E. and Antille, D. L. (2019). Soil and tillage research: Why still focus on soil compaction? *Soil and Tillage Research* 194: Article number: 104282. https://doi.org/10.1016/j.still.2019.05.028.

Bygden, G., Eliasson, L. and Wästerlund, I. (2003). Ruth depth, soil compaction and rolling resistance when using bogie tracks. *Journal of Terramechanics* 40(3): 179–190. https://doi.org/10.1016/j.jterra.2003.12.001.

Chamen, W. C. T., Watts, C. W., Leede, P. R. and Longstaff, D. J. (1992). Assessment of a wide span vehicle (gantry), and soil and cereal crop responses to its use in a zero-traffic regime. *Soil and Tillage Research* 24(4): 359–380. https://doi.org/10.1016/0167-1987(92)90119-V.

Chen, H. and Yang, Y. (2015). Effect of controlled traffic system on machine fuel saving in annual two crops region in North China Plain. *Soil and Tillage Research* 153: 137–144. https://doi.org/10.1016/j.still.2015.06.001.

Crossley, C. P., Kibiwot, V. N., Reynolds, A. J. and Rickson, R. J. (2001). Rut formation and rolling resistance on earth roads. *Journal of Agricultural Engineering Research* 78(1): 99–107. https://doi.org/10.1006/jaer.2000.0634.

Dalal, R. C., Wang, W., Robertson, G. P. and Parton, W. J. (2003). Nitrous oxide emissions from Australian agricultural lands and mitigation options: A review. *Soil Research* 41(2): 165–195. https://doi.org/10.1071/SR02064.

Douglas, J. T. and Crawford, C. E. (1989). *Effects of Wheel-Induced Compaction on Grass Yield and Nitrogen Uptake, 1988*. Departmental Note No. 19. Penicuik: Scottish Centre of Agricultural Engineering.

FAO. (2011). Energy-smart food for people and climate. Food and Agricultural Organisation of the United Nations. Available at: https://www.fao.org/3/i2454e/i2454e00.pdf (accessed January 2022).

Foley, J. P., Sandell, G. R., Szabo, P. M., Scobie, M. and Baillie, C. P. (2015). *Improving Energy Efficiency on Irrigated Australian Cotton Farms: Benchmarking Report. Final Report Cotton Research and Development Corporation. University of Southern Queensland, National Centre for Engineering in Agriculture Publication No. 1005371/1*. Toowoomba, QLD, Australia.

Friedrich, T. and Kassam, A. (2012). No-till farming and the environment: Do no-till systems require more chemicals? *Outlooks on Pest Management* 23(4): 153–157. https://doi.org/10.1564/23aug02.

Galambošová, J., Macák, M., Rataj, V., Antille, D. L., Godwin, R. J., Chamen, W. C. T., Žitňák, M., Vitázková, B., Ďuďák, J. and Chlpík, J. (2017). Field evaluation of controlled traffic farming in Central Europe using commercially available machinery. *Transactions of the ASABE* 60(3): 657–669. https://doi.org/10.13031/trans.11833.

Gasso, V., Sørensen, C. A. G., Oudshoorn, F. W. and Green, O. (2013). Controlled traffic farming: A review of the environmental impacts. *European Journal of Agronomy* 48: 66–73. https://doi.org/10.1016/j.eja.2013.02.002.

Godwin, R., Misiewicz, P., White, D., Smith, E., Chamen, T., Galambošová, J. and Stobart, R. (2015). Results from recent traffic systems research and the implications for future work. *Acta Technologica Agriculturae* 18(3): 57–63. https://doi.org/10.1515/ata-2015-0013.

Grogan, J., Morris, D. A., Searcy, S. W. and Stout, B. A. (1987). Microcomputer-based tractor performance monitoring and optimization system. *Journal of Agricultural Engineering Research* 38(4): 227–243. https://doi.org/10.1016/0021-8634(87)90091-6.

Hanna, M., Harmon, J. and Petersen, D. (2010). *Ballasting Tractors for Fuel Efficiency. Farm Energy Conservation and Efficiency Initiative*. Iowa State University Extension Publication No. PM2089g. Available at: https://dr.lib.iastate.edu/entities/publication/fa237a35-c8a0-4673-917c-e98d135d0957 (accessed January 2022).

Helsel, Z. R. (2019). Farm energy: Energy use and efficiency in pest control, including pesticide production, use, and management options. Available at: https://farm-energy.extension.org/energy-use-and-efficiency-in-pest-control-including-pesticide-production-use-and-management-options/ (accessed January 2022).

Höök, M. and Tang, X. (2013). Depletion of fossil fuels and anthropogenic climate change – A review. *Energy Policy* 52: 797–809. https://doi.org/10.1016/j.enpol.2012.10.046.

Horrocks, R. W. (2010). Overview of high-speed direct injection diesel engines. *Advanced Direct Injection Combustion Engine Technologies and Development* 2010: 3–60. https://doi.org/10.1533/9781845697457.1.3.

Howard, C. N., Kocher, M. F., Hoy, R. M. and Blankenship, E. E. (2013). Testing the fuel efficiency of tractors with continuously variable and standard geared transmissions. *Transactions of the ASABE* 56(3): 869–879. http://doi.org/10.13031/trans.56.10222.

Hoy, R. M., Rohrer, R., Liska, A., Luck, J. D. and Isom, L. (2014). Agricultural industrial advanced vehicle technology: Benchmarkstudy for reduced petroleum use. Adam Liska Papers No. 19. INL/EXT-14-33118. Lincoln: University of Nebraska; Idaho Falls, ID: Idaho National Laboratory, p. 83415. Available at: https://digitalcommons.unl.edu/cgi/viewcontent.cgi?article=1018&context=bseliska (accessed January 2022).

Hunt, D. R. (1986). *Engineering Models for Agricultural Production*. Westport, CT: The Avi Publishing, p. 260.

Hussein, M. A., Antille, D. L., Kodur, S., Chen, G. and Tullberg, J. N. (2021a). Controlled traffic farming delivers improved agronomic performance of wheat as a result of enhanced rainfall and fertilizer nitrogen use efficiency. *Acta Agriculturae Scandinavica, Section B – Soil and Plant Science* 71(5): 377–398. https://doi.org/10.1080/09064710.2021.1903984.

Hussein, M. A., Antille, D. L., Kodur, S., Chen, G. and Tullberg, J. N. (2021b). Controlled traffic farming effects on productivity of grain sorghum, rainfall and fertiliser nitrogen use efficiency. *Journal of Agriculture and Food Research* 3: Article number: 100111. https://doi.org/10.1016/j.jafr.2021.100111.

Inns, F. M. and Kilgour, J. (1978). *Agricultural Tyres*. London: Dunlop Limited, p. 70.

IPCC. (2021). AR6 climate change 2021: The physical science basis (full report). Available at: https://www.ipcc.ch/report/ar6/wg1/#FullReport (accessed January 2022).

Jungler, A. P. and Diotto, R. D. (2018). Sustentabilidade e Política Pública para Contaminantes no Brasil. *Research, Society and Development* 7(6): 1–18. https://doi.org/10.17648/rsd-v7i6.121.

Kesicki, F. (2010). The third oil price surge – What's different this time? *Energy Policy* 38(3): 1596–1606. https://doi.org/10.1016/j.enpol.2009.11.044.

Komandi, G. (1999). An evaluation of the concept of rolling resistance. *Journal of Terramechanics* 36(3): 159–166. https://doi.org/10.1016/S0022-4898(99)00005-1.

Kurjenluoma, J., Alakukku, L. and Ahokas, J. (2009). Rolling resistance and rut formation by implement tyres on tilled clay soil. *Journal of Terramechanics* 46(6): 267–275. https://doi.org/10.1016/j.jterra.2009.07.002.

Li, Y., Tullberg, J. N. and Freebairn, D. M. (2001). Traffic and residue cover effects on infiltration. *Soil Research* 39(2): 239–247. https://doi.org/10.1071/SR00017.

Li, Y. X., Tullberg, J. N. and Freebairn, D. M. (2007). Wheel traffic and tillage effects on runoff and crop yield. *Soil and Tillage Research* 97(2): 282–292. https://doi.org/10.1016/j.still.2005.10.001.

Li, Y. X., Tullberg, J. N., Freebairn, D. M. and Li, H. W. (2009). Functional relationships between soil water infiltration and wheeling and rainfall energy. *Soil and Tillage Research* 104(1): 156–163. https://doi.org/10.1016/j.still.2008.10.023.

Luhaib, A. A. A., Antille, D. L., Tullberg, J. N., Hussein, M. A. and Chen, G. (2017). *Effect of Controlled Traffic Farming on Energy Saving in Australian Grain Cropping Systems*. ASABE Paper No. 1700583. St. Joseph, MI: American Society of Agricultural and Biological Engineers. https://doi.org/10.13031/aim.201700583.

Masters, B., Rohde, K., Gurner, N., Higham, W. and Drewry, J. (2008). Sediment, nutrient, and herbicide runoff from cane farming practices in the Mackay WhitSunday region: A field-based rainfall simulation study of management practices. *Mackay-WhitSunday*. Queensland, Australia: Queensland Department of Natural Resources and Water, Natural Resource Management Group. Available at: http://reefcatchments.com.au/files/2013/02/rainfallsimdnrm_report.pdf (accessed December 2021).

McAllister, M. (1983). Reduction in the rolling resistance of tyres for trailed agricultural machinery. *Journal of Agricultural Engineering Research* 28(2): 127–137. https://doi.org/10.1016/0021-8634(83)90082-3.

McHugh, A. D., Tullberg, J. N. and Freebairn, D. M. (2009). Controlled traffic farming restores soil structure. *Soil and Tillage Research* 104(1): 164–172. https://doi.org/10.1016/j.still.2008.10.010.

McPhee,J. E., Aird, P. L., Hardie, M. A. and Corkrey, S. R. (2015). The effect of controlled traffic on soil physical properties and tillage requirements for vegetable production. *Soil and Tillage Research* 149: 33–45. https://doi.org/10.1016/j.still.2014.12.018.

McPhee, J. E., Antille, D. L., Tullberg, J. N., Doyle, R. B. and Boersma, M. (2020). Managing soil compaction – A choice of low-mass autonomous vehicles or controlled traffic? *Biosystems Engineering* 195: 227–241. https://doi.org/10.1016/j.biosystemseng.2020.05.006.

McPhee, J. E., Braunsck, M. V., Garside, A. L., Reid, D. J. and Hilton, D. J. (1995). Controlled traffic for irrigated double cropping in a necessary arid tropical environment: Part 3. *Journal of Agricultural Engineering Research* 60(3): 191–199. https://doi.org/10.1006/jaer.1995.1013.

Melland, A. R., Antille, D. L. and Dang, Y. P. (2017). Effects of strategic tillage on short-term erosion, nutrient loss in runoff and greenhouse gas emissions. *Soil Research* 55(3): 201–214. https://doi.org/10.1071/SR16136.

Mitchell, R., Wilhelm, N., Fisher, P., Tullberg, J., Bluett, C., Pearl, D., Dimos, N., and Benjamin, C. (2020). On the right track: Controlled traffic in the low rainfall zone of South-Eastern Australia. Available at: https://grdc.com.au/resources-and-publications/all-publications/publications/2020/on-the-right-track-controlled-traffic-in-the-low-rainfall-zone-of-south-eastern-australia (accessed December 2021).

Moreda, G. P., Muñoz-García, M. A. and Barreiro, P. (2016). High voltage electrification of tractor and agricultural machinery – A review. *Energy Conversion and Management* 115: 117–131. https://doi.org/10.1016/j.enconman.2016.02.018.

Morrison, J. E., Jr. (1985). Machinery requirements for permanent wide beds with controlled traffic. *Applied Engineering in Agriculture* 1(2): 64–67. https://doi.org/10.13031/2013.26766.

Ngo-Cong, D., Antille, D. L., van Genuchten, M. T., Nguyen, H. Q., Tekeste, M. Z., Baillie, C. P. and Godwin, R. J. (2021). A modeling framework to quantify the effects of compaction on soil water retention and infiltration. *Soil Science Society of America Journal* 85(6): 1931–1945. https://doi.org/10.1002/saj2.20328.

Özkan, M. (2015). A comparative study on energy and exergy analyses of a CI engine performed with different multiple injection strategies at part load: Effect of injection pressure. *Entropy* 17(1): 244–263. https://doi.org/10.3390/e17010244.

Pang, S. N., Zoerb, G. C. and Wang, G. (1985). Tractor monitor based on indirect fuel measurement. *Transactions of the ASABE* 28(4): 0994–0998. https://doi.org/10.13031/2013.32375.

Pangnakorn, U., George, D. L., Tullberg, J. N. and Gupta, M. L. (2003). Effect of tillage and traffic on earthworm populations in a Vertosol in South-East Queensland. In: *Proceedings of the 16th International Soil and Tillage Research Organization* (vol. 1). Brisbane, Australia, 13–18 July 2003, pp. 881–885.

Pimental, D. (2009). Energy inputs in food crop production in developing and developed nations. *Energies* 2(1): 1–24. https://doi.org/10.3390/en20100001.

Raper, R. L., Bailey, A. C., Burt, E. C., Way, T. R. and Liberati, P. (1995). The effects of reduced inflation pressure on soil-tyre interface stresses and soil strength. *Journal of Terramechanics* 32(1): 43–51. https://doi.org/10.1016/0022-4898(95)00002-I.

Rodgers, D., McPhee, J., Aird, P. and Corkrey, R. (2018). Soil arthropod responses to controlled traffic in vegetable production. *Soil and Tillage Research* 180: 154–163. https://doi.org/10.1016/j.still.2018.03.002.

Schlosser, J. F., de Farias, M. S., Bertollo, G. M., Russini, A., Herzog, D. and Casali, L. (2020). Agricultural tractor engines from the perspective of Agriculture 4.0. *Revista Ciência Agronômica* 51(5): e20207716.

Sims, R., Flammini, A., Puri, M. and Bracco, S. (2015). *Opportunities for Agri-Food Chains to Become Energy-Smart*. Rome, Italy: Food and Agriculture Organization of the United Nations.

Smil, V. (1999). Nitrogen in crop production: An account of global flows. *Global Biogeochemical Cycles* 13(2): 647–662. https://doi.org/10.1029/1999GB900015.

Stirling, G. R. (2008). The impact of farming systems on soil biology and soilborne diseases: Examples from the Australian sugar and vegetable industries – The case for better integration of sugarcane and vegetable production and implications for future research. *Australasian Plant Pathology* 37(1): 1–18. https://doi.org/10.1071/AP07084.

Stroud, J. L. (2020). No-till farming systems in Europe. In: Dang, Y., Dalal, R. and Menzies, N. (Eds) *No-Till Farming Systems for Sustainable Agriculture*. Cham: Springer. https://doi.org/10.1007/978-3-030-46409-7_31.

Subramaniam, M. N., Hayes, C., Tomazic, D., Downey, M. and Bruestle, C. (2011). Pre-turbo aftertreatment position for large bore diesel engines - compact & cost-effective aftertreatment with a fuel consumption advantage. *SAE International Journal of Engines* 4(1): 106–116. https://doi.org/10.4271/2011-01-0299.

Thomasson, J. A., Baillie, C. P., Antille, D. L., Lobsey, C. R. and McCarthy, C. L. (2019). Autonomous technologies in agricultural equipment: A review of the state of the art. In: *ASABE Distinguished Lecture Series. Tractor Design. Agricultural Equipment Technology Conference*, 11–13 February 2019 (vol. 40). Louisville, KY. St. Joseph, MIC: ASABE, pp. 1–17. ASABE Publication No. 913C0119. https://doi.org/10.13031/913C0119.

Thomasson, J. A., Baillie, C. P., Antille, D. L., McCarthy, C. L. and Lobsey, C. R. (2018). *A Review of the State of the Art in Agricultural Automation. Part II: On-Farm Agricultural Communications and Connectivity*. ASABE Paper No. 1801590. St. Joseph, MI: American Society of Agricultural and Biological Engineers. https://doi.org/10.13031/aim.201801801590.

Tomić, M., Savin, L., Simikić, M., Kiss, F., Kešelj, K., Ivanišević, M., Ponjičan, O., Zoranović, M. and Sedlar, A. (2021). Effects of biodiesel on changes in IC engine performances: A long-term experiment with farm tractors. *Fuel* 292: Article number: 120300. https://doi.org/10.1016/j.fuel.2021.120300.

Tullberg, J., Antille, D. L., Bluett, C., Eberhard, J. and Scheer, C. (2018). Controlled traffic farming effects on soil emissions of nitrous oxide and methane. *Soil and Tillage Research* 176: 18–25. https://doi.org/10.1016/j.still.2017.09.014.

Tullberg, J. N. (2000). Wheel traffic effects on tillage draught. *Journal of Agricultural Engineering Research* 75(4): 375–382. https://doi.org/10.1006/jaer.1999.0516.

Tullberg, J. N. (2014). Chapter 3: Energy in crop production systems. In: Bundschuh, J. and Chen, G. (Eds). *Sustainable Energy Solutions in Agriculture. Series: Sustainable Energy Developments 8*. Boca Raton, FL: CRC Press Taylor & Francis Group, pp. 53–76.

Tullberg, J. N. and Hilton, D. J. (1985). Future developments in tractors and field machinery. *Journal of the Australian Institution of Agricultural Science* 51: 238–242.

Tullberg, J. N., Yule, D. F. and McGarry, D. (2007). Controlled traffic farming - From research to adoption in Australia. *Soil and Tillage Research* 97(2): 272–281. https://doi.org/10.1016/j.still.2007.09.007.

Umbers, A. (2017). GRDC farm practices survey report 2016. Available at: https://grdc.com.au/__data/assets/pdf_file/0033/289842/GRDC-Farm-Practices-Survey-Report-2016.pdf?utm_source=website&utm_medium=download_link&utm_campaign=pdf_download&utm_term=National&utm_content=Farm%20Practices%20Survey%20Report%202016 (accessed November 2021).

Warwick HRI. (2007). AC0401 direct energy use in agriculture: Opportunities for reducing fossil fuel inputs. Final Report to Defra. London, UK: Department for Environment, Food and Rural Affairs, p. 57. Available at: https://ukerc.rl.ac.uk/pdf/AC0401_Final.pdf (accessed October 2021).

Wong, J. Y. (2009). *Terramechanics and Off-Road Vehicle Engineering*. Amsterdam, The Netherlands: Elsevier Ltd, p. 463. https://doi.org/10.1016/C2009-0-00403-6.

Woods, J., Williams, A., Hughes, J. K., Black, M. and Murphy, R. (2010). Energy and the food system. *Philosophical Transactions of the Royal Society of London. Series B, Biological Sciences* 365(1554): 2991–3006. https://doi.org/10.1098/rstb.2010.0172.

Yule, D. F. (1995). Controlled traffic for broadacre dry land farming – Better than sliced bread. In: Yule, D. F. and Tullberg, J. N. (Eds) *Controlled Traffic for Broadacre Dryland Farming: Proceedings of the 1st National Controlled Traffic Conference*, Yeppoon, QLD, Australia (accessed 13–14 September 1995). Available at: https://www.actfa.net/actfa-conferences/1995-ctf-conference/ (accessed January 2022).

Yule, I. J., Kohnen, G. and Nowak, M. (1999). A tractor performance monitor with DGPS capability. *Computers and Electronics in Agriculture* 23(2): 155–174. https://doi.org/10.1016/S0168-1699(99)00029-0.

Zentner, R. P., Lafond, G. P., Derksen, D. A., Nagy, C. N., Wall, D. D. and May, W. E. (2004). Effects of tillage method and crop rotation on non-renewable energy use efficiency for a thin Black chernozem in the Canadian Prairies. *Soil and Tillage Research* 77(2): 125–136. https://doi.org/10.1016/j.still.2003.11.002.

Zoz, F. M. and Grisso, R. D. (2003). Traction and tractor performance. In: *ASAE Distinguished Lecture Series Tractor Design* (vol. 27). Paper No. 913C0403. St. Joseph, MIC: American Society of Agricultural Engineers.

Zoz, F. M. and Wiley, J. C. (1995). A theoretical basis for tractor ballasting recommendations. In: *Proceedings of the 5th North American ISTVS Conference/Workshop*. Sasketoon, SK, Canada, 10–12 May 1995. International Society for Terrain-Vehicle Systems.

Chapter 4

Efficient water management and irrigation on farms

K. Reardon-Smith and S. Mushtaq, Centre for Applied Climate Sciences, University of Southern Queensland, Australia; M. Scobie and J. Eberhard, Centre for Agricultural Engineering, University of Southern Queensland, Australia; and T. N. Maraseni, Centre for Sustainable Agricultural Systems, University of Southern Queensland, Australia

1 Introduction

Global food security depends on the sustainability of agricultural production systems and of the natural resources (soils, water, functional landscapes) that support and sustain these. Irrigated agriculture currently meets 45% of global food demand, but is also the biggest consumer of freshwater globally, being responsible for 70% of withdrawals from global river systems (de Wrachien et al., 2021). However, seasonal rainfall and irrigation water supply are becoming less predictable in many parts of the world due to climate change and increasing climate variability (Döll, 2002). Due to global population growth and increasing demand for food and freshwater, many countries face a future with less reliable supplies of fresh water - a situation with severe implications for agricultural productivity and profitability. As declining water security places both food security and the well-being of many farmers and agriculturally-dependent communities at increasing risk, the improved efficiency of water use by the irrigation industry is becoming critically important.

Addressing water security issues through the use of more water-efficient irrigation technologies, such as pressurised drip and sprinkler systems and precision irrigation systems, significantly increases the energy demand of

http://dx.doi.org/10.19103/AS.2022.0100.06

modern agriculture. This adds to the costs of production as well as increasing agriculture's contributions to greenhouse gas (GHG) emissions and climate change, thereby potentially increasing pressure on food, energy and water security. Managing the trade-offs between food production and water and energy use - the food–water–energy nexus (FAO, 2014; Pittock et al., 2015; FAO, 2021) - and between climate change adaptation and mitigation is a significant industry and policy challenge (Maraseni et al., 2009a; Maraseni et al., 2012a; Mushtaq et al., 2015a,b). While improved irrigation practices can offset environmental footprints in the short run, large-scale implementation of irrigated agriculture remains an issue, which may have unintended consequences such as the risk of being maladaptive (Barnett and O'Neill, 2010). Investments aimed at reducing energy dependency and enhancing natural resource sustainability while ensuring future food security are needed to buffer the negative impacts of conventional food production intensification practices on the environment (Mushtaq et al., 2009b).

In this chapter, a brief overview of the current state of irrigation water resource management, both supply- and demand-side, is presented; innovative technologies and practices designed to improve water use efficiency and water productivity, such as the agricultural yield achieved per unit of water available or applied, are discussed; the energy implications of these are considered; and the importance of integrated decision-making at farm, irrigation scheme and policy levels is highlighted. Increasingly and to the extent possible, such decisions must account for trade-offs between food production and water and energy use (i.e. effectively, environmental security) and between adaptation and mitigation policy and practices.

2 Water use for irrigation in agricultural production systems

2.1 Background

Globally, arable land - which can be used for agricultural production with little or no modification - is in limited supply and in many cases threatened by chronic degradation and expanding urbanisation. At the same time, the global human population continues to increase, placing greater pressure on agricultural regions to produce greater quantities of food to ensure global food security (Gomiero, 2016). These factors, interacting with climate change, land-use dynamics, globalisation and changing consumption patterns, are driving agricultural and hydrological intensification (D'Odorico et al., 2013; van Vliet et al., 2012; Meyfroidt et al., 2013; Davis et al., 2015; Plieninger et al., 2016).

Agricultural intensification typically involves both increased use of agro-chemicals (including inorganic fertilisers, other soil ameliorants, and pesticides) and expansion of irrigated agriculture (Maraseni and Cockfield, 2012). Irrigation is typically implemented in arable landscapes where agricultural production is primarily constrained by temporally and spatially variable rainfall and water availability. Relative to rain-fed agricultural systems, irrigation can significantly reduce the climate-related risk and increase the reliability, productivity and profitability of cropping and fodder/pasture production in such regions. Where sufficient good quality water supplies are available, irrigation also offers an opportunity to grow a greater range of crops and, in some regions, to also double or triple crop the same area of land, such that several crops can be grown and harvested over the course of a single year (e.g. rice cropping in the Lower Mekong River Delta (An-Vo et al., 2021) and northwest Bangladesh (Mainuddin et al., 2019, 2020)).

In many regions, climate change - experienced as warming and greater evaporative demand and rainfall variability - threatens to adversely impact both water availability and crop yields (Döll, 2002). For example, modelling of cotton cropping systems in eastern Australia has indicated that future climate change impacts may lead both to declining yields (an estimated 17% decline by 2050) and to an almost 50% increase in the amount of irrigation required to maintain adequate soil moisture levels (Williams et al., 2015). Agricultural intensification, resulting in the expansion of irrigated crops as well as increased biomass growth and transpiration rates with the addition of nutrients, may further increase pressure on available water resources (Liu et al., 2015).

2.2 Supply-side irrigation water management

Significant investment in water supply infrastructure (in-stream and off-stream storage and distribution systems) occurs globally to meet a range of regional socio-economic and environmental needs, including the supply of water for agricultural irrigation requirements. Meeting the needs of multiple users places water resources at risk of over-exploitation. Managing water resources to ensure equitable access to all users (including the environment) and continuity of supply requires effective evidence-based policy, legislation, governance, monitoring and forecasting (Khan et al., 2008). Efficiencies in supply-side water resource management also need to be considered depending on the identity and scale of the water source - including surface waters (in-stream and off-stream, natural and impounded) and groundwater aquifers - and the distribution system (canal or pipeline).

Major investors in water storage and distribution infrastructure are governments, international aid agencies and the private sector, either individually or in partnership. Such investments are often aimed at delivering increased water and food security, and, in developing economies, they may also have humanitarian objectives associated with agricultural development, such as poverty alleviation (Khair et al., 2015, 2019). However, there is increasing recognition of the need to also consider the risks associated with such projects, including environmental impacts such as altered in-stream flow and sediment transport regimes; groundwater depletion; loss of critical ecological services (Eekhout et al., 2020); contributions to GHG emissions (Deemer et al., 2016); and detrimental socio-economic outcomes which may eventually prove maladaptive to the agricultural production systems and communities they aim to serve (Khair et al., 2015, 2019; de Vito et al., 2017). The success of these investments can be hindered where there is a lack of scientific understanding of potential biophysical and socio-economic, policy and legal consequences – both short- and long-term – and a breakdown in trust between the partner organisations (Khan and Mushtaq, 2009; Mushtaq and Bundschuh, 2016; Mutambara et al., 2016).

Irrigators and irrigation communities are also important actors in the management of water resources in water-scarce regions. In a review of the performance of farmer-managed smallholder irrigation schemes in Africa, Mutambara et al. (2016) identified a range of obstacles facing the sustainability of such regional development schemes and the need to implement 'best practices'. These included broad farmer participation and involvement, capital cost recovery, land tenure reform, improved water management, institutional management, traditional knowledge and management systems, water markets and adaptation to increasing water scarcity in the face of climate change. In Australia, which is often cited as a leader in agricultural water policy, failures in supply-side water policy also continue to be apparent (Kiem, 2013; Alston et al., 2016; Grafton, 2019; Colloff et al., 2021). For example, strong stakeholder response to early versions of the catchment-wide transboundary Murray Darling Basin Plan – whose objectives are to equitably meet the water needs of producers, catchment communities and the environment – led to significant political pressure and an eventual weakening of the environmental provisions of 'the plan'. The adopted plan has subsequently proved to have significant shortcomings and appears incapable of delivering its desired outcomes. In reality, supply-side water policy is difficult and rarely wholly evidence-based (Loch and McIver, 2012; Productivity Commission, 2018; Colloff et al., 2021).

The role of irrigation communities and the pressure placed on policy and regulators are also evident in the context of the supply of energy for use in irrigated agriculture. In Spain, Alcon et al. (2017) argued for collective action by irrigators in negotiating the purchase of electricity to ensure economies of scale and security of water provision, particularly where high energy-demanding water distribution systems increase the risk of irregular water supply. However, as shown by Khair et al. (2015, 2019), such considerations do not necessarily lead to sustainable water use, especially where maladaptive policies operate in combination with a lack of governance. In the semi-arid highland region of upper Balochistan, the 'fruit basket' of Pakistan, government policies aimed at increasing agricultural production and reducing poverty have led to rapid widespread installation of tube wells and unregulated groundwater extraction, largely driven by the provision of electricity subsidies. Such subsidies (in combination with irregular electricity supply) significantly influence farmers' adoption of tube wells and pumping decisions. Droughts, high returns received from irrigated fruit and vegetable cultivation, inefficient irrigation practices and lack of effective groundwater governance also have an impact on farmers' decisions, resulting in excessive water use despite awareness that the resource on which they rely is at increasing risk (Khair et al., 2019). A more rational electricity subsidy policy is critically needed to ensure that current groundwater levels (and access to supply) are maintained (Khair et al., 2015). This situation is not unique to Pakistan: increased competition among water users and expansion of groundwater exploitation are contributing to chronic groundwater degradation in many regions of the world (Morris et al., 2003; Shah et al., 2003; Bierkens and Wada, 2019; Gleeson et al., 2020). Comprehensive user-informed sustainable groundwater management policies and strong institutional support are required to ensure the viability of local groundwater and production systems and the well-being of farmers (Mushtaq et al., 2013b).

2.3 Demand-side irrigation water management

Where supply is unable to meet demand, water over-extraction and over-use threaten both water resource availability and the viability of irrigated agricultural production systems. This places responsibility not only on policymakers but also on water users, especially irrigators, to ensure the efficient use of water resources. Decisions to constrain irrigation water use are often driven by regulation, costs and biophysical feedbacks. For example, access to water resources may be restricted through a water management policy that regulates and enforces entitlements (also termed quotas) (DAWE, 2020). In some instances, where market mechanisms are in place, a price is paid per unit of water used for irrigation, providing negative feedback constraining over-use as well as a price signal that ensures high-value water use (influencing cropping decisions; Jackson et al., 2010). Other negative feedbacks may be biophysical; for example, the application of water in excess of plant growth requirements may lead to problems such as waterlogging, ponding, soil salinity issues, runoff and nutrient imbalance – with potentially detrimental impacts on crop yield and hence the productivity and profitability of agricultural enterprises (Feike et al., 2017; Ansari et al., 2018; Byrareddy et al., 2020). In a water-constrained environment, water security concerns should also, in theory, act to drive the adoption of on-farm water use efficiency (WUE) and other water conservation measures (Khair et al., 2015, 2019; van Duinen et al., 2016).

3 Water use efficiency: innovation and adaptation

Security of water supply and access both drive, and are increasingly challenged by, expansion of irrigated agriculture, especially in combination with possible climate change impacts. The impacts of water over-use on both agricultural production systems and on local and regional water resources are well recognised, as is competition for increasingly scarce water resources between irrigation and other environmental and societal needs. Greater awareness of the need to ensure water security has led to significant innovation such as technological advances and improved on-farm crop water management practices. These are predominantly aimed at achieving efficiency of water use on farms while maintaining and increasing crop yields (Jagermeyr et al., 2016; Brett et al., 2019). They include engineering and technological innovation which has led to the modernisation and increasing automation of irrigation infrastructure with significant potential for water savings at the farm level. Advances in real-time control and optimisation of irrigation, and increased use of remote sensing techniques, wireless communication systems and more versatile sensors may further improve on-farm WUE over time (Koech and Langat, 2018).

Despite these advances, poor management and configuration of irrigation systems can still waste significant quantities of water, leading to significant trade-offs between food and water security (Fig. 1). A review of empirical research publications, conducted by Pereira and Marques (2017), concluded that farms are generally inefficient with respect to water use and there remains significant potential for improvement. In many situations, barriers to the

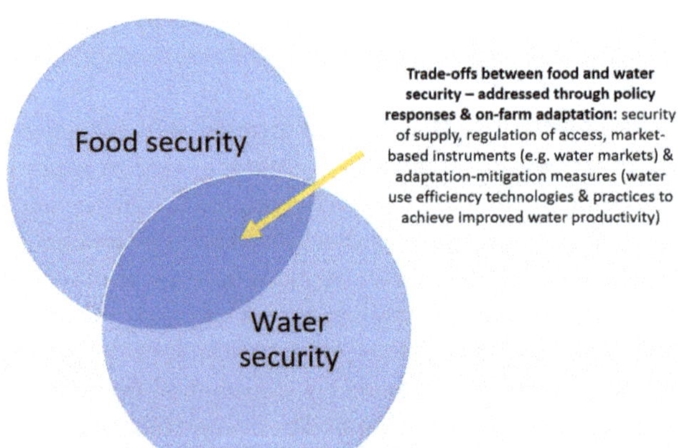

Figure 1 Food security is in part reliant on water security for enhanced agricultural production, although increased production can reduce water security with trade-offs that need careful management.

adoption of new agricultural technologies and practices remain, including lack of access to training, effective advisory and extension services, crop choice, localised irrigation systems, secure land tenure and credit opportunities. Any of these may constrain broadscale adoption and adaptation (Pereira and Marques, 2017).

Globally, it is estimated that up to 48% of agricultural water used could be saved through irrigation efficiency improvements in many of the world's river basins (Jagermeyr et al., 2016). Diversion of that saved water to support the expansion of irrigated agricultural land, in combination with adaptation to increase rain-fed yields in water-limited croplands, would significantly increase global food production. Jagermeyr et al. (2016) estimated that such strategies could reduce the water-related yield gap[1] by 62% and increase global agricultural production by 41%. However, in terms of water security, Koech and Langat (2018) stressed that such an outcome is effectively unsustainable in many regions. Instead, they argued that the adoption of water-efficient technologies and practices, which effectively increase water productivity and intensify production (i.e. achieve high levels of production using less water), must be accompanied by appropriate governance to prevent further expansion of irrigation and a net increase in total water consumption in water-limited environments. This includes regulations that limit water allocation and use and incentives for conservation. Feike et al. (2017) similarly argued that improved irrigation water productivity at the farm level may not significantly change or improve system-level over-use without policy intervention. This is an important policy consideration if irrigated agricultural systems are to remain sustainable in the long term (as discussed below).

3.1 Water productivity

Chartzoulakis and Bertaki (2015) discussed the importance of sustainable irrigation water use for agriculture in arid areas. The drive to develop policies aimed at WUE under scarcity conditions can achieve high levels of water productivity through better management, the uptake of new irrigation technologies and improved practices (e.g. soil moisture management, pest control, irrigation scheduling), leading to enhanced water efficiency. Other studies have looked at options for optimising irrigation water use to both maximise within-crop water productivity (i.e. the crop yield per unit of irrigation water applied) and also to potentially address increasing over-use of scarce surface water resources for irrigation and to minimise ecological degradation at the water resource system level. Increasing the efficiency of

1 The difference in crop yield between the highest achieved yield (with no water constraint) and the average yield with limited water availability (after Li et al., 2011)

on-farm water use requires decisions that consider a range of factors related to water management, including the irrigation system, the irrigation strategy, the method of scheduling irrigation (Jackson, 2009) and the economic implications of farm adoption decisions (Fernandez et al., 2020).

3.2 Irrigation practices

3.2.1 Irrigation management

Better management can increase the efficiency of irrigation, saving costs, energy and water resources. In order to ensure water use efficiency and address irrigation supply shortages, irrigators may practice irrigation scheduling based on either climatic data (such as cumulative pan evaporation or calculated from weather station parameters) or soil moisture-based measurements relative to crop requirements. Monitoring of crop water use directly through the plant in real-time is also possible but expensive and generally used only in research applications.

Irrigation scheduling decisions alone have been found to result in both yield benefits (up to 27.9%) and reduced irrigation energy consumption (up to 35.7%) associated with water savings in irrigated wheat cropping in Pakistan (Ansari et al., 2018). Such irrigation decisions may be based around the implementation of deficit irrigation practices designed to save water by subjecting the crop to periods of moisture stress with limited impact on crop yield. These have been found to be an effective adaptive response to water scarcity with the potential to deliver efficiency gains, water savings and benefits at the system level (Mushtaq and Moghaddasi, 2011; Shukr et al., 2021).

Shukr et al. (2021) compared the WUE of full irrigation and deficit irrigation practices across a range of irrigation delivery systems. These included gravity-fed furrow or surface irrigation and pressurised overhead sprinkler and sub-surface drip irrigation in cotton production systems in eastern Australia. For the pressurised irrigation systems, lint yield was greatest under full irrigation, while WUE was optimised under deficit irrigation at 60% of full irrigation for both systems; however, in the surface irrigation systems, crops were more likely to be exposed to periodic water stress when deficit irrigation was practised.

Similarly, Mushtaq and Moghaddasi (2011) compared three irrigation scheduling scenarios – 1) optimisation with full irrigation, 2) optimisation with deficit irrigation and 3) deficit irrigation without optimisation – using non-linear optimisation modelling with endogenous crop production and profit functions to evaluate the potential of deficit irrigation. This study indicated that optimisation with deficit irrigation could increase overall water use efficiency, resulting in water savings while maximising net returns. The authors concluded that deficit irrigation represents a cost-effective adaptive response that meets both climate and environmental objectives, especially where water saved through deficit irrigation is used to augment environmental flows and restore environmental balance.

In irrigated lowland rice production systems in Asia, alternate wetting and drying (AWD) water management involves irrigating rice fields to a desired depth, allowing them to partially drain then re-irrigating when the water falls to a certain level. This has been shown to reduce irrigation water input by up to 38% with no reduction in yield. Based on 'with and without' AWD comparisons (Lampayan et al., 2015), AWD can reduce water pumping costs and fuel consumption, thereby contributing to an increase in farmers' income – by 38% in Bangladesh, 32% in the Philippines and 17% in south Vietnam. Such information about the economics of irrigation management options can be critical in on-farm adoption decisions (Fernandez et al., 2020), while training in water-saving irrigation practices has been shown to increase the uptake of AWD practices amongst rice farmers in the Zhanghe irrigation system in China (Mushtaq et al., 2009a). In some instances, institutional policies and governance around water scarcity may also be required to ensure more widespread adoption (Mushtaq et al., 2009a).

3.2.2 Climate forecasting and irrigation decision making

Seasonal climate forecasts are available as a result of significant and ongoing advances in understanding climate drivers (Stone et al., 1996; Ali et al., 2018) and the role of interactions between the oceans, land, ice cover and atmosphere (Hansen et al., 2006). They hold significant value for agricultural decision-makers (e.g. Schepen et al., 2020). Forecasts at a range of time scales are produced by organisations such as the UK Meteorological Office and, in Australia, the Bureau of Meteorology with readily accessible online data (Northern Australia Climate Program, 2021; WMO, USQ and CIAT, 2021) and through climate service providers (Cobon et al., 2021).

In regions subject to significant climatic variability, variability in rainfall, runoff and recharge of both surface water and groundwater resources affect irrigated crop production. Advance warning of likely climatic conditions for the upcoming irrigated cropping season can provide decision-makers with an opportunity to better plan and manage their available water resources.

No seasonal climate forecast will be 100% accurate given the dynamic and complex nature of global climate systems and limitations in modelling capacity. However, bio-economic modelling shows that the use of seasonal climate forecast information (even at a forecast skill level of 60%) in irrigation scheduling decisions in a sugarcane farming system in the Burdekin catchment, eastern Australia, has significant potential to increase water use efficiency, save both water and energy and deliver improved crop yield and profitability (An-Vo et al., 2019; Cobon et al., 2019).

Similarly, water allocation forecasts, including the forecast price of irrigation water where water market mechanisms are in place, that incorporate seasonal climate information can inform cropping and irrigation water management decisions (Mushtaq et al., 2012; Nguyen-ky et al., 2018). In the Coleambally irrigation area (CIA) in southeastern Australia, such decisions might include changing the combination of winter crop planted, abandoning irrigation for certain summer crops, temporary purchase or sale of water on the water market and deficit irrigation scheduling, enabling irrigators to avoid losses that may have otherwise been incurred. Without such forecast information, irrigators may overestimate the amount of water they will be able to access throughout the season (i.e. seasonal water allocations) leading to poorly informed cropping plans and over-commitment of resources and inputs. This information was found to be of particular value in periods of relative water scarcity, when allocations, hence access to irrigation water volumes, were constrained at well below irrigators' legal entitlements (Mushtaq et al., 2012).

3.3 Water-efficient irrigation technologies

3.3.1 Precision irrigation

Advances in agricultural technologies have the potential to increase both crop productivity and water and energy efficiency in irrigated agricultural systems, especially in arid and semi-arid regions (McClintock, 2009; Mushtaq et al., 2013a). This is the case even in more humid regions such as northern Europe where supplementary irrigation enables more effective/efficient use of rainfall and the greater control afforded by the technologies reduces the risk of productivity losses and environmental impacts (including ponding, deep drainage, nitrate leaching) associated with conventional irrigation (Daccache et al., 2015).

3.3.2 Pressurised irrigation systems

Surface and sub-surface drip irrigation systems were developed over 60 years ago for use in annual row cropping systems. Since then, drip irrigation has undergone continued evolution in terms of materials used, installation and management, and its use has expanded into a range of cropping systems, including perennial crops. This has resulted in improved yield, crop quality and water productivity, and significant savings in water use and agronomic costs (weed control, fertilisation, tillage) (Ayars et al., 2015).

Similarly, sprinkler irrigation systems have undergone significant development since the early iterations of solid set and side roll systems (which are still in use) towards large mobile irrigation machines including centre pivot and lateral move irrigators. These systems can apply accurate water volumes to row and broadacre crops and pastures. While uniformity issues can be a problem in windy conditions, the pressure and volume controls generally increase the performance. Well-designed systems are able to meet crop needs without exceeding the soil's surface infiltration rate, thereby reducing the potential for ponding and runoff.

High-pressure irrigation systems continue to be used in some regions. These systems offer the ability to irrigate small and disaggregated parcels of land by moving the irrigator to new areas when necessary. However, their performance is heavily impacted by strong crosswinds and their energy operating costs make them less attractive to farmers looking for more water and energy-efficient irrigation systems (Maraseni et al., 2012a,b; Mushtaq et al., 2013a).

3.3.3 Decision support and management tools

Technology-assisted decision support and irrigation management is an area of recent innovation with the potential to improve on-farm crop water management to further improve water efficiency. Such technologies have the capacity for efficient automated farm monitoring and are increasingly incorporated into decision support tools and automated irrigation management systems (Khan et al., 2018).

Wireless sensor networks (WSNs) incorporate a wide range of sensors that directly interact with the physical environment, remotely sensing and monitoring soil parameters, like temperature, moisture and fertility, and providing real-time data to inform management decision-making based on crop requirements for water and fertiliser. Technology platforms such as the internet of things (IoT), which can retrieve real-time data and integrate them to determine optimal irrigation volumes, are increasingly proposed for 'smart' water management. Kamienski et al. (2018, 2019) reported on the SWAMP project, which developed and applied IoT-based methods and approaches for smart water management in precision irrigation systems across a range of crops, climates and countries. Initial results from these pilot applications are promising but further work is recommended to ensure scalability and reduce computational requirements (Kamienski et al., 2019). Other trials have shown that the application of IoT technologies can save up to 26% of water when using temperature data and up to 34% when temperature, humidity and soil moisture data inputs are used (Gloria et al., 2020).

Advances in IoT and WSN technologies, and the increasing availability of low-cost sensors that can be connected to nodes, mean that such irrigation management and agriculture monitoring systems are becoming increasingly affordable. García et al. (2020) reviewed the current state of the art regarding smart irrigation systems (nodes, wireless technologies), including the parameters that are monitored in irrigation systems regarding water quantity and quality, soil characteristics and weather conditions, and the challenges and best practices for the implementation of such sensor-based irrigation systems.

The increased capacity and availability of satellite technology are also beginning to feed into on-farm management decision-making. For example, a dedicated web GIS and satellite-based irrigation advisory system can now provide decision-makers with near real-time information (within 24–36 h) on crop canopy development (leaf area index, albedo and soil cover) derived from high-resolution (20 m) multispectral satellite images. In combination with in-situ agro-meteorological data and the calculation of crop evapotranspiration, such systems can provide timely site-specific recommendations for irrigation scheduling decisions, enabling growers to better control the irrigation process and improve the overall efficiency of irrigation systems. The crop heterogeneity captured by high-resolution images also enables a better understanding of the variability of soil texture and fertility, plant nutrition and irrigation system performance. In many cases, this has resulted in reduced water use and increased irrigation efficiency (Vuolo et al., 2015).

3.4 Adoption of water use efficiency irrigation innovations

In addition to improved technologies and practices to increase agricultural water use efficiencies and water productivity, it is also essential that key barriers to on-farm adoption are recognised and addressed, as discussed above. These include critical barriers in terms of financial capacity and skill in effectively implementing innovations. To be effective in facilitating widespread adoption, extension programmes should also (i) acknowledge profit as an over-riding strategic objective in farm decision-making; (ii) consider water as both a key input to farming systems and a core factor in achieving sustainability (i.e. addressing the 'water–energy–food nexus'); (iii) adopt an adaptive framework that accommodates both short- and long-term goals; (iv) promote innovations that contribute to meeting policy outcomes and (v) provide evidence of reduced production risks (Smidt et al., 2016). It is also the case that, across the world's irrigated cropping systems, many innovations are unlikely to be equally accessible. Limitations in terms of irrigation infrastructure, internet access and technical capacity to install, maintain and manage such systems, amongst other constraints, may also play a critical role (see the case studies below).

3.4.1 Case study 1: Measurement and assessment is the first step to improvement and optimisation

In many parts of the world, evaporation from open water storage reduces agricultural water availability, potentially limiting crop production. In the northeastern parts of India, water storage ponds are a common aspect across the landscape. While these ponds have multiple uses, they offer a good source of irrigation water at various times of the year (Jana et al., 2012). When investigating solutions to mitigate losses at any point in the irrigation system (water source, transfer or application system), it is important to critically measure and assess the system. Schmidt et al. (2019) investigated losses in several open storage ponds in the states of West Bengal and Bihar in India. Hourly and weekly water storage level data were collected and analysed using purpose-built software able to separate the seepage loss from the evaporative loss. While there are long hot summers in India, the major mechanism of loss was in fact seepage through the bed and banks of the storage pond rather than through evaporation from the water surface. Around 2.7 mm per day was being lost to evaporation compared with 4.9 mm per day lost to seepage. This assessment allows quantification of the costs and economic returns of measures to mitigate this loss.

3.4.2 Case study 2: Optimising irrigation system performance can lead to reduced costs

A lychee grower on the Sunshine Coast in Queensland, Australia, undertook both an irrigation system assessment and a pump evaluation (Eberhard et al., 2013). The Irrigation Pump Evaluation and Reporting Tool (IPERT) online assessment was used to collect physical measurements (motor speed (rpm), pump speed (rpm), suction, pressure and discharge rate, and electricity used) as well as other information about the irrigation system and electricity tariffs. The overall efficiency of the motor and pump was calculated to be 27.2% (motor efficiency 87.5% and pump efficiency 33.4%). Using the electricity tariff information, the cost of using this pump was estimated to be approximately AU$168 per million litres.

The grower then invested $6500 on a new direct drive pump, motor, filters, suction line and fittings to improve the pumping efficiency and reduce irrigation costs. The combined efficiency of the new system was higher at 40.7% (motor efficiency 87.5%, pump efficiency 49%). This reduced pumping costs to AU$ 113/Ml, a saving of $55/Ml pumped. Based on the grower's average annual pumping hours, the payback period for this investment was approximately 2.7 years.

This case study highlights the value of measurement in making informed decisions to reduce both energy use and costs, thereby improving economic performance and minimising the environmental impact of irrigation systems (Eberhard et al., 2013).

3.4.3 Case study 3: Without proper training and management, technology alone will not produce results

In the past 30 years, Vietnam has developed into the second-highest coffee-producing nation in the world. Coffee agronomy has been widely supported by the private sector, and the need for irrigation in the dry season is well accepted. However, the volume and timing (scheduling) of irrigation is not based on measurement or understanding of crop water requirements. For many years, the standard recommendation for coffee irrigation in the Central Highlands was 800 l/tree per irrigation. Usually, three or four irrigations were applied from January to April before the onset of the monsoon (Byrareddy et al., 2020).

Buon Ma Thout, in the Dak Lak province in the Central Highlands, is a significant coffee-growing region. Farmers intercrop robusta coffee with black pepper and, increasingly, fruit and nut trees including durian, avocado and macadamia. The trees also provide shade for optimum coffee bean production (Vaast et al., 2006). Irrigation water can be accessed from tube wells, open dug-wells and surface water streams, and often the resource is shared with neighbouring farmers. The most popular practice is traditional irrigation by hand using a hosepipe to fill a basin at the base of each tree. This practice takes hundreds of hours each season and the volume applied to each tree is highly variable.

A number of farmers have been involved in projects funded by development agencies or coffee certification programmes. One such project provided a case study of farmers with underground water mains and laterals for a drip irrigation system designed to deliver 20 l/h, coupled with monitoring and control. One system had a fertigation plant to allow the application of nutrients through the irrigation system. The other had a system of biophysical monitoring of soil moisture and weather to aid in scheduling irrigation applications. Farmers could start and stop the irrigation remotely using their mobile phone.

Moving from the traditional hosepipe and basin method to fertigation-enabled precision automated drip irrigation in a single season was a very large step. Some of the farmers already had a good understanding of how to use fertigation and were happy with the automated control and use of the mobile phone for starting and stopping irrigations. In addition to savings in labour, the drip irrigation system provided the ability to apply small amounts of water more regularly to better match the crop water demand and reduce the agronomic and environmental impacts of over-irrigation. Other farmers continued to apply their traditional understanding by starting the irrigation system and allowing it to run until the bund was filled. This is not the best practice for this type of technology. The lower flow rate from drip emitters enables significantly more water to be applied to the crop before the bund is full, thereby increasing the likelihood of deep percolation of irrigation water through the soil profile down to below the crop root zone.

This case study highlights the risks associated with poor management, even with high-performance precision irrigation systems, and the need for training alongside access to any new technology.

4 Challenges

4.1 Water-energy trade-offs

While adaptation through the adoption of innovations is critical to buffering the effects of climate variability and change on both food and water security, it is unlikely to completely negate them (Jagermeyr et al., 2016) and may, in fact, contribute to further change. For instance, in Australia, government policies and community perceptions of the irrigation sector generally promote the conversion from gravity-fed to pressurised water use efficient irrigation systems as a way of reducing water consumption. However, optimising water savings through the adoption of pressurised irrigation systems may have unintended resource and environmental consequences due to the increase in energy consumption and GHG emissions (Jackson et al., 2010).

Irrigation systems, apart from those that are entirely gravity-fed, generally require pumps to access, distribute and/or pressurise water. Conventional pumping systems require electrical energy usually derived from fossil fuels (diesel generators or a coal- or gas-fired electricity grid). Significant energy is also used to manufacture, transport and operate the majority of other cropping farm inputs such as seed, agri-chemicals, fertilisers and equipment (e.g. Maraseni et al., 2009b). Increased energy consumption has implications for on-farm costs of production and also, potentially, for the level of production-related GHG emissions (Mushtaq et al., 2013a; Maraseni et al., 2021).

Modern irrigation technologies may save significant volumes of water and offer an effective way to adapt to increasing climate variability and hence manage risks to food and water security. They may also result in potential conflicts between climate change mitigation and adaptation policies (Mushtaq et al., 2013a). Without appropriate intervention, such trade-offs will likely drive cascading impacts on environmental conditions and functions (de Vito et al., 2017), with potentially negative consequences for food production and security

(Fig. 2). In the context of increasing global concerns regarding climate change, such trade-offs point to the need to simultaneously optimise both water and sustainable energy use in irrigated cropping systems (Maraseni et al., 2012a,b; Mushtaq et al., 2013a; Hafeez et al., 2014; Tarjuelo et al., 2015).

Technological advances in recent decades have improved WUE of irrigation equipment and water productivity of irrigated agriculture, as well as significant on-farm infrastructure investment such as in Australia's Murray Darling Basin, to address water shortages (Mushtaq et al., 2013a). However, such on-farm infrastructure investment has the potential to also increase energy consumption and generate considerable GHG emissions, suggesting potential conflict in terms of national climate change mitigation and adaptation policies.

For example, the introduction of a price on carbon (a mitigation policy) will have financial implications for the adoption of new water-efficient irrigation technologies (an adaptation measure) and may limit the extent to which these technologies are taken up (Maraseni et al., 2012b). On the other hand, in combination with increasing water scarcity, higher fossil fuel prices are likely to lead to more energy-efficient farming and increased use of renewable energy and energy-efficient technologies that are both cost-effective and significantly reduce irrigation-related GHG emissions. Ultimately, decision-makers need to balance the need to ensure productivity and sustainability in the face of a rapidly growing world population and increasing demand for food and fibre and the need to reduce carbon dioxide emissions from fossil fuel use in agriculture,

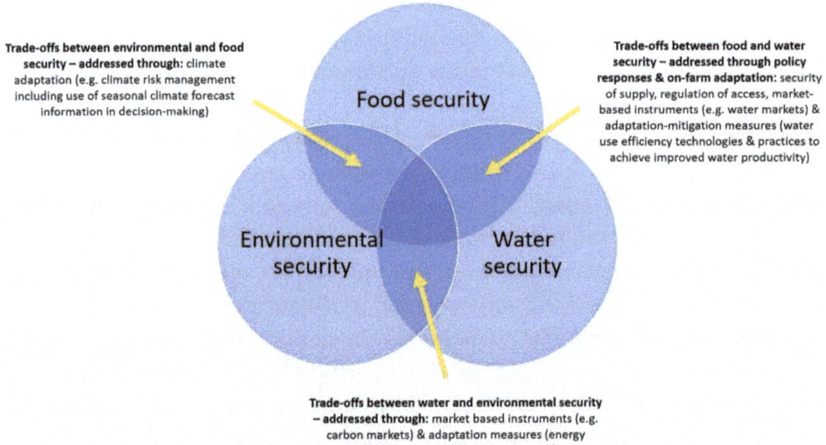

Figure 2 Cascading negative consequences on food security occur where increased fossil fuel energy is used to achieve water security in irrigated agricultural systems and where environmental consequences are not effectively mitigated.

which contributes some 22% of global carbon dioxide emissions (FAO, 2011; Bundschuh et al., 2014).

Jackson et al. (2010, 2011) examined water and energy patterns of crop production on farms in two irrigated areas of Australia, one using surface water and the other groundwater. The aim was to better understand the risk, uncertainty and sensitivity associated with the link between water and energy consumption at the farm scale. The study found that irrigation water and energy needs were sensitive to the water source. Converting from surface to pressurised systems in the surface water supplied region resulted in a 10–66% reduction in water application, accompanied by an increase in energy consumption of up to 163%. Conversely, in the groundwater-dependent region, increased WUE resulted in a 12–44% decline in energy consumption due to the reduced amount of water being pumped. In order to optimise energy and water use, it was recommended that pressurised irrigation systems be used in areas requiring extraction of groundwater, while efficient gravity-based methods, coupled with good management practices, be promoted in surface water supplied areas (Jackson et al., 2010). These recommendations were supported by modelling that investigated the impact of climatic (evapotranspiration and rainfall) and technical (irrigation system efficiency, pump efficiency, suction and discharge head) factors on water and energy use for a given irrigation method (Jackson et al., 2011). Overall, an irrigation system that minimises water and energy consumption, as well as GHG emissions, will reduce risk and enhance resilience at the farm level (Jackson et al., 2011).

4.2 Integrated analysis of trade-offs (water and energy use and greenhouse gas emission implications)

Increasingly complex energy–agriculture relationships require an in-depth understanding of water and energy trade-offs. For example, analysing the complex energy, water and economic dynamics across a selection of major rice-growing countries, Mushtaq et al. (2009b) showed that trade-offs exist between yield and energy inputs. Overall, higher yields can be achieved with higher levels of energy input, but improved energy efficiency may be evident where enhanced technology and improved farm management practices are adopted. Similarly, water-saving irrigation practices may contribute to greater water productivity with economic and energy benefits associated with reduced pumping (Mushtaq et al., 2009b). However, even 'soft options' for increasing water productivity to address the problem of increasing water scarcity (such as water re-use or recycling) have potential economic and environmental trade-offs, requiring additional investment and potential for increased GHG emissions. For example, water re-use may not only contribute significantly to water productivity but may also have significant energy use and GHG implications. In the Upper Pampanga River integrated irrigation system (UPRIIS) in the Philippines, Maraseni et al. (2010) found that water re-use increased water productivity by an estimated 21% but, due to the increase in pumping requirements compared to gravity-fed canal irrigation without re-use, GHG emissions also increased by almost 150%. The study concluded that water re-use should only be a priority in areas where water scarcity is a serious issue (Maraseni et al., 2010).

Applying an integrated assessment framework to explore trade-offs associated with the adoption of more water-efficient but energy-intensive irrigation technologies, Maraseni et al. (2012a) investigated water and environmental security in five case study systems that had converted from traditional surface irrigation to pressurised sprinkler or drip irrigation systems. Trade-offs in terms of fuel and energy-related emissions were apparent in all instances except where older energy inefficient hand-shift and roll-line sprinkler irrigation systems were replaced. This suggested, when implementing on-farm infrastructure investment policy, that priority should be given to replacing such inefficient and energy-intensive sprinkler irrigation systems (Maraseni et al., 2012a,b; Mushtaq et al., 2013a). The results of the study also indicated that beneficial changes in the use of agricultural machinery and agri-chemicals were apparent where newer pressurised sprinkler or drip irrigation equipment was adopted, with additional energy efficiencies achieved through the use of precision techniques such as fertigation and application of selective herbicides through the irrigation system. Such 'win–win' situations, rather than trade-offs, are not necessarily expected. The authors argue that such an integrated evidence-based approach to policy development and strategic decision-making is critical for the prioritisation of investments on both economic and environmental grounds and to ensure that conflict in designing national climate change mitigation and adaptation policies is avoided (Maraseni et al., 2012a,b; Mushtaq et al., 2013a, 2015a,b).

Similarly, modernisation of irrigation systems in Spain is delivering increased WUE and productivity as well as improved operation and management of irrigation systems and working conditions for farmers (Tarjuelo et al., 2015). Here, open-channel gravity-based surface irrigation systems are being replaced with pressurised distribution networks and irrigation systems. However, there are also increased costs due to the investment in, and increased energy demand for, the new technologies. Such trade-offs between water conservation and energy consumption point to the need to balance water and energy use to minimise detrimental financial and environmental impacts. The economic, social and environmental viability of the irrigation modernisation process should be assessed on a case-by-case basis to ensure the appropriate design and management of irrigation systems. Communications and capacity building, as well as the integration of associated advisory services and

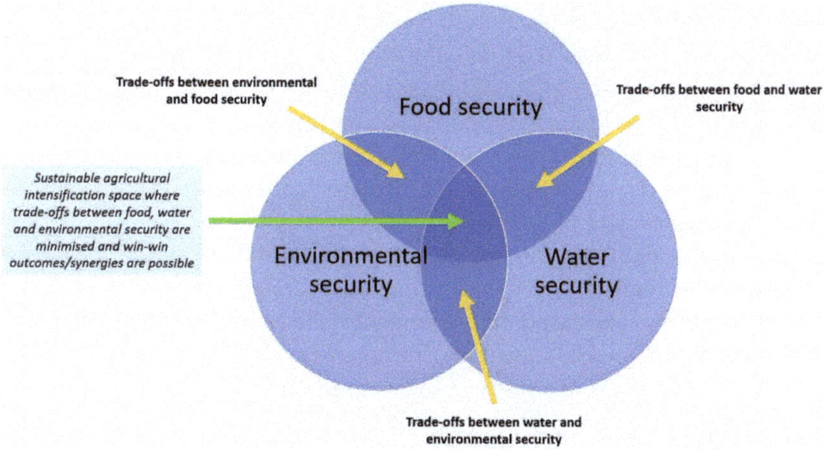

Figure 3 Sustainable intensification of irrigated agricultural production systems requires integrated system-level considerations of food, water and environmental security.

web-based GIS platforms, will also aid in ensuring the real-time transfer and sharing of information that enable farmers to access feedback on their irrigation management decisions in terms of consumption of water, energy and other production inputs (Tarjuelo et al., 2015).

4.3 Sustainable intensification

Strategies to meet the challenge of increasing agricultural yields per hectare to address increasing demand from a growing global population, while also avoiding over-exploitation and other impacts on land, water and energy resources, require solutions that simultaneously optimise water use, reduce energy consumption and improve crop quality and yields, while minimising potential conflicts and trade-offs between economic, food production and environmental values (Gadanakis et al., 2015). Termed 'sustainable intensification', such strategies also aim to reduce the contribution and exposure of agriculture to the social, economic and environmental impacts resulting from climate change and increasing climate variability, including climate extremes (Fig. 3).

At the farm scale, conservation agriculture-based sustainable intensification (CASI) practices[2] have been implemented in rice-based cropping systems on the Eastern Gangetic Plains of South Asia. Compared to traditional agricultural management practices, they have increased productivity and profitability;

2 CASI encourages the adoption of conservation agriculture principles for crop management including mechanised crop establishment with zero tillage, intensification through crop diversification/rotations, and more efficient crop irrigation and nutrient management. In small studies across parts of South Asia, these have been shown to reduce water, energy and labour requirements while maintaining or increasing crop productivity (Gathala et al., 2020).

reduced production costs; increased gross margins and reduced water, energy and labour requirements and GHG emissions. Benefits include increased crop productivity (up to 10% of yield), water productivity (up to 19%) and energy inputs per unit of production (up to 26%). Energy demand was reduced by up to 11%, labour requirements up to 50% and production costs by up to 22%. Overall, improved management led to increased gross margins of 12–32% and reduced GHG emissions by 10–17% (Gathala et al., 2020).

In these systems, enhanced implementation and uptake of CASI were evident where there was effective community and government support as part of a community-based implementation framework that (i) connected with external stakeholders including researchers, extension agents, financial institutions and agro-industries; and (ii) engaged with and provided evidence of success to policymakers and governments. Such an implementation framework that effectively supports farmers in implementing improved management practices has significant potential to assist smallholders in transitioning to low-energy, sustainable food production with positive outcomes in terms of the food–energy–water nexus (Gathala et al., 2020).

At a regional planning scale, Maraseni et al. (2021) conducted an integrated trade-offs analysis of the energy–water–food nexus, land use and climate change for a major irrigated cropping area in the southern Murray Darling Basin of Australia. They applied a future-looking model to explore whether a regional strategy could optimise the use of land and water resources and achieve both increased crop production and reduced GHG emissions while sustaining economic outcomes. This analysis showed that (i) up to 50% reduction in GHG emissions from irrigated crop production is possible without compromising total gross margins when surplus water and land available is reallocated to environmental planting; and (ii) further emissions reduction without trade-offs is achievable where carbon pricing and/or water market mechanisms are used to ensure higher value water use. Integrated approaches such as this can contribute to the development of government emissions reduction policy with the enhanced potential to achieve balance, equity and overall sustainability across multiple policy areas (Maraseni et al., 2021).

4.3.1 Alternative energy sources

As discussed earlier, increasing the uptake of pressurised irrigation systems to improve the efficiency of irrigation systems often results in substantially increased energy demand. Together with rising energy costs, this could challenge the profitability of farming and also increase the contribution of agriculture to GHG emissions and climate change (García et al., 2018). Sustainable and less costly energy alternatives include the use of solar energy for pumping irrigation water. Solar photovoltaics (PV), wind power, small hydro and bioenergy have become increasingly affordable options to conventional fossil fuel electricity generation (including off-grid diesel generators). The development of control systems able to synchronise renewable power availability with the energy demand for water pumping in different sectors of irrigation networks also provides enhanced capacity to manage the key variables of such systems. Many successful commercial solar-powered water pumping systems have been installed in the last two decades. For example, in an olive orchard in southern Spain, a solar PV system provided enough water to satisfy crop requirements throughout the irrigation season and avoided GHG emissions of 1.2 t CO_2-equivalent (García et al., 2018). In all cases, it is important that the use of such technologies is supported by an understanding of the daily soil water balance and irrigation energy demand of the crop, the hydraulic behaviour of the irrigation network and the variable power production of solar and wind systems (Scobie et al., 2018).

4.3.1.1 Integrated technological advances

Autonomous off-grid systems - for example, combining solar PV with battery storage for irrigating tomatoes (Yahyaoui et al., 2017) - are now widely used. Correct management of such installations ensures efficient crop irrigation by balancing energy and water requirements and extends the lifetime of components, particularly the batteries. Such systems may use fuzzy logic inside the control system to firstly evaluate the water volume needed by the crop during the vegetative cycle using a detailed model of crop evapotranspiration and irrigation frequency for the specific site and crop characteristics. Based on this and the energy availability, a control algorithm determines the switching of relays which connect the solar panels, inverter, batteries and water pumps. The control algorithm fulfils the specified objectives by considering criteria related to the water volume needed to irrigate the crop, the safe and cost-effective operation of the batteries and the operating periods of the pump (Yahyaoui et al., 2017).

4.4 Integrated evidence-based agricultural, water and energy policy development

4.4.1 Unintended consequences

The importance of managing access to, and use of, water resources has long been recognised by legislators around the world, especially in regions subject to significant seasonal and climatic variability in water availability and demand. There has been a necessary focus on promoting greater WUE in water-scarce areas, but this has potential for unintended consequences such as 'increased ex-post water consumption'. The paradox is that, while improved irrigation efficiency may reduce crop water use, it may also result in increased water consumption.

Berbel et al. (2018) conducted a micro-economic analysis of the relationship between irrigation efficiency, water demand and water pricing conducted to better understand the effects of water-saving investments and increased irrigation efficiency on water use and consumption. They found that, while greater efficiency significantly reduced in-crop water use (delivering increased water productivity), there was little impact on overall water consumption regardless of water cost due to a 'rebound effect' resulting from irrigated area expansion, crop-mix changes and market forces (Berbel et al., 2018). Such a situation - where a technology designed to enhance the efficiency of using a natural resource and lower consumption results in increased consumption of the resource - is referred to as a Jevons' paradox. Policymakers need to account for users' behavioural responses to avoid such unintended perverse consequences (Sears et al., 2018).

Policymakers also need to better understand the trade-offs and synergies between climate change, energy and water sector policies in order to ensure that agricultural water savings are effectively rewarded, GHG emissions are penalised and agricultural production systems are economically viable, environmentally sound and able to operate in a low-carbon, water-constrained economy (Mushtaq et al., 2015a,b). For example, the Australian Government 'National Plan for Water Security' (McCormick, 2007) aimed to address the challenge of increasing water scarcity through significant on-farm infrastructure investment to facilitate the adoption of new water-efficient pressurised irrigation systems. However, conversion to these systems had significant potential to increase on-farm energy consumption and GHG emissions, thereby creating potential conflicts in terms of mitigation and adaptation policies (Maraseni et al., 2012a). Similarly, Jobbins et al. (2015) looked at three case studies in Morocco, where farmers had adopted drip irrigation, through the lens of the 'water–energy–food nexus'. Despite the conceptually apparent benefits of implementing these technical options, water and energy efficiency did not necessarily result in reduced consumption of either water or energy. This suggests that caution is required in applying the nexus concept in policy formulation and that the risks of unintended consequences, including exacerbating other dimensions of poverty and inequality when policies are focused on issues of scarcity and efficiency, need to be recognised (Jobbins et al., 2015).

Effective and sustainable management of water resources on extended timescales necessitates a comprehensive understanding of the complex interactions between the physical, agricultural and socio-economic drivers of water use (Smidt et al., 2016; Pham et al., 2020, 2021). Pham et al. (2020, 2021) argued for integrated policy development that addresses the range of cross-sectoral drivers impacting the sustainability of an irrigated cropping system including regional population dynamics, land and water governance and adoption of water use efficient technologies and practices. Unless these align, they can exacerbate water use and constrain the sustainability of agricultural production systems, resulting in a range of socio-economic and environmental impacts. Overall, water policy and management fall into the realm of complex decision-making, and need to be informed by the challenge of minimising water–energy–food trade-offs while also aiming to ensure the sustainability (and profitability) of agricultural production (Smidt et al., 2016; Maraseni et al., 2021).

4.4.2 Water governance, adaptation and mitigation strategies

Climate change is expected to significantly impact the agriculture, water and energy sectors. Developing effective and complementary mitigation and adaptation policies is a significant challenge to policymakers (Mushtaq et al., 2015a,b). For example, in the water sector, climate adaptation approaches need to be evaluated in the context of climate change mitigation policies to ensure that both economic and environmental outcomes are optimised. Policies for improved water management, especially those focussing on adaptation to better manage climate impacts (such as conversion to new water-efficient irrigation technologies), may lead to increased GHG emissions, indicating a potential conflict. Climate change mitigation measures aimed at reducing GHG contributions may in turn influence access to, and management of, water resources for irrigation (Dubreuil et al., 2013). There is often a fine balance between either minimising the trade-off between climate adaptation and mitigation approaches or creating a win–win situation (Mushtaq et al., 2015a,b; Maraseni et al., 2021).

Unfortunately, in the majority of countries, policy and development planning sit within sectoral silos and coordination between different government levels and sectors can be a key barrier to improvements in integrated water policy development and its implementation (Barbosa et al., 2016, 2017; Khair et al., 2019). Such institutional structures fail to account for the inter-relatedness of water, energy, and food production or for the critical role of the environment in the provision of vital ecological services and natural resources. In many countries, this is often accompanied by a lack of both political will and strong institutional and governance capacity. Policy development should also be informed by, and responsive to, local/regional issues, while its effective implementation requires a level of local authority and responsibility (Hering et al., 2015).

Greater coordination between and within government departments, the private sector and other stakeholders as well as effective governance arrangements are needed to ensure sustainable intensification of agricultural production systems and the resilience of agricultural landscapes, as well as agriculture's ongoing contributions to the United Nation's overarching Sustainable Development Goals (SDGs) such as eliminating poverty (SDG #1), zero hunger (#2), clean water and sanitation (#6), affordable and clean energy (#7), climate action (#13) and life on land (#15) (UN, 2015).

5 Acknowledgements

This work was supported by the University of Southern Queensland through the German Federal Ministry for the Environment, Nature Conservation and Nuclear Safety (BMUB)'s International Climate Initiative funded project: *Applying seasonal climate forecasting and innovative insurance solutions to climate risk management in the agriculture sector in SE Asia*.

6 Where to look for further information

- Food and Agriculture Organization of the United Nations (2014). The Water-Energy-Food Nexus A new approach in support of food security and sustainable agriculture. https://www.fao.org/3/bl496e/bl496e.pdf.
- United Nations (2014), UN World Water Development Report https://www.unwater.org/water-facts/water-food-and-energy/.
- US Department of Energy (2014). The Water-Energy Nexus: Challenges and Opportunities. Overview and Synthesis. https://www.energy.gov/sites/default/files/2014/07/f17/Water%20Energy%20Nexus%20Executive%20Summary%20July%202014.pdf.

7 References

Alcon, F., García-Bastida, P. A., Soto-García, M., Martínez-Alvarez, V., MartinGorriz, B. and Baille, A. (2017). Explaining the performance of irrigation communities in a water-scarce region. *Irrigation Science* 35(3), 193–203.

Ali, M., Deo, R. C., Downs, N. J. and Maraseni, T. N. (2018). Multi-stage committee based extreme learning machine model incorporating the influence of climate parameters and seasonality on drought forecasting. *Computers and Electronics in Agriculture* 152, 149–165.

Alston, M., Whittenbury, K., Western, D. and Gosling, A. (2016). Water policy, trust and governance in the Murray-Darling Basin. *Australian Geographer* 47(1), 49–64.

Ansari, R., Liaqat, M. U., Khan, H. I. and Mushtaq, S. (2018). Energy efficiency analysis of wheat crop under different climate- and soil-based irrigation schedules. *Proceedings of* the Multidisciplinary Digital Publishing Institute 2(5), 184.

An-Vo, D.-A., Mushtaq, S., Reardon-Smith, K., Kouadio, L., Attard, S., Cobon, D. and Stone, R. (2019). Value of seasonal forecasting for sugarcane farm irrigation planning. *European Journal of Agronomy* 104, 37–48.

An-Vo, D.-A., Radanielson, A. M., Mushtaq, S., Reardon-Smith, K. and Hewitt, C. (2021). A framework for assessing the value of seasonal climate forecasting in key agricultural decisions. *Climate Services* 22, 100234.

Ayars, J. E., Fulton, A. and Taylor, B. (2015). Subsurface drip irrigation in California–Here to stay? *Agricultural Water Management* 157, 39–47.

Barbosa, M. C., Alam, K. and Mushtaq, S. (2016). Water policy implementation in the state of São Paulo, Brazil: Key challenges and opportunities. *Environmental Science and Policy* 60, 11–18.

Barbosa, M. C., Mushtaq, S. and Alam, K. (2017). Integrated water resources management: Are river basin committees in Brazil enabling effective stakeholder interaction? *Environmental Science and Policy* 76, 1–11.

Barnett, J. and O'Neill, S. (2010). Maladaptation. *Global Environmental Change* 20(2), 211–213.

Berbel, J., Gutiérrez-Martín, C. and Expósito, A. (2018). Impacts of irrigation efficiency improvement on water use, water consumption and response to water price at field level. *Agricultural Water Management* 203, 423–429.

Bierkens, M. F. P. and Wada, Y. (2019). Non-renewable groundwater use and groundwater depletion: A review. *Environmental Research Letters* 14(6), 063002.

Brett, P, McCarthy, A., McCarthy, C., Long, D., Gillies, M., Foley, J. and Baillie, C. (2019). Advancing automation in the agricultural working environment. *Proceedings of the Conference of the Australian Society of Sugar Cane Technologists*, Toowoomba, Queensland, Australia, 30 April, pp. 14-20.

Bundschuh, J., Chen, G. and Mushtaq, S. (2014). Towards a sustainable energy technologies-based agriculture. In: *Sustainable Energy Solutions in Agriculture: Sustainable Energy Developments*, Taylor & Francis (CRC Press), London, United Kingdom. Pp. 3-15.

Byrareddy, V., Kouadio, L., Kath, J., Mushtaq, S., Rafiei, V., Scobie, M. and Stone, R. (2020). Win-win: Improved irrigation management saves water and increases yield for robusta coffee farms in Vietnam. *Agricultural Water Management* 241, 106350.

Chartzoulakis, K. and Bertaki, M. (2015). Sustainable water management in agriculture under climate change. *Agriculture and Agricultural Science Procedia* 4, 88–98.

Cobon, D., Jarvis, C., Reardon-Smith, K., Guillory, L., Pudmenzky, C., Nguyen-Huy, T., Mushtaq, S. and Stone, R. (2021). Northern Australia Climate Program–Supporting adaptation in rangeland grazing systems through more targeted climate forecasts, improved drought information and an innovative extension program. *The Rangeland Journal* 43(3), 87–100.

Cobon, D. H., Darbyshire, R., Crean, J., Kodur, S., Simpson, M. and Jarvis, C. (2019). Valuing seasonal climate forecasts in the Northern Australia beef industry. *Weather, Climate, and Society* 12(1), 3–14.

Colloff, M J., Grafton, R. Q. and Williams, J. (2021). Scientific integrity, public policy and water governance in the Murray-Darling Basin, Australia. *Australasian Journal of Water Resources*, 25(2), 121–140 . doi: 10.1080/13241583.2021.1917097.

Daccache, A., Knox, J. W., Weatherhead, E. K., Daneshkhah, A. and Hess, T. M. (2015). Implementing precision irrigation in a humid climate–Recent experiences and on-going challenges. *Agricultural Water Management* 147, 135–143.

Davis, J., O'Grady, A. P., Dale, A., Arthington, A. H., Gell, P. A., Driver, P. D., Bond, N., Casanova, M., Finlayson, M., Watts, R. J., Capon, S. J., Nagelkerken, I., Tingley, R.,

Fry, B., Page, T. J. and Specht, A. (2015). When trends intersect: The challenge of protecting freshwater ecosystems under multiple land use and hydrological intensification scenarios. *Science of the Total Environment* 534, 65–78.
DAWE (2020). *Australian Water Markets*. Australian Department of Agriculture, Water and Environment (DAWE). Available at: https://www.agriculture.gov.au/water/markets.
Deemer, B. R., Harrison, J. A., Li, S., Beaulieu, J. J., DelSontro, T., Barros, N., Bezerra-Neto, J. F., Powers, S. M., dos Santos, M. A. and Vonk, J. A. (2016). Greenhouse gas emissions from reservoir water surfaces: A new global synthesis. *Bioscience* 66(11), 949–964.
de Vito, R., Portoghese, I., Pagano, A., Fratino, U. and Vurro, M. (2017). An index-based approach for the sustainability assessment of irrigation practice based on the water-energy-food nexus framework. *Advances in Water Resources* 110, 423–436.
de Wrachien, D, Schultz, B. and Goli, M. B. (2021). Impacts of population growth and climate change on food production and irrigation and drainage needs: A world-wide view. *Irrigation and Drainage* 2021, 1–15.
D'Odorico, P., Bhattachan, A., Davis, K. F., Ravi, S. and Runyan, C. W. (2013). Global desertification: Drivers and feedbacks. *Advances in Water Resources* 51, 326–344.
Döll, P. (2002). Impact of climate change and variability on irrigation requirements: A global perspective. *Climatic Change* 54(3), 269–293.
Dubreuil, A., Assoumou, E., Bouckaert, S., Selosse, S. and Maïzi, N. (2013) Water modeling in an energy optimization framework–The water-scarce middle east context. *Applied Energy* 101, 268–279.
Eberhard, J., McHugh, A., Scobie, M., Schmidt, E., McCarthy, A., Uddin, M. J., McKeering, L. and Poulter, R. (2013). *Improving Irrigation Efficiency through Precision Irrigation in South East Queensland*. Project Report, National Centre for Engineering in Agriculture, University of Southern Queensland.
Eekhout, J P. C., Boix-Fayos, C., Pérez-Cutillas, P. and de Vente, J. (2020). The impact of reservoir construction and changes in land use and climate on ecosystem services in a large Mediterranean catchment. *Journal of Hydrology* 590, 125208.
FAO (2011). Energy smart-food for people and climate, *United Nations Food and Agricultural Organisation*, Food and Agriculture Organization. Available at: http://www.fao.org/3/i2454e/i2454e.pdf.
FAO (2014). The water-energy-food nexus–A new approach in support of food security and sustainable agriculture. *United Nations Food and Agricultural Organisation*, Food and Agriculture Organization. Available at: http://www.fao.org/3/bl496e/bl496e.pdf.
FAO (2021). Water–energy–food nexus. *United Nations Food and Agricultural Organisation*, Food and Agriculture Organization. Available at: http://www.fao.org/land-water/water/watergovernance/waterfoodenergynexus/en/.
Feike, T., Khor, L. Y., Mamitimin, Y., Ha, N., Li, L., Abdusalih, N., Xiao, H. and Doluschitz, R. (2017). Determinants of cotton farmers' irrigation water management in arid Northwestern China. *Agricultural Water Management* 187, 1–10.
Fernández, J. E., Alcon, F., Diaz-Espejo, A., Hernandez-Santana, V. and Cuevas, M. V. (2020). Water use indicators and economic analysis for on-farm irrigation decision: A case study of a super high density olive tree orchard. *Agricultural Water Management* 237, 106074.
Gadanakis, Y, Bennett, R., Park, J. and Areal, F. J. (2015). Improving productivity and water use efficiency: A case study of farms in England. *Agricultural Water Management* 160, 22–32.

García, A. M., Fernández García, I., Camacho Poyato, E., Montesinos Barrios, P. and Rodríguez Díaz, J. A. (2018). Coupling irrigation scheduling with solar energy production in a smart irrigation management system. *Journal of Cleaner Production* 175, 670–682.

García, L., Parra, L., Jiminez, J. M., Lloret, J. and Lorenz, P. (2020). IoT-based smart irrigation systems: An overview on the recent trends on sensors and IoT systems for irrigation in precision agriculture. *Sensors* 20(4), 1042.

Gathala, M. K., Laing, A. M., Tiwari, T. P., Timsina, J., Islam, M. S., Chowdhury, A. K., Chattopadhyay, C., Singh, A. K., Bhatt, B. P., Shrestha, R., Barma, N. C. D., Rana, D. S., Jackson, T. M. and Gerard, B. (2020). Enabling smallholder farmers to sustainably improve their food, energy and water nexus while achieving environmental and economic benefits. *Renewable and Sustainable Energy Reviews* 120, 109645.

Gleeson, T., Cuthbert, M., Ferguson, G. and Perrone, D. (2020). Global groundwater sustainability, resources, and systems in the Anthropocene. *Annual Review of Earth and Planetary Sciences* 48(1), 431–463.

Glória, A., Dionisio, C., Simoes, G., Cardosa, J. and Sebastiao, P. (2020). Water management for sustainable irrigation systems using internet-of-things. *Sensors* 20(5), 1402.

Gomiero, T (2016). Soil degradation, land scarcity and food security: Reviewing a complex challenge. *Sustainability* 8(3), 281.

Grafton, R. Q. (2019). Policy review of water reform in the Murray–Darling Basin, Australia: The "do's" and "do nots". *Australian Journal of Agricultural and Resource Economics* 63(1), 116–141.

Hafeez, M., Bundschuh, J. and Mushtaq, S. (2014). Exploring synergies and tradeoffs: Energy, water, and economic implications of water re-use in rice-based irrigation systems. *Applied Energy* 114, 889–900.

Hansen, J. W., Challinor, A., Ines, A., Wheeler, T. and Moron, V. (2006). Translating climate forecasts into agricultural terms: Advances and challenges. *Climate Research* 33, 27–41.

Hering, J. G., Sedlak, D. L., Tortajada, C., Biswas, A. K., Niwagaba, C. and Breu, T. (2015). WATER. Local perspectives on water. *Science* 349(6247), 479–480.

Jackson, T. (2009). *An Appraisal of the On-Farm Water and Energy Nexus in Irrigated Agriculture*, A doctoral thesis submitted to Charles Sturt University, Australia.

Jackson, T M., Ahmad Hanjra, M. A., Khan, S. and Hafeez, M. M. (2011). Building a climate resilient farm: A risk based approach for understanding water, energy and emissions in irrigated agriculture. *Agricultural Systems* 104(9), 729–745.

Jackson, T. M., Khan, S. and Hafeez, M. (2010). A comparative analysis of water application and energy consumption at the irrigated field level. *Agricultural Water Management* 97(10), 1477–1485.

Jägermeyr, J., Gerten, D., Schaphoff, S., Heinke, J., Lucht, W. and Rockström, J. (2016). Integrated crop water management might sustainably halve the global food gap. *Environmental Research Letters* 11(2), 025002.

Jana, S. K., Palanisami, K. and Das, A. (2012). A study on tank irrigation productivity in the dry zones of West Bengal. *Indian Journal of Agricultural Economics* 67, 1–13.

Jobbins, G., Kalpakian, J., Chriyaa, A., Legrouri, A. and El Mzouri, E. H. (2015). To what end? Drip irrigation and the water–energy–food nexus in Morocco. *International Journal of Water Resources Development* 31(3), 393–406.

Kamienski, C., Soininen, J. P., Taumberger, M., Dantas, R., Toscano, A., Salmon Cinotti, T., Filev Maia, R. and Torre Neto, A. (2019). Smart water management platform: IoT-based precision irrigation for agriculture. *Sensors* 19(2), 276.

Kamienski, C, Soininen, J.-P., Taumberger, M., Fernandes, S., Toscano, A., Cinotti, T. S., Maia, R. F. and Neto, A. T. (2018). Swamp: An IOT-based smart water management platform for precision irrigation in agriculture. *Global Internet of Things Summit (GIoTS)*, Bilbao, Spain, pp. 1–6.

Khair, S. M., Mushtaq, S. and Reardon-Smith, K. (2015). Groundwater governance in a water-starved country: Public policy, farmers' perceptions, and drivers of tubewell adoption in Balochistan, Pakistan. *Groundwater* 53(4), 626–637.

Khair, S. M., Mushtaq, S., Reardon-Smith, K. and Ostini, J. (2019). Diverse drivers of unsustainable groundwater extraction behaviour operate in an unregulated water scarce region. *Journal of Environmental Management* 236, 340–350.

Khan, R., Ali, I., Zakarya, M., Ahmad, M., Imran, M. and Shoaib, M. (2018). Technology-assisted decision support system for efficient water utilization: A real-time testbed for irrigation using wireless sensor networks. *IEEE Access* 6, 25686–25697.

Khan, S. and Mushtaq, S. (2009). Regional partnerships to assist public–private investments in irrigation systems. *Agricultural Water Management* 96(5), 839–846.

Khan, S., Mushtaq, S., Ahmad, A. and Hafeez, M. (2008). Trade-off analysis for restoring environmental flows through irrigation demand management. *Australasian Journal of Water Resources* 12(1), 1–20.

Kiem, A S. (2013). Drought and water policy in Australia: Challenges for the future illustrated by the issues associated with water trading and climate change adaptation in the Murray-Darling Basin. *Global Environmental Change* 23(6), 1615–1626.

Koech, R. and Langat, P. (2018). Improving irrigation water use efficiency: A review of advances, challenges and opportunities in the Australian context. *Water* 10(12), 1771.

Lampayan, R. M., Rejesus, R. M., Singleton, G. R. and Bouman, B. A. M. (2015). Adoption and economics of alternate wetting and drying water management for irrigated lowland rice. *Field Crops Research* 170, 95–108.

Li, X., Waddington, S. R., Dixon, J., Joshi, A. K. and de Vicente, M. C. (2011). The relative importance of drought and other water-related constraints for major food crops in South Asian farming systems. *Food Security* 3(1), 19–33.

Liu, Y., Pan, Z., Zhuang, Q., Miralles, D. G., Teuling, A. J., Zhang, T., An, P., Dong, Z., Zhang, J., He, D., Wang, L., Pan, X., Bai, W. and Niyogi, D. (2015). Agriculture intensifies soil moisture decline in Northern China. *Scientific Reports* 5, 11261.

Loch, A. and McIver, R. (2012). The Murray-Darling Basin plan and public policy failure: A transaction costs analysis. *Sustainable Irrigation and Drainage* 168, 481–494.

Mainuddin, M, Alam, M., Maniruzzaman, Md., Islam, T., Kabir, J., Hasan, M., Scobie, M. and Schmidt, E. (2019). *Irrigated Agriculture in the Northwest Region of Bangladesh*. CSIRO, Canberra, Australia.

Mainuddin, M., Maniruzzaman, Md., Alam, M. M., Mojid, M. A., Schmidt, E. J., Islam, M. T. and Scobie, M. (2020). Water usage and productivity of Boro rice at the field level and their impacts on the sustainable groundwater irrigation in the North-West Bangladesh. *Agricultural Water Management* 240, 106294.

Maraseni, T., An-Vo, D.-A., Mushtaq, S. and Reardon-Smith, K. (2021). Carbon smart agriculture: An integrated regional approach offers significant potential to increase profit and resource use efficiency, and reduce emissions. *Journal of Cleaner Production* 282, 124555.

Maraseni, T. N. and Cockfield, G. (2012). Including the costs of water and greenhouse gas emissions in a reassessment of the profitability of irrigation. *Agricultural Water Management* 103, 25–32.

Maraseni, T. N., Maroulis, J. and Cockfield, G. (2009a). An analysis of Australia's Carbon Pollution Reduction Scheme. *International Journal of Environmental Studies* 66(5), 591–603.

Maraseni, T N., Mushtaq, S. and Maroulis, J. (2009b). Greenhouse gas emissions from rice farming inputs: A cross-country assessment. *The Journal of Agricultural Science* 147(2), 117–126.

Maraseni, T. N., Mushtaq, S., Hafeez, M. and Maroulis, J. (2010). Greenhouse gas implications of water re-use in the Upper Pampanga River Integrated Irrigation System, Philippines. *Agricultural Water Management* 97(3), 382–388.

Maraseni, T. N., Mushtaq, S. and Reardon-Smith, K. (2012a). Climate change, water security and the need for integrated policy development: The case of on-farm infrastructure investment in the Australian irrigation sector. *Environmental Research Letters* 7(3), 034006.

Maraseni, T. N., Mushtaq, S. and Reardon-Smith, K. (2012b). Integrated analysis for a carbon- and water-constrained future: An assessment of drip irrigation in a lettuce production system in eastern Australia. *Journal of Environmental Management* 111, 220–226.

McClintock, A. (2009). Investment in irrigation technology: An application of real options analysis. *53rd Australian Agricultural and Resource Economics Society Conference*, Cairns, Australia, 11–13 February 2009, 47934.

McCormick, B. (2007). *National Plan for Water Security*, Parliament of Australia. Available at: https://www.aph.gov.au/About_Parliament/Parliamentary_Departments/Parliamentary_Library/pubs/rp/BudgetReview_2007-_2008/National_Plan_for_Water_Security.

Meyfroidt, P., Lambin, E. F., Erb, K.-H. and Hertel, T. W. (2013). Globalization of land use: Distant drivers of land change and geographic displacement of land use. *Current Opinion in Environmental Sustainability* 5(5), 438–444.

Morris, B L., Lawrence, A. R. L., Chilton, P. J. C., Adams, B., Calow, R. C. and Klinck, B. A. (2003). *Groundwater and Its Susceptibility to Degradation: A Global Assessment of the Problem and Options for Management*. United Nations Environment Programme, Nairobi, Kenya. 126 pp.

Mushtaq, S. and Bundschuh, J. (2016). An innovative systematic approach to internalize external costs of salinization in major irrigated systems. *Groundwater for Sustainable Development* 2–3, 16–26.

Mushtaq, S., Chen, C., Hafeez, M., Maroulis, J. and Gabriel, H. (2012). The economic value of improved agrometeorological information to irrigators amid climate variability. *International Journal of Climatology* 32(4), 567–581.

Mushtaq, S., Khan, S., Hafeez, M. and Hanjra, M. A. (2009a). Does reliability of water resources matter in the adoption of water-saving irrigation practices? A case study in the Zhanghe irrigation system, China. *Water Policy* 11(6), 661–679.

Mushtaq, S., Maraseni, T. N., Maroulis, J. and Hafeez, M. (2009b). Energy and water tradeoffs in enhancing food security: A selective international assessment. *Energy Policy* 37(9), 3635–3644.

Mushtaq, S., Maraseni, T. N. and Reardon-Smith, K. (2013a). Climate change and water security: estimating the greenhouse gas costs of achieving water security through investments in modern irrigation technology. *Agricultural Systems* 117, 78–89.

Mushtaq, S., Reardon-Smith, K., Stone, R. and Syed, M. K. (2013b). A blueprint for sustainable groundwater management in Balochistan, Pakistan. In: *Free Flow: Reaching Water Security Through Cooperation*, United Nations Educational, Scientific and Cultural Organization, Paris, France.

Mushtaq, S, Maraseni, T. N. and Reardon-Smith, K. (2015a). Trade-offs and synergies between water and energy use in rural Australia. In: *Climate, Energy and Water: Managing a Complex Trinity*, Pittock et al. (Eds), Cambridge University Press, New York, USA.

Mushtaq, S., Maraseni, T. N., Reardon-Smith, K., Bundschuh, J. and Jackson, T. (2015b). Integrated assessment of water-energy-GHG emissions tradeoffs in an irrigated lucerne production system in eastern Australia. *Journal of Cleaner Production* 103, 491–498.

Mushtaq, S. and Moghaddasi, M. (2011). Evaluating the potentials of deficit irrigation as an adaptive response to climate change and environmental demand. *Environmental Science and Policy* 14(8), 1139–1150.

Mutambara, S., Darkoh, M. B. K. and Atlhopheng, J. R. (2016). A comparative review of water management sustainability challenges in smallholder irrigation schemes in Africa and Asia. *Agricultural Water Management* 171, 63–72.

Nguyen-ky, T., Mushtaq, S., Loch, A., Reardon-Smith, K., An-Vo, D.-A., Ngo-Cong, D. and Tran-Cong, T. (2018). Predicting water allocation trade prices using a hybrid Artificial Neural Network-Bayesian modelling approach. *Journal of Hydrology* 567, 781–791.

Northern Australia Climate Program (2021). Climate outlook review. Available at: https://www.nacp.org.au/forecasts/climate_outlook.

Pereira, H. and Marques, R. C. (2017). An analytical review of irrigation efficiency measured using deterministic and stochastic models. *Agricultural Water Management* 184, 28–35.

Pham, Y., Reardon-Smith, K., Mushtaq, S., Deo and Deo, R. C. (2020). Feedback modelling of the impacts of drought: A case study in coffee production systems in Viet Nam.*Climate Risk Management* 30, 100255.

Pham, Y. H., Reardon-Smith, K., and Deo, R. C. (2021). Evaluating drought management strategies: application of system dynamics modelling for sustainable crop production. *Journal of Environmental Management* 292, JEMA-D-21-00184R1.

Pittock, J., Hussey, K. and Dovers, S. (2015). *Climate, Energy and Water: Managing a Complex Trinity*. Cambridge University Press, New York, USA.

Plieninger, T., Draux, H., Fagerholm, N., Bieling, C., Bürgi, M., Kizos, T., Kuemmerle, T., Primdahl, J. and Verburg, P. H. (2016). The driving forces of landscape change in Europe: A systematic review of the evidence. *Land Use Policy* 57, 204–214.

Productivity Commission (2018). *Murray-Darling Basin Plan: Five-Year Assessment*, Final Report, Commonwealth of Australia.

Schepen, A, Everingham, Y. and Wang, Q. J. (2020). An improved workflow for calibration and downscaling of GCM climate forecasts for agricultural applications–A case study on prediction of sugarcane yield in Australia. *Agricultural and Forest Meteorology* 291, 107991.

Schmidt, E., Sugden, F., Scobie, M., Mainuddin, M., Leder, S., Mishra, R., Thapa, B., Bastakoti, R. and da Silva, S. (2019). *Improving Water Use for Dry Season Agriculture by Marginal and Tenant Farmers in the Eastern Gangetic Plains (ACIAR Final Report)*, Centre for Agricultural Engineering, University of Southern Queensland.

Scobie, M., Schmidt, E., Bowtell, L., Hill, A. and Garmendia, A. (2018). *Feasibility of Solar PV Irrigation in Queensland, Reports 1-3*. Queensland Department of Energy and Water Supply, Brisbane, Queensland, Australia.

Sears, L., Caparelli, J., Lee, C., Pan, D., Strandberg, G., Vuu, L. and Lin Lawell, C.-Y. (2018). Jevons' paradox and efficient irrigation technology. *Sustainability* 10(5), 1590.

Shah, T., Roy, A. D., Qureshi, A. S. and Wang, J. (2003). Sustaining Asia's groundwater boom: An overview of issues and evidence. *Natural Resources Forum* 27(2), 130–141.

Shukr, H. H., Pembleton, K. G., Zull, A. F. and Cockfield, G. J. (2021). Impacts of effects of deficit irrigation strategy on water use efficiency and yield in cotton under different irrigation systems. *Agronomy* 11(2), 231.

Smidt, S J., Haacker, E. M. K., Kendall, A. D., Deines, J. M., Pei, L., Cotterman, K. A., Li, H., Liu, X., Basso, B. and Hyndman, D. W. (2016). Complex water management in modern agriculture: Trends in the water-energy-food nexus over the High Plains Aquifer. *Science of the Total Environment* 566-567, 988–1001.

Stone, R. C., Hammer, G. L. and Marcussen, T. (1996). Prediction of global rainfall probabilities using phases of the Southern Oscillation Index. *Nature* 384(6606), 252–255.

Tarjuelo, J. M., Rodriguez-Diaz, J. A., Abadía, R., Camacho, E., Rocamora, C. and Moreno, M. A. (2015). Efficient water and energy use in irrigation modernization: lessons from Spanish case studies. *Agricultural Water Management* 162, 67–77.

UN (2015). *Global Sustainable Development Report*. United Nations (UN), New York, USA.

Vaast, P., Bertrand, B., Perriot, J.-J., Guyot, B. and Génard, M. (2006). Fruit thinning and shade improve bean characteristics and beverage quality of coffee (Coffea arabica L.) under optimal conditions. *Journal of the Science of Food and Agriculture* 86(2), 197–204.

van Duinen, R., Filatova, T., Jager, W. and van der Veen, A. (2016). Going beyond perfect rationality: Drought risk, economic choices and the influence of social networks. *The Annals of Regional Science* 57(2–3), 335–369.

van Vliet, N, Mertz, O., Heinimann, A., Langanke, T., Pascual, U., Schmook, B., Adams, C., Schmidt-Vogt, D., Messerli, P., Leisz, S., Castella, J.-C., Jørgensen, L., Birch-Thomsen, T., Hett, C., Bech-Bruun, T., Ickowitz, A., Vu, K. C., Yasuyuki, K., Fox, J., Padoch, C., Dressler, W. and Ziegler, A. D. (2012). Trends, drivers and impacts of changes in swidden cultivation in tropical forest-agriculture frontiers: A global assessment. *Global Environmental Change* 22(2), 418–429.

Vuolo, F., D'Urso, G., De Michele, C., Bianchi, B. and Cutting, M. (2015). Satellite-based irrigation advisory services: A common tool for different experiences from Europe to Australia. *Agricultural Water Management* 147, 82–95.

Williams, A., White, N., Mushtaq, S., Cockfield, G., Power, B. and Kouadio, L. (2015). Quantifying the response of cotton production in eastern Australia to climate change. *Climatic Change* 129(1–2), 183–196.

WMO, USQ, and CIAT (2021). *DeRISK SE Asia*, World Meteorological Organization (WMO), University of Southern Queensland (USQ), International Centre for Tropical Agriculture (CIAT). Available at: https://www.deriskseasia.org/ viewed 2806/2021.

Yahyaoui, I., Tadeo, F. and Segattoa, M. V. (2017). Energy and water management for drip-irrigation of tomatoes in a semi-arid district. *Agricultural Water Management* 183, 4–15.

Part 2

On-farm renewable energy heat and power generation

Chapter 5

Assessing and modelling the costs of on-farm distributed renewable energy systems

Julio Pombo-Romero, University of Vigo, Spain

1 Introduction

Modern farming is an increasingly energy-intensive activity. As a consequence, the cost of energy is one of the most important inputs of agrarian production and has a highly relevant impact on the profitability of farms. Furthermore, the impact of the significant use of energy inputs goes beyond the sustainability of the farms and also affects the environmental and socio-economic conditions of rural communities. Almost 32% of global end-use energy consumption and 22% of all greenhouse gas (GHG) emissions (including non-energy-related methane and nitrous oxide) are related to the global food supply chain. On-farm activities behind the 'farm gate' are responsible for around one-fifth of these total energy inputs to the food sector and two-thirds of total GHG emissions based on the high global warming potentials of the non-CO_2 gases (FAO, 2011).

Energy consumption on farms is greatly related to the mechanization and intensification of agriculture that has taken place over the last century and has relied heavily on fossil fuels (Capuano, 2018). As a result, agricultural

http://dx.doi.org/10.19103/AS.2022.0100.07

dependence on fossil fuels is a cause for concern because of the negative impacts on environmental, economic and social sustainability (Pimentel and Pimentel, 2007). In this context, the 'decarbonization' of the farming sector is gaining momentum as a strategic goal of the different stakeholders, which include farmers, investors and public authorities.

Alongside improvements in the energy efficiency of the agrarian processes, the generalization of the use of on-farm distributed generation-renewable energy systems (DG-RES) is the most relevant action that can be taken to decarbonize the sector. DG-RES are particularly well suited to agricultural holdings as they rely on the availability of natural resources such as solar irradiation, wind and biomass that are often in abundance on farmlands. Furthermore, the economic viability of DG-RES is becoming possible for an increasingly large proportion of farms, due to the availability of reliable technological solutions and reductions in the cost of many critical components of the systems such as solar photovoltaic (PV) and solar water heating panels. As a result, nowadays, farmers have access to a broad range of technologies, business models, information and suppliers of on-farm DG-RES suited to producing electricity, heat or biofuels using resources available on the farm itself.

Despite the many potential benefits, on-farm DG-RES often involve high up-front costs to design, install and build the systems together with specialized operation and maintenance procedures to operate them efficiently. Furthermore, solar and wind DG-RES are variable with outputs depending on the availability of the resource at any given time. DG-RES are also susceptible to technical failure and degradation over their lifetime. Depending on such factors and other relevant variables – particularly the cost of capital – the resulting costs of energy from DG-RES are not straightforward to assess. After considering all the factors, for some projects, the resulting cost is not competitive with incumbent solutions. In fact, in order to facilitate the introduction of on-farm DG-RES, it is particularly relevant to allow farmers to produce accurate assessments of the cost of the resulting energy to enable comparison with conventional energy supplies, as securing a competitive cost of energy is the single most important reason for adopting these green technologies (Pombo-Romero and Regueiro-Ferreira, 2016).

Producing accurate assessments of the cost of energy of on-farm DG-RES is relevant not only for farmers at the 'project level' but also for governments, developmental agencies, researchers and other stakeholders who require methods and tools to produce accurate assessments of the costs and benefits of on-farm DG-RES. Impacts that go beyond the farm are included such as environmental and socio-economic impacts on rural communities. There is increased interest and effort in producing and improving such methodologies. These efforts can result in proposals of new approaches and models to suit the requirements of the different stakeholders and consider the nature and impacts of the costs involved. Most methods and tools used to assess the costs of on-farm DG-RES are common to any other application of DG-RES.

DG-RES include solar, wind and biomass which can be directly connected to the energy demand centres of a farm including heating systems or electrical equipment (Gonzalez-Longatt and Fortoul, 2005). This flexibility and the variety of alternatives allows many combinations of technologies, off-take alternatives and financing options that ultimately determine the cost of on-farm DG. Furthermore, a comprehensive cost analysis should assess the impact of uncertainty regarding not only the performance of the system but also the regulatory and political events that can affect the costs at some unknown moment in the future.

In this chapter, the most relevant methodologies and related research activities when assessing and modelling the costs of on-farm DG-RES are presented and discussed. In addition, a case study on the estimation of the costs of a solar PV irrigation system was included to show the application of some of the most relevant methodologies.

2 Private costs of on-farm distributed generation-renewable energy systems

Costs can be measured in different ways and from different perspectives to provide a range of insights that can be useful for specific stakeholders (Annex 1). The cost of an on-farm RES project can be examined from the perspective of the farmer who will be mainly interested in the cost of the energy obtained. Or it can be analysed from a broader perspective that will also take into account externalities such as environmental impacts and the use of public infrastructure. More generally, one can distinguish between a private perspective of costs, which refers to those that are paid for by those directly involved in the use or production of the goods, and social costs that refer to the use of inputs that are available in the public domain. Both approaches are often complementary, and practitioners and researchers often combine methodologies and concepts from both when producing their analyses.

The first approach is oriented towards estimating the 'private cost' of a system, focusing exclusively on the economic resources that the promoters of the project must allocate to it. This approach is intended to support decision-making based on financial criteria, namely the return on investment, which tends to be what investors, including farmers, prioritize when considering new projects. When assessing the private costs of DG-RES, the methods are normally those used for investment appraisal, based on discounted cash flows (Allan et al., 2015).

As in any other sort of project or investment, the cost assessment of DG-RES often starts by estimating the capital expenditure (CAPEX) and the operational expenditure (OPEX) that must be paid in order to build and operate the system. CAPEX includes the hardware such as solar panels or biomass boilers and the cost of engineering, procurement and construction (EPC). The other main part of the budget is related to OPEX which includes items that are necessary during the operational life of the systems such as the maintenance and the cost of fuel, if required. The costs of the different elements of CAPEX and OPEX are obtained from the market, so estimation does not require a particular set of tools. Nevertheless, as the market for DG-RES systems is not completely mature, the CAPEX and OPEX costs of a specific type of DG-RES can vary significantly depending on the location and the supplier.

Once the 'spot' prices of the inputs are known, a number of methodologies can be used to assess the resulting cost of energy from the private perspective, including the basic cost of energy (BCE), the techno-economic analysis (TEA) and the levelized cost of energy (LCOE).

2.1 Basic cost of energy

BCE provides the first approach to assess the cost of the systems in terms of cost per unit of energy (for example in $/kWh). The method consists of the simple addition of all the costs (CAPEX and OPEX) undertaken by the promoter in planning, installing, operating and maintaining the system. The resulting total cost over the operational lifespan of the system is then divided by the expected total amount of electricity, heat or fuels to be produced. This technique has many limitations as important factors such as the time value of capital and the riskiness of the project are not taken into consideration. Nevertheless, many researchers use this approach particularly in the context of life cycle cost analysis (Kannan et al., 2006).

2.2 Techno-economic analysis

TEA combines process modelling, engineering and construction design and economic evaluation in order to assess the viability of a system or compare alternative systems. In the context of on-farm DG-RES, TEA consists of using economic assessment and process modelling in order to simultaneously analyse several alternatives that can produce energy, for example, in order to compare the use of a diesel generator versus a solar PV system to power irrigation water pumps. TEA normally includes the time value of money and elements to assess the riskiness of the project and the impact on the expected costs and benefit elements that are considered as key in any economic appraisal.

Time value of money and riskiness must be included in valuations because rational agents prefer projects whose expected returns will be obtained earlier in time and with less uncertainty. This means that the cost of capital increases when projects have longer payback times and greater uncertainty over the results as if the riskiness is high, then consequently, the expected return must be high.

The modern portfolio theory provides the framework for the economic analysis of risk (Markowitz, 1991), and the capital asset pricing model (CAPM) allows for the relation of this riskiness and the cost of capital of every single project (Sharpe, 1964). Then, alongside OPEX and CAPEX, any investment such as an on-farm DG-RES has an 'opportunity cost' as the money used there could have been invested instead in alternative assets and produced a return. The higher the opportunity cost is, the higher the riskiness of the project, so an appropriate 'discount factor' must be estimated to calculate the cost of the project in real terms. The discount factor that is used for a project is consistent with its weighted average cost of capital (WACC), that is, the level of return that investors and debt providers require from a project in a specific context and which is the basis of a TEA's valuation and cost calculations.

TEA employs the most widely used valuation techniques in on-farm DG-RES based on discounted cash flow (DCF) methods. The financial literature presents a variety of DFC approaches that can be applied to on-farm DG-RES depending on the characteristics of the project (Fernandez, 2007). The investment appraisal based on DCF is normally focused on the calculation of either the net present value (NPV) and/or the internal rate of return (IRR). NPV is the result of the difference between the summation of the present value of future cash inflows and cash outflows from the project discounted at a given discount rate.

$$NPV = \sum_{t=1}^{n} \frac{\text{Cash inflow } s_t - \text{Cash outflow } s_t}{(1+r)^t}, \tag{1}$$

where r is the discount rate and n is the number of years of lifespan expectancy.

If the value of r used for this discount rate is WACC, then a positive NPV indicates that the project has the expected capacity to create value and can be considered for execution. The other key indicator of DCF is the IRR which is the discount rate at which NPV = 0.

$$0 = NPV = \sum_{t=1}^{n} \frac{\text{Cash inflow } s_t - \text{Cash outflow } s_t}{(1+IRR)^t}. \tag{2}$$

If the value of IRR estimated for a project is lower than its WACC, then the NPV is negative, and the project should not be considered appropriate for investment purposes.

In the case of TEA for DG-RES, the outputs of the models consist not only of financial indicators, such as the NPV and the IRR, but also of a number of indicators for each technological alternative including the net energy ratio. This indicates whether or not more energy is produced than consumed in a specific production process and is particularly relevant when analysing biofuels (Fore et al., 2011; Gittinger, 1982). Cost components are normally broken down, and a sensitivity analysis, with respect to the most relevant variables, is often presented as a result of the TEA. A classic reference for this methodology in the context of farming is the World Bank handbook[1] (Gittinger, 1982).

In the case of on-farm DG-RES, researchers are using TEA to evaluate the use of solar irrigation and its suitability over incumbent alternatives in different configurations and countries. Most of the published studies on solar PV-powered irrigation systems present positive NPV, indicating the suitability of the solution in a wide range of contexts (Carrêlo et al, 2020; Sarkar and Ghosh, 2017; Niajalili et al, 2017).

The heating and cooling of greenhouses (Sethi and Sharma, 2008) and livestock buildings (Barbari et al, 2012) are also particularly well suited to on-farm DG-RES applications. The use of small on-farm wind turbines to produce electricity has also been considered using TEA, with positive returns in contexts where the cost of alternatives is expensive (Nouni et al, 2007) or where there are government feed-in tariffs available (Xydis, 2015). In the case of biomass, specific TEA methodology applications have been developed in order to include elements such as system configuration and the type and quality of the feedstock. The objective of TEA biomass analysis is, in most cases, the comparison of the different alternatives in terms of cost of energy obtained (Unrean et al, 2018; Carpenter et al, 2014; Abdelhady et al, 2018; Mana et al, 2021). A particularly tailored tool is the biomass logistic model developed by the Idaho National Laboratory that allows analysts to improve estimations of the cost of delivered feedstock (Cafferty et al., 2013), which is a highly relevant element in order to assess the cost of the energy produced by biomass systems.

Another active area of research in on-farm DG-RES that uses TEA is 'agro-photovoltaics' where agricultural production is carried out on the ground below a PV installation (Goetzberger and Zastrow, 1982) Therefore, both energy and crops can be produced on the same area of land. The aim of agro-PV TEA analysis is to identify what type of crops and practices can better co-exist with the on-farm DG-RES (Schindele et al., 2020).

TEA research on DG-RES presents a large variation in the estimated cost of energy, depending on the availability of local providers and the abundance of the energy resource. These differences indicate the difficulty of extrapolating the TEA results from DG-RES projects from different locations and times so any comparison of results available in the literature must be carried out considering these factors.

[1] Available at http://documents1.worldbank.org/curated/en/584961468765021837/pdf/multi0page.pdf.

2.3 Levelized cost of energy

While TEA allows for the calculation of a large variety of indicators, LCOE is considered as the standard metric for the assessment of the cost of producing energy at the system level (Short et al., 1995). Estimating LCOE allows promoters and investors to establish the break-even level of the projects by comparing the resulting cost of energy with other alternatives (Hansen, 2019). LCOE is also popular among policy makers to design effective supporting schemes (Ouyang and Lin, 2014). Its calculation consists of estimating the total cost of constructing and operating an electricity-generating plant divided by the expected total electricity production. The future costs and the production are 'discounted' using a discount rate that reflects the cost of capital and the riskiness of the project.

$$LCOE = \frac{\sum_{t=1}^{n} \frac{CAPEX_t + OPEX_t + FuelEX_t}{(1+r)^t}}{\sum_{t=1}^{n} \frac{EnGen_t}{(1+r)^t}}, \quad (3)$$

where n is the number of years of lifespan of the system, $FuelEX_t$ is the fuel expenditure in year t, $EnGen_t$ is the amount of energy produced in year t, and r is the discount rate appropriate for the project.

Despite the widespread use of LCOE, a number of shortcomings have been identified (Nissen and Harfst, 2019). Consequently, and given the importance of LCOE calculations in all types of analyses, several research areas have been established with the objective of improving the accuracy. These are focused on at least four different aspects (Shen et al., 2020): investment-related cost (Geissmann and Ponta, 2017), operational-related cost (Rhodes et al., 2017; Khatib, 2016), plant performance and the uncertainty and risk-related costs (Nissen and Harfst, 2019; Heck et al., 2016).

Location is considered an important investment-related element to determine the investment costs that must be included in LCOE, as the different levels of prices of CAPEX elements can vary significantly among different countries. In order to introduce this factor, geographically based methodologies have been included in LCOE calculations (Rhodes et al., 2017; Heck et al., 2016; Ondraczek et al., 2015; Wang et al., 2011; Ioannou et al., 2017; Lucheroni and Mari, 2018). Alongside location, other investment-related costs that have been considered relevant for DG-RES LCOE calculations include the existence of subsidies (Ondraczek et al., 2015; Orioli and Di Gangi, 2015; Rahman et al., 2018) and the specific business model (Bruck et al., 2018).

The need to introduce operational-related costs into DG-RES LCOE estimations arises mainly because of the different variable generation profiles of the systems. Thus, some DG-RES must be complemented with a backup or energy storage system in case electricity is needed but the renewable resource is not available. Some analysts include the cost of balancing this energy (Graham, 2018; Ueckerdt et al., 2013). Plant performance can also affect the LCOE as the reliability of the system can affect the energy output and, consequently, the denominator of the equation. In this case, the LCOE calculations have been improved by using metrics that introduce failure probabilities based on historical track records (Topper et al., 2019; Tu et al., 2019).

The introduction of uncertainty in LCOE models in order to improve cost assessment is another active area of research. In this regard, techniques including sensitivity analysis and Monte Carlo simulations allow for the introduction of elements such as the project risk (Geissmann and Ponta, 2017; Ioannou et al., 2017; Heck et al., 2016) and the volatility of inputs in LCOE estimations (Geissmann and Ponta, 2017; Reichenberg et al., 2018). Risk assessment techniques also allow for the identification of which input parameters are the most influential in relation to the riskiness of the results of the project (Obi et al., 2017). In particular, the variable nature of renewable energy production introduces a high level of uncertainty that hampers an accurate estimation of the output over the life of the system. In response, LCOE calculations are being modified to include assumptions and models that account for that uncertainty. This can be done by introducing, among other quantitative techniques, scenario-based and probabilistic methods (Rhodes et al., 2017; Heck et al., 2016).

A limitation of LCOE, that has not been fully addressed, is the limited capacity of existing models to include the impact of the institutional framework and specifically the impact of policies and policy change on the future cost of energy. These issues have played a critical role in the economic performance of on-farm DG-RES on many occasions (Alonso et al., 2016), so LCOE calculations should include them where possible.

In the context of on-farm DG-RES, LCOEs have been estimated for a number of applications including PV and wind energy as well as biomass and biogas for heat and electricity production from manure and other on-farm organic wastes and residues. There is a wide variation in the resulting energy cost depending on the technology, the system configuration, the location and other factors with LCOE ranging from \$0.04/kWh to \$186/kWh (Table 1).

Another indicator for DG-RES being evaluated is energy return on energy invested (EROI). This is not related to costs so is not included in this chapter. In essence, it is a check that the energy involved in material extraction, manufacturing, transporting and installing a specific renewable energy

Table 1 Review of on-farm DG-RES private costs selected to cover some of the more relevant DG-RES technologies available

Technology	Country/region	LCOE (USD/unit)	IRR (%)	Reference
PV irrigation	Mediterranean	$0.05–0.2*	10–28	Carrêlo et al. (2020)
PV irrigation	Bangladesh	$0.18/kWh	12.95	Sarkar and Ghosh (2017)
PV-greenhouse	Algeria	$0.068/kWh		Mostefaoui and Amara (2019)
PV-greenhouse	Greece	$0.1/kWh*		Vourdoubas (2020)
Wind	India	$0.1–186/kWh	-	Nouni et al. (2007)
Biomass	Egypt	$0.0633–0.1055/kWh		Abdelhady et al. (2018)
Biofuel (ethanol)	Vietnam	$1.19/L		Diep et al. (2015)
Biogas	Jordan	$0.055–0.071/kWh	11–17	Jarrar et al. (2020)
Biogas	Italy	$0.2–0.3/kWh*	22.9	Pantaleo et al. (2013)
Agro-photovoltaic	Germany	$0.1/kWh		Schindele et al. (2020)

* Converted from Euro at March 2021 exchange rate of €1–USD 1.18.

Table 2 A selection of international LCA studies of various on-farm DG-RES technologies

Technology	Country	Remarks	Reference
Biogas	Germany	Greenhouse gas emissions can be reduced with anaerobic digestion of liquid manure.	Lansche and Müller (2012)
Biogas	Spain	The environmental profile of the biogas system turned out to be highly dependent on the selection of system boundaries and the allocation method.	Lijó et al. (2014)
Biofuel (ethanol)	International (review)	Lignocellulosic bioethanol has lower environmental impacts in most of the categories and a positive energy balance compared with gasoline or first-generation bioethanol produced from maize and sugarcane.	Morales et al. (2015)
Biogas	Europe (review)	The type of feedstock, e.g. maize, green crop or animal manure, is a determining factor for the environmental impacts of biogas systems.	Hijazi et al. (2016)
Biomass	Sweden	Biomass used for heat and power reduces GHG emissions considerably compared to burning fossil fuels but has higher acidifying emissions.	Kimming et al. (2011)
Biofuel	Sweden	Biofuels produced from cereal straw had only 32–39% of the impact on reducing GHGs (kg CO_2-eq) compared to short-rotation *Salix*-based systems.	Ahlgren et al. (2008)
PV irrigation	South Africa	Replacing grid electricity with solar photovoltaic-generated electricity can lead to a 34% reduction in the GHG emissions per tonne of irrigated maize (though this depends on the share of coal-, oil- and gas-fired power plants supplying the grid).	Wettstein et al. (2017)
PV irrigation	Spain/Italy	The energy payback time of PV irrigation systems ranges from 1.94 to 5.25 years, depending on the capacity of the system.	Todde et al. (2018)
PV irrigation	China	Benefits from the avoidance of CO_2 emission reductions are much higher than lifetime carbon emissions.	Yang et al. (2014)
Solar PV and solar thermal	Iran	Using solar PV and solar water heating panels improves all energy-environmental indicators of greenhouse strawberry production.	Hosseini-Fashami et al. (2019)

technology such as a wind turbine is less than the energy that this technology will produce over its lifetime of operation.[2]

3 Social costs of energy

'Private cost' analysis allows farmers to assess the costs, expected returns and uncertainties that are related to building and operating on-farm DG-RES. Nevertheless, a comprehensive cost assessment should include not only the 'private costs' that are incurred directly by the farmer/investors but also all the different direct and indirect impacts that producing and operating these systems will have on society as a whole. Thus, a second, and often complementary, approach is oriented to estimating the 'social cost' of the systems. In this case, a widened focus is applied in order to include costs that are not necessarily assumed directly by the investors including environmental impacts, externalities, the use of public goods and any other element that will be 'consumed' as a result of installing such systems (Strantzali and Aravossis, 2016). In this case, the targeted audience of the results of these analyses tend to be policy makers and public institutions, including academics, that seek to understand all the implications that producing and introducing a 'disruptive' technology may have on a wide number of aspects.

The different methodologies available for estimating the social costs of energy are included in the socio-economic analysis. This type of analysis is based on the assumption that any enterprise or project has ramifications that affect not only the promoters or 'shareholders' but also the rest of society. In the case of natural resources being involved, such is the case of on-farm DG-RES, the environment is also affected. These considerations are normally not present in financial appraisal methods such as LCOE or TEA, so specific cost metrics and methodologies must be employed. A useful and clear conceptual framework of socio-economic analysis relevant to assessing the costs of on-farm DG-RES (Wang et al., 2011) showed that the two most frequent approaches used are cost–benefit analysis (CBA) and life cycle assessment (LCA).

3.1 Cost–benefit analysis

CBA identifies and estimates the cost and benefits of projects in relation to any change in welfare that would result from its execution. In order to do this, CBA practitioners not only identify and compare the private costs and benefits of the project (for example by carrying out TEA) but also include those impacting on the overall economic and natural environment. Therefore, it is possible to

[2] See here for more details and examples of EROI relating to renewable energy technologies https://www.sciencedirect.com/topics/engineering/energy-return-on-investment.

establish the social cost or net benefit to be expected from project execution. Central to most CBA analysis is the assumption that by reducing project analysis to 'private costs' and market price mechanisms, many elements can be missed because factors such as the existence of imperfect markets, the consumption of public goods or the impact of information asymmetry are not included in the cost calculation. By producing investment appraisal from the perspective of public policy and societal costs, CBA can address these limitations (Pearce, 1998).

For this reason, CBA methodology includes measures of the gains and losses of individuals in monetary terms, irrespective of their direct involvement in the project. By aggregating the results, an estimation of the net social gain or loss of undertaking the project is possible. The results are economic indicators that are complementary to those estimated by TEA, including the economic net present value (ENPV) and the economic internal rate of return (EIRR), which are the 'social' versions of NPV and IRR, respectively. On this basis, it is possible to design and implement effective regulation and policy actions to correct market failures. For example, in the case of on-farm DG-RES, it often occurs that the estimated NPV is negative while the ENPV is positive as a result of positive externalities of the projects, for example, because of the creation of economic opportunities in the vicinity of the project that are not captured by its promoters. In situations such as these, public authorities may introduce subsidies to compensate the promoters for the negative NPV thus allowing an increase in the number of projects executed. The amount and impact of these subsidies can be determined by using CBA (Johansson and Kriström, 2019).

In order to be useful, CBA requires the social, economic, political and institutional context in which the project will be produced to be taken into consideration. For example, the installation of a biofuel refinery may affect the price paid to the farmers for the feedstock grown in the area which in turn may reduce the competitiveness of any other productive activity that relies on such inputs, such as for cattle feed. CBA practitioners and researchers have to deal with the difficulties of identifying all these indirect impacts and to assign a monetary value to them, particularly when considering impacts on non-tradable goods. In these cases, preferences often have to substitute monetary values by using techniques such as multi-criteria decision aid (Zhou et al., 2006).

CBA is particularly well suited to assessing the environmental cost–benefit of DG-RES. These types of studies are normally carried out in eight steps (Hanley and Spash, 1996):

1 identification of gainers and/or losers;
2 identification of impacts (job creation, landscape destruction);
3 identification of the impacts with economic significance;
4 physical identification and schedule of relevant impacts;
5 monetary valuation of impacts;

6 estimation of a discount rate and cash flows;
7 estimation of NPV and IRR; and
8 sensitivity analysis.

Apart from the difficulties of identifying and assigning an economic value to non-tradable goods, the estimation of an adequate discount rate is particularly challenging for practitioners and particularly relevant for the results of the valuation.

One of the most relevant issues dealt with by using CBA is the determination of the overall impacts of DG-RES versus centralized production, a particularly important issue for energy planners. In this case, some elements of each option are particularly relevant including, in the case of DG-RES, the problem of wind and solar variability and the need to adapt the electricity grid. In the case of more centralized renewable electricity generation, the costs of transmission and distribution losses should be included (Holtmeyer et al., 2013).

3.2 Life cycle assessment

Among the different techniques that can be used to assess the social costs and benefits of DG-RES, LCA is probably the most widely used (Hauschild et al., 2018). All the inputs, including the use of tradable and non-tradable resources, and direct and indirect impacts related to the installation and operation of the system, are quantified during its full life cycle. While LCA is particularly focused on assessing environmental impacts, other elements such as geographical and trade effects are also taken into consideration.

The LCA approach is based on the assumption that the different stakeholders involved in the development of a project cannot limit their assessment to the elements which are directly related only to themselves. Thus, the costs and benefits of a project must be taken into consideration, irrespective of who is going to pay for, or benefit from, them and when they will happen. In order to support a comprehensive appraisal of projects, LCA allows for the estimation of environmental performance indicators, including energy intensity and energy payback time.

The standardization of the LCA methodological framework in ISO-14000 series has provided a common handbook that allows for a rich variety of on-farm DG RES analyses (Arvanitoyannis, 2008). The LCA methodology requires the production of a detailed analysis of each process involved and the obtention of data relevant to energy and water consumption, GHG emissions and other externalities in each of the steps of construction, operation, maintenance and disposal of the systems. The difficulty of obtaining data for such a comprehensive set of processes is a particularly challenging element of LCA studies that can hamper its application to small and stand-alone projects (Finnveden et al., 2009).

LCA methodology is normally complemented by life cycle costing (LCC) analysis in order to enable an economic assessment of a project. LCC assessment takes into consideration all types of costs, not only those that are relevant for investors such as LCOE and TEA (Miah et al., 2017). Thus, while LCA focuses on the evaluation of the material and energy flows involved in the project, LCC permits the translation of such flows into monetary units. LCC analysis typically starts by defining the different cost stages, so an estimation of total costs can be carried out. The estimations and information obtained can then be used as inputs for risk/sensitivity analysis.

The results obtained with LCA and LCC are frequently geographically dependent because key variables to assess DG-RES projects, such as resource availability or the cost of materials, can vary considerably between different locations. In order to include this aspect, many authors use geographic information systems (GIS).

LCA has been used to analyse on-farm DG-RES for a wide range of applications and different countries (Table 2).

4 Risk analysis methods

Regardless of the focus of the cost analysis (whether private or social) and of the methodology used, modellers must consider how to introduce uncertainty when producing their assessments. Uncertainty is inherent to the estimated value of cost and returns for on-farm DG-RES as they will continue over decades of operation.

Analysis of uncertainty in cost calculations is often carried out by producing a sensitivity analysis that consists of varying the values of variables and quantifying their influence on the main indicators of the project, such as LCOE or GHG emissions. Sensitivity analysis is a deterministic approach that fails to recognize key factors to understand uncertainty, including endogeneity among variables (when missing from a model) and, crucially, the probability of the occurrence of significant events. In order to include these issues, cost models with probabilistic methods have been proposed that allow for the introduction of factors such as probability distributions and correlations between variables (Hasan et al., 2019). These probabilistic approaches require the selection of probability distributions for uncertain input factors based on observed cases such as data records or surveys.

The most frequently used probabilistic approach to assess risk and uncertainty in the cost of DG-RES is the Monte Carlo method (MCM). It consists of assigning a random value to the variable containing uncertainty and then running the model and recording the result. The process is repeated in order to obtain a set of results that can be used to assess the probability of the evolution of the variable to follow a given path. MCM can be applied in order to consider simultaneous variations in several parameters of a project and their impact on the expected cost and return. For example, the MCM allows simultaneous changes in parameters that present uncertainty, such as operation and maintenance costs, electricity generation and variability, and obtaining in this manner a probability distribution function for the value of IRR instead of a single value.

If the probability distribution of uncertain variables is not available, non-parametric methods can be used to introduce uncertainty into the models. A commonly used non-parametric model is quantile regression (QR), which is particularly suited to financial risk modelling (Koenker and Bassett Jr., 1978). QR is an extension of ordinary regression models that allow for the assessment of the relationship between the independent variables and the dependent variable when the conditions for regression analysis are not met.

5 Case study

LCOE is a metric that is widely used to assess the cost of DG-RES and to evaluate the economic sustainability of new projects from the 'private' perspective. A case study from the Duero Valley, Spain, is presented that consists of a cost assessment for the opportunity to introduce a solar PV system to power water pumps in an irrigation system in order to substitute for having a connection to the electricity grid. The decision rule in this case consists of comparing the calculated LCOE of the solar system with the price of the electricity that is currently being charged to the farmer by the electricity company (P_e). If LCOE < P_e, then the project should be executed.

In this case, P_e is already known and was calculated at $96.7 MWh[3] on the basis of the analysis of the bills charged by the electricity company over the last 5 years. To quantify the LCOE, estimations are required for the expected energy production of the DG-RES over the expected lifespan of the system (t), CAPEX ($), OPEX ($) and the discount rate r (see eq. 3).

5.1 Estimating energy production

The first variable that affects the energy production of a PV system is its capacity, normally expressed in kilowatt-peak (kWp). This variable is determined during the design of the irrigation system as a function of the expected energy demand of the pumps and the configuration of the system itself.

Another element needed to estimate energy production is the solar radiation available at the location of the system expressed in annual peak

3 Euros were used in this case study: €1 = USD 1.18 in March 2021.

Table 3 Parameters for the technical description of an on-farm solar PV-powered irrigation system on-farm

Parameter and symbol	Value	Unit
Capacity of the system (*S*)	125	kWp
Solar radiation (*Sun*)	2540	Peak sun hours
System efficiency (*Sys_eff*)	78%	
Lifespan	25	Years
Annual panel degradation (*Deg*)	0.8%	W/yr
Availability factor (*Av*)	97%	% of total time

sun hours.[4] There are a number of GIS-based tools that produce accurate estimations of solar radiation for specific geographical coordinates (see Annex 2). In this case, the SISIFO simulation tool was used (see https://www.sisifo.info/en/default). The system configuration refers to the power efficiency of the panels and inverters and also to the existence or not of a tracking mechanism to optimize the orientation of the solar panels towards the movement of the sun across the sky. These elements are introduced in the model using factors accounting for losses of efficiency. As the system will have production pauses caused by technical reasons, an 'availability ratio' must also be included. Since the availability ratio is a cause of uncertainty, a probabilistic model can also be used to estimate its value. Finally, the lifespan of the system and the degradation rate of the solar panels were also estimated, based on the information provided by the manufacturer. Thus, the parameters that describe the production of the system are outlined in Table 3.

In order to account for the progressive loss of production capacity of the PV cells, a linear degradation factor was introduced in the estimation of the future energy production which is then reduced each year according to its value. The energy yield (*Ey*), expressed in MWh, was calculated for year *t*:

$$Ey_t = S \times Sun \times Sys\_eff \times Av \times \left(1 - Deg\right)^t. \quad (4)$$

The resulting *Ey* is presented for the first 6 years of the life of the system in Table 4. Nevertheless, the estimation of electricity generation over the entire operational life of the system (25 years) was used in the calculations.

5.2 Estimating the capital expenditure

The capital expenditure of a solar PV system is mostly paid for during the planning and construction phase, when intensive use of human and material resources is required to build the system. The most costly contribution to CAPEX

4 One 'peak sun hour' is equivalent to 1000 W/m² of the sun's radiation collected in 1 h.

Table 4 Annual electricity (MWh) expected to be generated by the solar PV system over the first 6 years of operation

Year	1	2	3	4	5	6
Ey (MWh)	238.30	236.39	234.50	232.63	230.76	228.92

is by the solar panels, followed by the EPC activities. The EPC budget includes contractors hired to design and install the different components. PV systems for irrigation also require civil engineering works to be carried out in order to prepare the location where the systems will be installed. In this case study, a sun tracker was also included in order to improve the efficiency of the system. The frequency converters have a significant impact on CAPEX because their lifespans are shorter than the rest of the system and must be replaced every 10–15 years. Monitoring systems and the cost of taxes and permits are included in the estimation of CAPEX. In this case, the cost of the site was considered to be zero as it was provided by the project owner (though it could be included as an opportunity cost if considered to be of significance). Inverters are not required for this system configuration as the resulting installation will not be connected to the general grid. The economic value of the aforementioned elements was provided by the suppliers at market prices (Table 5). The summation of these costs for each year *t* is $CAPEX_t$.

5.3 Estimating operation expenditure

In order to assess the LCOE of the project, OPEX of the system has to be estimated. In this case study, OPEX will be paid at a constant rate during the entire operational lifetime, measured as $/MWh produced. The calculation of OPEX is inherently more uncertain than that of CAPEX, especially if fuel is required to operate the system, as is also the case for bioenergy plants. Uncertainty over the values of OPEX-related variables can be assessed in the model by using probabilistic methods. In order to introduce the expected inflation, the calculation of future inflows and outflows of cash must be adjusted. In this case study, the variables and the values included for the first 6 years in the LCOE calculation are outlined in Table 6.

5.4 Estimating the discount rate

LCOE provides a measure of the cost of energy that takes into consideration the cost of capital presented as a discount factor *r*. The discount factor is normally the weighted average cost of capital that is calculated based on the proportion of debt and equity used to fund the investment. In this case, the project will be executed based exclusively on equity, so the discount rate will be equal to

Table 5 Components of the capital expenditure (CAPEX $) of a 125 kWp solar photovoltaic system designed to power water pumps for an irrigation scheme

	Capital cost	
Year	1	12
PV modules	$51 625	
Variator Inversor	$14 160	
Sun tracker	$15 930	
Monitoring	$5310	
Replacement frequency converter		$15 250
Engineering, procurement, construction	$49 265	
Civil engineering works	$17 700	
Taxes and permits	$8850	
Total	$CAPEX_1 = \$162\,480$	$CAPEX_{12} = \$17\,958$

Table 6 Components of the operational expenditure (OPEX $/MWh generated annually - see Table 4) of a 125 kWp solar photovoltaic system designed to power water pumps for an irrigation scheme

Year	1	2	3	4	5	6
Maintenance	$18.2	$18.5	$18.9	$19.3	$19.7	$20
Management	$9.1	$9.3	$9.4	$9.6	$9.8	$10
Monitoring & communications	$8.9	$9	$9.2	$9.4	$9.6	$9.8
Security	$2.3	$2.3	$2.4	$2.4	$2.5	$2.5
Insurance	$3	$3	$3.1	$3.1	$3.2	$3.3
Total/MWh	$41.3	$42.1	$43	$43.8	$44.7	$45.6
Annual total	$9842	$9958	$10 076	$10 195	$10 316	$10 438
Inflation	2%					

the cost of equity. Among the different methods available to estimate the cost of equity, the CAPM model is the most widely used that calculates the cost of equity (k_e) of a project (Sharpe, 1964):

$$k_e = r_f + \beta \times (R_M - r_f), \quad (5)$$

where r_f is the risk-free rate (normally the return of government bonds of the country where the project is based with a maturity equivalent to the lifespan of the system), β is the systematic risk of the project (which can be obtained using comparables (Damodaran, 2002)), and R_M is the return of the risky asset of the country (normally the historical return of its stock market).

The estimated values of each of these variables for the solar PV case study project located in Spain are presented in Table 7.

5.5 Estimation of levelized cost of energy

Using the estimated discount factor, the summation of the expected energy yield discounted at *r* over the lifespan of the project is

$$\sum_{t=1}^{25} \frac{EnGen_t}{(1+r)^t} = 2297.09 \text{ MWh}. \tag{6}$$

It was assumed the CAPEX will be paid for in 2 years, most of it being spent during construction in year 1, with the remainder in year 12 when inverters are expected to be replaced. The resulting value of CAPEX discounted at the rate *r* is

$$CAPEX_1 + \frac{CAPEX_{12}}{(1+r)^{12}} = \$162480 + \$5949 = \$168790.$$

Regarding the OPEX, cash flows have been estimated and discounted for each of the 25 years of operation, adding up a total of \$108 995. Therefore, applying eq. 3 to the values estimated for the cost of the project:

$$LCOE = \frac{\$168790 + \$108995}{2297 \text{ MWh}} = \$120.9/\text{MWh}$$

The estimated LCOE from the solar PV system is higher than the current price of the energy obtained from the electricity grid (\$96.7/MWh), so in this case, the decision should be to not invest. Nevertheless, a sensitivity analysis can

Table 7 Parameters included in the calculation of the cost of equity (eq. 5) for use in predicting the LCOE for a 125 kWp solar PV project to power irrigation pumps on a farm in Spain

Variable	Value	Description
Risk-free rate (r_f)	1.072%	Yield of Spanish 25 year bond
Systematic risk of the project (β)	1.23	For solar assets in Spain
Return for risky asset (R_M)	8.04%	Historic annual return of Spanish stocks
Discount rate/cost of equity (k_e)	9.64%	Calculated for solar PV irrigation project in Spain

produce relevant information to support or revise this decision. In particular, a project investor may consider

1 to fund part of the project with debt in order to reduce *r* and consequently the LCOE of the project and/or
2 to change the configuration of the system in order to increase the efficiency.

5.6 Sensitivity analysis

A sensitivity analysis can estimate the LCOE for different levels of debt and the efficiency of the system. The original scenario in this case study was based on a level of debt for the project of 0% and a 78% efficiency. Increasing the levels of these variables gives a reduction of LCOE. So, the objective of the sensitivity analysis is to assess if it may be possible to reduce LCOE below the $96.7/MWh tariff currently being paid by the farmer to the power company for grid electricity (P_e) (Table 8).

Increasing the level of debt used to fund the project to 75% or more will give a lower LCOE that can become less than the retail price being paid for electricity from the grid. Increasing the efficiency of the system can also contribute to reducing the LCOE to competitive levels but at a lower rate than increasing the level of debt. However, increasing the system efficiency will probably require an increase in CAPEX that will result in an increase in LCOE. This possibility can be assessed with another sensitivity analysis. On the basis of the cost analysis presented, potential investors in solar PV to power the irrigation pumps can have access to a comprehensive set of information that

Table 8 Sensitivity analysis of the LCOE ($/MWh) of solar PV in relation to the proportion of debt used to fund the CAPEX (base case = 0%) and the efficiency of the system (base case = 78%). LCOEs that are lower than the current electricity charge of $96.7/MWh ($P_e$) (as shown in italics) indicate scenarios where changing to a solar PV system is economically viable

		Efficiency of the system					
		79%	80%	81%	82%	83%	84%
% of debt	50%	$104.1	$103.4	$102.8	$102.1	$101.5	$100.8
	55%	$102.6	$101.9	$101.3	$100.6	$100	$99.4
	60%	$101.1	$100.5	$99.8	$99.2	$98.6	$98
	65%	$99.6	$99	$98.4	$97.8	$97.2	*$96.6*
	70%	$98.1	$97.5	$96.9	*$96.3*	*$95.8*	*$95.2*
	75%	*$96.7*	*$96.1*	*$95.5*	*$94.9*	*$94.4*	*$93.9*
	80%	*$95.3*	*$94.7*	*$94.1*	*$93.6*	*$93*	*$92.5*
	85%	*$93.8*	*$93.3*	*$92.7*	*$92.2*	*$91.7*	*$91.2*

will help decide whether or not to execute the project based on which option has the highest potential to produce a financial return.

6 Conclusion

The decarbonization of the agrarian sector depends in part on the introduction of on-farm DG-RES. To generate clean heat energy, electricity and liquid biofuels on farms requires significant investments and involves financial risks and uncertainties. To deal with this, decision-makers such as investors, public authorities and farmers should have access to reliable information, methodologies and tools that allow them to assess the financial suitability of the projects.

The introduction of on-farm DG-RES can have impacts beyond the farm by affecting the economy and the environment of local communities. In this regard, researchers can play a key role by producing reliable, accurate and practical data sets and methodologies to support on-farm DG RES assessments at both private and public levels. This will facilitate the adoption of the most suitable technologies for a specific case and reduce the high level of uncertainty inherent in technology substitution processes. Higher levels of adoption and satisfaction based on properly assessed investment decisions can also accelerate the adoption of on-farm DG-RES, thereby further improving technical solutions and reducing their costs.

The existing methodologies and models used to assess the cost of on-farm DG-RES projects are largely based on those used for project appraisal in general. They include TEA, LCOE, CBA and LCA. These models are flexible frameworks that allow researchers to propose modifications in order to better reflect any peculiarities of on-farm DG RES in assessing the cost of the energy generated. These contributions are focusing on a wide range of issues ranging from how to assess the impact of feedstock price variations into bioenergy production costs to identifying what is the optimal level of government support to encourage deployment of new on-farm DG-RES projects.

On-farm DG-RES technologies continue to evolve so new analysis and research is needed on their effectiveness. Researchers should allow for the replacement of up-to-date data sets on the performance of state-of-the art systems working in real-world conditions. Data sets are key to facilitate accurate estimations of energy generation systems, operational costs and technical risks that are the basis of reliable cost valuations.

The production of reliable models and the application of advanced simulation tools can increase the quality of cost estimations and reduce the uncertainty of investments. In particular, probabilistic and computer-based methods are continuously evolving and provide an opportunity to apply the latest advances to improve cost analysis of on-farm DG-RES. Research is based on clear definitions of the technologies so that the information, indicators and results can be directly comparable. To assess social and environmental impacts, efforts are particularly relevant to reveal the cost of GHG emissions (and the benefits from avoiding them). International efforts are oriented towards integrating this cost in market prices, for example by implementing carbon costs. By providing farmers, policy makers and society as a whole with reliable tools to accurately assess the cost of on-farm DG-RES and to support the decarbonization of the farming sectors, researchers have the opportunity to make a significant contribution not only to scientific knowledge but also to the on-going efforts for sustaining the planet.

7 Annex 1: Focus and methodologies for on-farm distributed renewable energy systems

Table A1 Impacts that are most commonly referenced in the literature when assessing the cost of on-farm DG-RES projects and the type of methodology that is best suited to estimating its value (Strantzali and Aravossis, 2016)

Criteria	TEA	LCOE	CBA	LCA
Technical				
Efficiency	✓	✓	✓	✓
Reliability	✓	✓	✓	✓
Resource availability	✓	✓	✓	✓
Installed capacity	✓	✓	✓	✓
Safety			✓	✓
Lifespan	✓	✓	✓	✓
Economic				
CAPEX	✓	✓	✓	✓
OPEX	✓	✓	✓	✓
Cost of energy	✓	✓	✓	✓
Cost of capital	✓	✓		
Payback	✓	✓		
IRR	✓	✓		
Life cycle cost			✓	✓
NPV	✓	✓		
Environmental				
CO_2 emissions			✓	✓
Land use			✓	✓
Impacts on ecosystems			✓	✓
NO_x emissions			✓	✓
SO_x emissions			✓	✓
Noise			✓	✓

Criteria	TEA	LCOE	CBA	LCA
Social				
Job creation			✓	✓
Social acceptability			✓	✓
Social benefits			✓	✓
Visual impact			✓	✓
Local development			✓	✓
Impacts on health			✓	✓
Income from jobs			✓	✓

8 Where to look for further information

The estimation of the costs for the CAPEX and OPEX can be obtained from reliable and updated sources including:

- The U.S. National Renewable Laboratory (NREL) (https://www.nrel.gov/analysis/).
- The International Renewable Energy Agency (IRENA) (https://www.irena.org/costs).
- The Danish Energy Agency (https://ens.dk/en/our-responsibilities/global-cooperation/levelized-cost-energy-calculator).

Solar radiation levels for a specific location can be obtained from these GIS-based tools:

Tool	Developer	Web site
ArcGIS	Environmental Systems Research Institute	www.arcgis.com
Solargis	Solargis	www.solargis.com
SISIFO	Instituto de Energía Solar	www.sisifo.info
National Solar Radiation Database (NSRDB)	National Renewable Energy Laboratory	https://nsrdb.nrel.gov/
Global Solar Atlas	The World Bank	www.globalsolaratlas.info

LCA software that can be used for on-farm DG-RES assessment:

Software	Developer	Web site
Bousted	Bousted Consulting	www.boustead-consulting.co.uk

Software	Developer	Web site
Euklid	Frauenhofer Institut	www.ivv.fhg.de
GaBi	PE Europe GmbH	www.gabi-software.com
LCAmanager	SIMPPLE	www.simpple.com
OpenLCA	GreenDeltaTC	www.greendeltatc.com
REGIS	Sinum AG	www.sinum.com
SALCA	Agroscope Reckenholz-Tänikon	www.agroscope.admin.ch
SimaPro	Pré-consultants	www.pre.nl
TEAM	Ecobilan-Pricewaterhouse Coopers	www.pwcglobal.com
WISARD	Pricewaterhouse Coopers	www.ecobilan.eu
WRATE	UK Enviromental Agency	www.enviroment-agency.gov.uk/wrate

9 References

Abdelhady, S., Borello, D. and Shaban, A. (2018). Techno-economic assessment of biomass power plant fed with rice straw: Sensitivity and parametric analysis of the performance and the LCOE. *Renewable Energy* 115, 1026–1034. https://doi.org/10.1016/j.renene.2017.09.040.

Ahlgren, S., Baky, A., Bernesson, S., Nordberg, Å., Norén, O. and Hansson, P.-A. (2008). Future fuel supply systems for organic production based on Fischer–Tropsch diesel and dimethyl ether from on-farm-grown biomass. *Biosystems Engineering* 99(1), 145–155.

Allan, G., Eromenko, I., Gilmartin, M., Kockar, I. and McGregor, P. (2015). The economics of distributed energy generation: A literature review. *Renewable and Sustainable Energy Reviews* 42, 543–556. https://doi.org/10.1016/j.rser.2014.07.064.

Alonso, P. M., Hewitt, R., Pacheco, J. D., Bermejo, L. R., Jiménez, V. H., Guillén, J. V., Bressers, H. and de Boer, C. (2016). Losing the roadmap: Renewable energy paralysis in Spain and its implications for the EU low carbon economy. *Renewable Energy* 89, 680–694.

Arvanitoyannis, I. S. (2008). ISO 14040: Life cycle assessment (LCA)–Principles and guidelines. In *Waste Management for the Food Industries*, pp. 97–132. Amsterdam: Elsevier.

Barbari, M., Leso, L. and Simonini, S. (2012). Evaluating thermal performance of a photovoltaic-integrated roofing system for livestock buildings: Comparison with conventional coverings. *The Ninth International Livestock Environment Symposium (ILES IX); International Conference of Agricultural Engineering-CIGR-AgEng. Agriculture and Engineering for a Healthier Life*, Valencia, Spain, 8–12 July 2012.

Bruck, M., Sandborn, P. and Goudarzi, N. (2018). A levelized cost of energy (LCOE) model for wind farms that include power purchase agreements (PPAs). *Renewable Energy* 122, 131–139.

Cafferty, K. G., Muth, D. J., Jacobson, J. J. and Bryden, K. M. (2013). Model based biomass system design of feedstock supply systems for bioenergy production. *ASME 2013 International Design Engineering Technical Conferences and Computers and Information in Engineering Conference*.

Capuano, L. (2018). *International Energy Outlook 2018 (IEO2018)*, Vol. 21. Washington, DC: US Energy Information Administration (EIA).

Carpenter, D., Westover, T. L., Czernik, S. and Jablonski, W. (2014). Biomass feedstocks for renewable fuel production: A review of the impacts of feedstock and pretreatment on the yield and product distribution of fast pyrolysis bio-oils and vapors. *Green Chemistry* 16(2), 384–406.

Carrêlo, I. B., Almeida, R. H., Narvarte, L., Martinez-Moreno, F. and Carrasco, L. M. (2020). Comparative analysis of the economic feasibility of five large-power photovoltaic irrigation systems in the Mediterranean region. *Renewable Energy* 145, 2671-2682. https://doi.org/10.1016/j.renene.2019.08.030.

Damodaran, A. (2002). Relative valuation. In: *Investment Valuation*.New York: John Wiley & Sons.

Diep, N. Q., Sakanishi, K., Nakagoshi, N., Fujimoto, S. and Minowa, T. (2015). Potential for rice straw ethanol production in the Mekong delta, Vietnam. *Renewable Energy* 74, 456–463. https://doi.org/10.1016/j.renene.2014.08.051.

FAO (2011). *Energy-Smart Food for People and Climate*. Rome: Food and Agricultural Organisation of the United Nations.

Fernandez, P. (2007). Valuing companies by cash flow discounting: Ten methods and nine theories. *Managerial Finance* 33(11), 853–876.

Finnveden, G., Hauschild, M. Z., Ekvall, T., Guinée, J., Heijungs, R., Hellweg, S., Koehler, A., Pennington, D. and Suh, S. (2009). Recent developments in life cycle assessment. *Journal of Environmental Management* 91(1), 1–21.

Fore, S. R., Porter, P. and Lazarus, W. (2011). Net energy balance of small-scale on-farm biodiesel production from canola and soybean. *Biomass and Bioenergy* 35(5), 2234–2244. https://doi.org/10.1016/j.biombioe.2011.02.037.

Geissmann, T. and Ponta, O. (2017). A probabilistic approach to the computation of the levelized cost of electricity. *Energy* 124, 372–381.

Gittinger, J. P. (1982). *Economic Analysis of Agricultural Projects*. Baltimore, MD: John Hopkins University Press.

Goetzberger, A. and Zastrow, A. (1982). On the coexistence of solar-energy conversion and plant cultivation. *International Journal of Solar Energy* 1(1), 55–69.

Gonzalez-Longatt, F. and Fortoul, C. (2005). Review of the distributed generation concept: Attempt of unification. *International Conference on Renewable Energies and Power Quality (ICREPQ 05)*, pp. 16–18.

Graham, P. (2018). *Review of Alternative Methods for Extending LCOE Calculations to Include Balancing Costs*. Camberra: CSIRO.

Hanley, N. and Spash, C. (1996). Cost benefit analysis and the environment. Environmental Values 5(2).

Hansen, K. (2019). Decision-making based on energy costs: Comparing levelized cost of energy and energy system costs. *Energy Strategy Reviews* 24, 68–82.

Hasan, K. N., Preece, R. and Milanović, J. V. (2019). Existing approaches and trends in uncertainty modelling and probabilistic stability analysis of power systems with renewable generation. *Renewable and Sustainable Energy Reviews* 101, 168–180. https://doi.org/10.1016/j.rser.2018.10.027.

Hauschild, M. Z., Rosenbaum, R. K. and Olsen, S. I. (2018). *Life Cycle Assessment*. Cham: Springer.

Heck, N., Smith, C. and Hittinger, E. (2016). A Monte Carlo approach to integrating uncertainty into the levelized cost of electricity. *The Electricity Journal* 29(3), 21–30.

Hijazi, O., Munro, S., Zerhusen, B. and Effenberger, M. (2016). Review of life cycle assessment for biogas production in Europe. *Renewable and Sustainable Energy Reviews* 54, 1291–1300. https://doi.org/10.1016/j.rser.2015.10.013.
Holtmeyer, M. L., Wang, S. and Axelbaum, R. L. (2013). Considerations for decision-making on distributed power generation in rural areas. *Energy Policy* 63, 708–715. https://doi.org/10.1016/j.enpol.2013.07.087.
Hosseini-Fashami, F., Motevali, A., Nabavi-Pelesaraei, A., Hashemi, S. J. and Chau, K. (2019). Energy-life cycle assessment on applying solar technologies for greenhouse strawberry production. *Renewable and Sustainable Energy Reviews* 116, 109411. https://doi.org/10.1016/j.rser.2019.109411.
Ioannou, A., Angus, A. and Brennan, F. (2017). Stochastic prediction of offshore wind farm LCOE through an integrated cost model. *Energy Procedia* 107, 383–389.
Jarrar, L., Ayadi, O. and Al Asfar, J. (2020). Techno-economic aspects of electricity generation from a farm based biogas plant. *Journal of Sustainable Development of Energy, Water and Environment Systems* 8(3), 476–492.
Johansson, P. and Kriström, B. (2019). Welfare evaluation of subsidies to renewable energy in general equilibrium: Theory and application. *Energy Economics* 83, 144-155. https://doi.org/10.1016/j.eneco.2019.06.024.
Kannan, R., Leong, K. C., Osman, R., Ho, H. K. and Tso, C. P. (2006). Life cycle assessment study of solar PV systems: An example of a 2.7 kWp distributed solar PV system in Singapore. *Solar Energy* 80(5), 555–563.
Khatib, H. (2016). A review of the IEA/NEA Projected Costs of Electricity - 2015 edition. *Energy Policy* 88, 229–233. https://doi.org/10.1016/j.enpol.2015.10.030.
Kimming, M., Sundberg, C., Nordberg, A., Baky, A., Bernesson, S., Norén, O. and Hansson, P. A. (2011). Life cycle assessment of energy self-sufficiency systems based on agricultural residues for organic arable farms. *Bioresource Technology* 102(2), 1425-1432. https://doi.org/10.1016/j.biortech.2010.09.068.
Koenker, R. and Bassett Jr., G. (1978). Regression quantiles. *Econometrica* 46(1), 33–50.
Lansche, J. and Müller, J. (2012). Life cycle assessment of energy generation of biogas fed combined heat and power plants: Environmental impact of different agricultural substrates. *Engineering in Life Sciences* 12(3), 313–320.
Lijó, L., González-García, S., Bacenetti, J., Fiala, M., Feijoo, G., Lema, J. M. and Moreira, M. T. (2014). Life cycle assessment of electricity production in Italy from anaerobic co-digestion of pig slurry and energy crops. *Renewable Energy* 68, 625–635. https://doi.org/10.1016/j.renene.2014.03.005.
Lucheroni, C. and Mari, C. (2018). Risk shaping of optimal electricity portfolios in the stochastic LCOE theory. *Computers and Operations Research* 96, 374 –385.
Mana, A. A., Allouhi, A., Ouazzani, K. and Jamil, A. (2021). Feasibility of agriculture biomass power generation in Morocco: Techno-economic analysis. *Journal of Cleaner Production* 295, 126293. https://doi.org/10.1016/j.jclepro.2021.126293.
Markowitz, H. M. (1991). Foundations of portfolio theory. *The Journal of Finance* 46(2), 469–477.
Miah, J. H., Koh, S. C. L. and Stone, D. (2017). A hybridised framework combining integrated methods for environmental life cycle assessment and life cycle costing. *Journal of Cleaner Production* 168, 846–866.
Morales, M., Quintero, J., Conejeros, R. and Aroca, G. (2015). Life cycle assessment of lignocellulosic bioethanol: Environmental impacts and energy balance. *Renewable*

and Sustainable Energy Reviews 42, 1349–1361. https://doi.org/10.1016/j.rser.2014.10.097.

Mostefaoui, Z. and Amara, S. (2019). Renewable energy analysis in the agriculture-greenhouse farms: A case study in the Mediterranean region (Sidi bel abbes, Algeria). *Environmental Progress and Sustainable Energy* 38(3), e13029.

Niajalili, M., Mayeli, P., Naghashzadegan, M. and Poshtiri, A. H. (2017). Techno-economic feasibility of off-grid solar irrigation for a rice paddy in Guilan province in Iran: A case study. *Solar Energy* 150, 546–557. https://doi.org/10.1016/j.solener.2017.05.012.

Nissen, U. and Harfst, N. (2019). Shortcomings of the traditional "levelized cost of energy" [LCOE] for the determination of grid parity. *Energy* 171, 1009–1016. https://doi.org/10.1016/j.energy.2019.01.093.

Nouni, M. R., Mullick, S. C. and Kandpal, T. C. (2007). Techno-economics of small wind electric generator projects for decentralized power supply in India. *Energy Policy* 35(4), 2491–2506. https://doi.org/10.1016/j.enpol.2006.08.011.

Obi, M., Jensen, S. M., Ferris, J. B. and Bass, R. B. (2017). Calculation of levelized costs of electricity for various electrical energy storage systems. *Renewable and Sustainable Energy Reviews* 67, 908–920. https://doi.org/10.1016/j.rser.2016.09.043.

Ondraczek, J., Komendantova, N. and Patt, A. (2015). WACC the dog: The effect of financing costs on the levelized cost of solar PV power. *Renewable Energy* 75, 888–898.

Orioli, A. and Di Gangi, A. (2015). The recent change in the Italian policies for photovoltaics: Effects on the payback period and levelized cost of electricity of grid-connected photovoltaic systems installed in urban contexts. *Energy* 93, 1989–2005.

Ouyang, X. and Lin, B. (2014). Levelized cost of electricity (LCOE) of renewable energies and required subsidies in China. *Energy Policy* 70, 64–73.

Pantaleo, A., Gennaro, B. D. and Shah, N. (2013). Assessment of optimal size of anaerobic co-digestion plants: An application to cattle farms in the province of bari (italy). *Renewable and Sustainable Energy Reviews* 20, 57–70. https://doi.org/10.1016/j.rser.2012.11.068.

Pearce, D. (1998). Cost benefit analysis and environmental policy. *Oxford Review of Economic Policy* 14(4), 84–100.

Pimentel, D. and Pimentel, M. H. (2007). *Food, Energy, and Society*. Boca Raton, FL: CRC Press.

Pombo-Romero, J. and Regueiro-Ferreira, R. (2016). RESFARM: Report on analysis on survey data. https://doi.org/10.13140/RG.2.2.33469.77287.

Rahman, S. M., Spalding-Fecher, R., Haites, E. and Kirkman, G. A. (2018). The levelized costs of electricity generation by the CDM power projects. *Energy* 148, 235–246.

Reichenberg, L., Hedenus, F., Odenberger, M. and Johnsson, F. (2018). The marginal system LCOE of variable renewables–Evaluating high penetration levels of wind and solar in Europe. *Energy* 152, 914–924. https://doi.org/10.1016/j.energy.2018.02.061.

Rhodes, J. D., King, C., Gulen, G., Olmstead, S. M., Dyer, J. S., Hebner, R. E., Beach, F. C., Edgar, T. F. and Webber, M. E. (2017). A geographically resolved method to estimate levelized power plant costs with environmental externalities. *Energy Policy* 102, 491–499.

Sarkar, M. N. I. and Ghosh, H. R. (2017). Techno-economic analysis and challenges of solar powered pumps dissemination in Bangladesh. *Sustainable Energy Technologies and Assessments* 20, 33–46. https://doi.org/10.1016/j.seta.2017.02.013.

Schindele, S., Trommsdorff, M., Schlaak, A., Obergfell, T., Bopp, G., Reise, C., Braun, C., Weselek, A., Bauerle, A., Högy, P., Goetzberger, A. and Weber, E. (2020).

Implementation of agrophotovoltaics: Techno-economic analysis of the price-performance ratio and its policy implications. *Applied Energy* 265, 114737. https://doi.org/10.1016/j.apenergy.2020.114737.

Sethi, V. P. and Sharma, S. K. (2008). Survey and evaluation of heating technologies for worldwide agricultural greenhouse applications. *Solar Energy* 82(9), 832–859.

Sharpe, W. F. (1964). Capital asset prices: A theory of market equilibrium under conditions of risk. *The Journal of Finance* 19(3), 425–442.

Shen, W., Chen, X., Qiu, J., Hayward, J. A., Sayeef, S., Osman, P., Meng, K. and Dong, Z. Y. (2020). A comprehensive review of variable renewable energy levelized cost of electricity. *Renewable and Sustainable Energy Reviews* 133, 110301. https://doi.org/10.1016/j.rser.2020.110301.

Short, W., Packey, D. J. and Holt, T. (1995). *A Manual for the Economic Evaluation of Energy Efficiency and Renewable Energy Technologies*. Golden, CO: NREL.

Strantzali, E. and Aravossis, K. (2016). Decision making in renewable energy investments: A review. *Renewable and Sustainable Energy Reviews* 55, 885–898. https://doi.org/10.1016/j.rser.2015.11.021.

Todde, G., Murgia, L., Carrelo, I., Hogan, R., Pazzona, A., Ledda, L. and Narvarte, L. (2018). Embodied energy and environmental impact of large-power stand-alone photovoltaic irrigation systems. *Energies* 11(8), 2110.

Topper, M. B. R., Nava, V., Collin, A. J., Bould, D., Ferri, F., Olson, S. S., Dallman, A. R., Roberts, J. D., Ruiz-Minguela, P. and Jeffrey, H. F. (2019). Reducing variability in the cost of energy of ocean energy arrays. *Renewable and Sustainable Energy Reviews* 112, 263–279. https://doi.org/10.1016/j.rser.2019.05.032.

Tu, Q., Betz, R., Mo, J., Fan, Y. and Liu, Y. (2019). Achieving grid parity of wind power in China –Present levelized cost of electricity and future evolution. *Applied Energy* 250, 1053–1064. https://doi.org/10.1016/j.apenergy.2019.05.039.

Ueckerdt, F., Hirth, L., Luderer, G. and Edenhofer, O. (2013). System LCOE: What are the costs of variable renewables? *Energy* 63, 61–75.

Unrean, P., Fui, B. C. L., Rianawati, E. and Acda, M. (2018). Comparative techno-economic assessment and environmental impacts of rice husk-to-fuel conversion technologies. *Energy* 151, 581–593.

Vourdoubas, J. (2020). Co-production of vegetables and electricity in agricultural greenhouses in Crete, Greece. Is it feasible? *Journal of Agriculture and Life Sciences* 7(1).

Wang, X., Kurdgelashvili, L., Byrne, J. and Barnett, A. (2011). The value of module efficiency in lowering the levelized cost of energy of photovoltaic systems. *Renewable and Sustainable Energy Reviews* 15(9), 4248–4254.

Wettstein, S., Muir, K., Scharfy, D. and Stucki, M. (2017). The environmental mitigation potential of photovoltaic-powered irrigation in the production of South African maize. *Sustainability* 9(10), 1772.

Xydis, G. (2015). A wind energy integration analysis using wind resource assessment as a decision tool for promoting sustainable energy utilization in agriculture. *Journal of Cleaner Production* 96, 476–485. https://doi.org/10.1016/j.jclepro.2013.11.030.

Yang, J., Olsson, A., Yan, J. and Chen, B. (2014). A hybrid life-cycle assessment of CO2 emissions of a PV water pumping system in China. *Energy Procedia* 61, 2871–2875. https://doi.org/10.1016/j.egypro.2014.12.326.

Zhou, P., Ang, B. and Poh, K. (2006). Decision analysis in energy and environmental modelling: An update. *Energy* 31(14), 2604–2622.

Chapter 6

Energy-smart innovation and renewable energy systems on farms: an overview

Ralph E. H. Sims, Massey University, New Zealand

1 Introduction

Food production 'behind the farm gate' requires reliable inputs of energy to power tractors and machinery; run irrigation pumps; provide heat for drying, livestock housing and greenhouses; and supply electricity for refrigeration, lighting, running motors and so on. Whether for cropping, fruit and vegetables, pastoral and intensive livestock production, aquaculture or fishing, reliable energy inputs are essential. Farm forest production is less energy intensive so has not been included in this analysis.

On-farm energy demand accounts for around one-fifth of the total energy consumption in the whole agri-food supply chain (Fig. 1). This sector also emits around one-third of all global anthropogenic greenhouse gas emissions, including indirect upstream emissions from fossil fuel extraction and refining (Crippa et al., 2021) and indirect embedded emissions from the manufacture of fertiliser, agri-chemicals, machinery, buildings and so on (Sims et al., 2015; Tubiello et al., 2020).

Global end-use energy demand is more than 300 EJ/year, mainly met by fossil fuels of which nearly one-third of annual demand is consumed by the agri-food supply chain. The more intensive farm and food processing

http://dx.doi.org/10.19103/AS.2022.0100.23

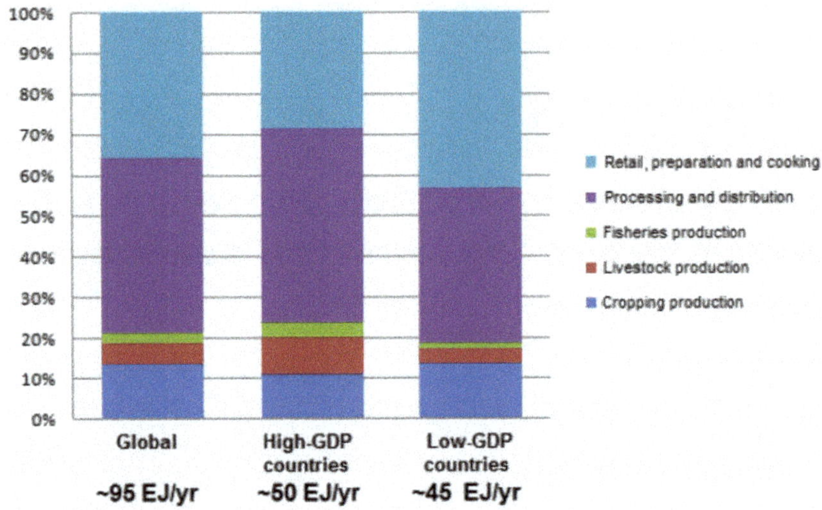

Figure 1 Shares of direct energy inputs throughout the global agri-food supply chain with differences shown between the top 50 GDP countries and the remaining developing countries (FAO, 2011).

systems in the top 50 high-GDP countries consume around 35 GJ/capita/year, whereas for the lower-GDP countries (where subsistence farming is common and a greater share of the total energy input is used for domestic cooking), the agri-food sector consumes a significantly lower 8 GJ/capita/year (FAO, 2011).

The total energy required to bring food to the table is around 15% of total consumer energy supply in the United States, about 20% in the United Kingdom, over 30% in New Zealand, where the economy depends on food production for exporting a wide range of processed products, and as much as 55% in low-GDP countries. Here, around 10% of the total is for on-farm production and transport of products over short distances, 15% is for processing and 75% is for cooking using traditional biomass in the form of fuelwood, charcoal and animal dung.

Renewable energy sources, including from hydro, wind, biomass, geothermal, solar and, to a small degree, ocean energy, are being used more each year to aid the transition to a less carbon-intensive and more sustainable energy-based system. In addition to large-scale renewable energy installations for electricity grid connection, combined heat and power plants and industrial heat loads of hundreds of MW capacity, small-scale systems are being employed to provide heat and power at the farm scale. The production and use of biofuels and hydrogen energy carriers for transport continue to evolve, with particular use targeted for heavy trucks, aviation and shipping that are difficult to electrify. However, they could also have future applications on farms for fuelling tractors and machinery, grain drying, generating electricity using engine generating sets and fuel cells and so on.

This chapter focuses on the energy demands of farms large enough to supply supermarket retail companies and does not cover subsistence farming. Whether small- or large scale, farm enterprises will usually have access to modern post-harvest treatments and food storage facilities that also require electricity or fossil fuel inputs (e.g. for a cool chain for transporting milk, fish, fresh fruit and vegetables or for cereal grain drying and storage).

With regard to energy demand, energy intensity and direct and indirect fossil fuel dependence, there are major differences between subsistence farming in developing countries, large corporate, industrialised food production systems and other categories of farm enterprises in between (Table 1). Production intensity, labour demand, investment capital availability and type of market supplied also vary.

Some land owners can benefit when large-scale renewable energy projects are developed on their land because they have exceptional resources, such as high wind speeds, a waterway suitable for hydro power generation or above-average levels of solar radiation. The land also should be situated relatively close to the electricity demand to minimise transmission costs.

This chapter provides an introduction and overview to Part 2 of this publication by outlining the range of renewable energy technologies and systems suitable for use at the on-farm scale. Such systems can also be used for food processing plants whether located on the farm or more centrally at a larger scale. Energy efficiency and conservation opportunities to reduce on-farm energy demand can result in lower capital investments in renewable energy systems of smaller capacity. Energy efficiency details are covered in Part 1 so only a brief overview is presented here. It precedes a section outlining recent trends in renewable energy systems including large-scale wind, hydro, bioenergy and solar farm developments on agricultural land. In the following sections, greater emphasis is placed on utilising distributed renewable energy resources and developing systems at the farm scale. Policies and measures to encourage the deployment of distributed energy systems in rural locations are discussed with real-world case studies used to illustrate practical options that other farmers could implement.

2 The use of energy efficiency to help decarbonise food production systems

There are usually opportunities to reduce energy demand on farms at all scales. Introducing more energy-efficient technologies and systems can help to reduce fossil fuel dependence. Displacing the remaining fossil fuels with local,

Table 1 A qualitative assessment of typical 'small'- and 'large'-scale farms and fisheries to provide an indication of the range of farming systems around the world (FAO, 2011)

Scale of producer	Overall input intensity	Human labour units	Animal power use	Fossil fuel dependence	Capital availability	Majorfood markets	Gnergy intensity
Subsistence level	Low	I-2	Common	Zero	Micro-finance	Own use	Low
Smali family unit	Low	2-3	Possible	Low/medium	Limited	Local fresh/process/own use	Low to high?
	High	2-3	Rarely	Medium/high	Limited	Local fresh/regional process/ own use	Low to high?
Small business	Low	3–10	Rarely	Medium/high	Medium	Local/regional/export	Low to high?
	High	3–10	Never	High	Medium	Local/regional/export	Low to high?
Large corporate	High	10–50	Never	High	Good	Regional process/export	Low to high?

low-carbon, renewable energy sources is then a cheaper proposition because the total energy demand is lower. New business models can be applied.

Many opportunities exist to improve energy intensity by lowering the energy inputs per unit of food produced (MJ/kg), for example by improving the operating performance of tractors and field machinery, reducing natural gas demands during fertiliser manufacture, through more precise application of fertilisers and agri-chemicals, reducing heat losses through better designed heated greenhouses and controls and avoiding water losses. Currently, a large proportion of water applied to crops by irrigation schemes is wasted due to leaks and inefficient application methods. Improvements from monitoring soil moisture levels and applying only when needed can save both water and energy used for pumping. The overall aim should be using less energy inputs to produce similar or greater amounts of food per unit of land area and water input.

Energy-saving measures can assist the profitability of individual enterprises, including for example using boat and motor designs and hull coatings that give lower fuel consumption when used for capture fishing. Energy efficiency can result in either direct savings owing to changes in technology or behaviour or indirect savings arising as a co-benefit from agro-ecological farming practices or from social change such as a reduced intake of animal protein. Besides containing costs and reducing greenhouse gas emissions from lower fossil fuel demand, energy efficiency measures can also help to make food production less vulnerable to any future interruptions in fossil fuel supplies. Reducing energy intensity depends both on farmers making behavioural changes and on the development and deployment of new practices and technologies that can improve energy efficiency at little or no cost.

Historically, energy costs have been a small component of the total operating costs for many food production enterprises. Therefore, policies and incentives to reduce energy demand have not been strong. However, as energy costs increase and more businesses set targets to reduce their carbon footprints, renewed interest in improving energy efficiency will gain win-win benefits.

Where significant capital investment in modern equipment is required to adopt improved energy-efficient technologies, this can be a constraint (Flammini et al., 2019). Examples include precision farming systems, irrigation monitoring and soil moisture sensing, boat propeller designs, LED lighting, advanced heat exchangers and heat pumps and variable-speed electric motors. A balance needs to be sought between energy efficiency savings and current and projected energy costs to achieve an acceptable payback period.

Energy reduction strategies across the diverse range of agri-food management options are complex and can involve trade-offs. For example, agro-ecological farming practices can be energy efficient and also achieve good yields and benefit livelihoods. Any methods used to reduce energy inputs that also lower productivity (e.g., by simply cutting back the amount of fertiliser applied rather than optimising the frequency, time and accuracy of application) are rarely beneficial and should be avoided. Food production systems with high overall inputs do not necessarily have high energy intensities, especially when they result in increased productivity. Conversely, lower yields from low input systems can result in relatively high energy intensities.

For smaller-scale family farms, there may be a case for increasing both direct and indirect energy inputs over time in order to improve productivity and water use efficiency. Energy conservation and efficiency measures can be implemented in several ways at all stages along the agri-food chain, including on the farm (see the chapters in Part 1 for details).

For both large and small farming systems, any means of avoiding losses and deterioration post-harvest should be encouraged as the energy embedded in the food production system will be wasted. In addition, the growing competition for land and water will be reduced, given that globally we fail to consume one-third of all the food produced.

Any improvements in energy efficiency bear the risk of a 'rebound effect' that can occur when lower energy bills result and thus encourage additional energy purchases in other areas. For example, a fishing company that saves fuel by more efficient operation of the boats might use the money saved to purchase larger and faster outboard motors that have higher fuel consumption. The energy intensity of the fish catch (MJ/kg fish) would increase although the time spent on the water would decline. The scale of rebound effects and their duration are the subject of much debate, although it is a common response to reducing energy demand and should therefore be taken into account when estimating potential overall energy savings.

3 Trends in renewable energy

Globally, renewable energy provides around 10% of total primary energy (IEA, 2019). Energy from combusting traditional biomass used for cooking and heating adds a further 4–5%. Almost 170 countries have renewable electricity policies and targets in place, around 70 have renewable energy transport fuel policies and over 25 have renewable energy heating and cooling policies (REN21, 2021).

Electricity represents around 17% of total global final energy use with renewable electricity producing over 26% of total power generation in 2019 (IEA, 2020). Wind power has now reached over 650 GW installed capacity globally with solar photo-voltaic (PV) systems at a similar level. Both continue to grow exponentially as their costs continue to decline. Almost half of the renewable energy use is captured for large-scale centralised electricity generation through hydro-power plants, wind turbine farms, bioenergy power systems (including co-generation of heat), high-temperature geothermal fields, solar PV systems and a few tidal range plants (with wave and ocean current technologies still under early development).

Heating and cooling services meet over half of the total final energy demand with renewables providing over 10% of the total. Geothermal, solar thermal and the combustion of biomass and biogas can provide useful heat at high to low temperatures. Thus, renewable energy to meet annual global heating demand has a similar total energy value as that used for electricity generation.

Conversely, whereas transport fuels (including for tractors and off-road vehicles) represent almost one-third of total final energy consumption, renewables provide only 3.3% of that share. Bioethanol and biodiesel have long been blended with gasoline and diesel at a large commercial scale in several countries. Biofuels can also be produced and used directly on farms similar to biogas (see sections below). Renewable electricity is used to power an increasing number of battery and hybrid road vehicles as well as trains, boats and even tractors. Green hydrogen (produced from the electrolysis of water using renewable electricity) has the potential to provide fuel for large boats and aircraft in the future and possibly also at a smaller scale.

Around one-third of global end-use energy demand from coal, oil and gas (95 EJ) is consumed annually to produce, process and bring to the table the world's food supply (Fig. 1). Direct energy inputs are essential on modern farms for harvesting, irrigation, storage and also along the supply chain for food processing, transport, retailing and cooking. Indirect energy inputs needed for synthetic fertiliser and machinery manufacture, for example, are important but are not included in the analysis shown in Fig. 1. Most of the agri-food sector's energy demand is met by the combustion of fossil fuels (Sims et al., 2015). In addition, human labour, animal power, the combustion of traditional biomass (such as fuelwood) to provide heat and small shares of renewable heat and electricity contribute to the total energy demand.

The financial cost, volatile prices and greenhouse gas emissions resulting from using oil, coal and natural gas on farms raise concerns about future energy security, production costs, competitiveness and affordable food prices for consumers (OECD, 2017). Emissions of black carbon (a short-lived climate forcer that arises mainly from diesel engines, brick making, charcoal production and fuelwood combustion in cookstoves) are also of growing concern (Sims et al., 2017).

On-farm energy and water inputs will need to increase over the coming decades to avert climate-related impacts while simultaneously attempting to increase agricultural productivity. However, increasing fossil fuel dependence has to be avoided. Decarbonising energy sources, improving water use efficiency and making a transition to a more sustainable food supply system through a combined water-energy-food nexus approach are urgent (IRENA, 2016). Decentralised energy systems that depend on renewable energy will have a key role to play. In addition to large-scale, commercial energy systems, renewable energy sources can be used at the small scale for utilisation by individual farms or in rural communities in distributed energy systems (see below).

The capital costs for solar PV systems have decreased by around 90% in the past decade as have the costs for lithium-ion batteries (from around US$400/kWh capacity to US$100/kWh), and wind turbines are also around 40% cheaper. The levelised cost of on-shore wind power is currently around US$50–100/MWh and solar PV power is around US$60–160/MWh, so both are becoming competitive with coal- and gas-fired power plants in many regions, especially where a price on carbon emissions has been implemented. The total capacity of solar and wind power installed during recent years has exceeded previous projections and presently provides around 7% of the total electricity supply.

Concentrated solar power (CSP) plants (that use mirrors to focus the solar energy to reach high temperatures for driving steam turbines) are increasing in regions with high solar intensity. They have the beneficial characteristic of heat being stored in liquids and so power can be generated to meet peak demands such as in the evenings (SolarPACES, 2021). Small-scale CSP plants suitable for distributed energy mini-grids are evolving but not yet sufficiently reliable or commercially viable (Giovannelli, 2015).

Bioenergy is also making a growing contribution to provide electricity and heat as co-generation and to provide space heating in buildings and industrial process heat. Liquid biofuels for transport fuels have grown at a slower pace.

In 2019, over US$301 billion was invested in new renewable power, heat and transport systems having grown annually over the past decade (REN21, 2020). The total installed capacity for renewable electricity technologies reached 2588 GW (Table 2).

3.1 Large-scale renewable power generation and land use

Wind farms and solar farms to supply power into the national grid can be constructed on farmland as a revenue earner for the landowner. In specific areas, landowners may also extract geothermal heat sources to provide a range of low- and high-temperature applications (including for electricity generation) depending on the source and if the location is not too far from the demand to make the cost of energy transmission non-viable. Geothermal heat can also be used for industrial applications such as prawn production, the drying of sawn timber, paper pulp production and food processing but is rarely used at the farm scale as few landowners have access to suitable resources. Therefore, geothermal heat is not included here.

Table 2 Installed global power capacities for renewable energy technologies in 2019, annual energy demands for renewable heating and transport fuels and annual rate of increase

Electricity generation technology	**Installed capacity (GW)**	**Increase from 2018 (%)**
Hydropower	1150	1.32
Wind power	651	10.15
Solar PV	627	22.46
Bioenergy	139	6.10
Geothermal	13.9	5.30
Concentrating solar power	6.2	10.71
Ocean energy and tidal range	0.5	0
Heating fuels	**Thermal energy (TJ/yr)**	**Increase from 2018 (%)**
Modern bioenergy	14100	1.43
Solar hot water	1400	0
Geothermal direct heat	1021000[a]	9.63
Transport fuels	**Litres/yr (billion)**	**Increase from 2018 (%)**
Ethanol	114	2.70
Biodiesel (fatty acid methyl esters)	47	14.63
Biodiesel (hydro-treated vegetable oil)	6.5	8.33

[a] https://www.geothermal-energy.org/pdf/IGAstandard/WGC/2020/01018.pdf

Biomass plantations have been grown as feedstock for heat and power plants as well as for liquid biofuels. For example, many hectares of short rotation willow (*Salix*) trees were planted in Sweden in the 1990s to be harvested every 3–4 years for bioenergy heat and power applications and then the trees were allowed to coppice (regrow from their cut stems). The Swedish biomass industry continues to expand (Dunphy, 2018). However, a similar approach in the United Kingdom failed when the pioneering 10-MW Yorkshire ARBRE gasification power plant became insolvent (Piterou et al., 2008). Many farmers had been contracted for 16 years to grow and supply the willow biomass, but the plantations then became a liabilvity.

Growing crops specifically for biomass on agricultural land (such as sugarcane in Brazil and corn in United States for bioethanol) has resulted in the 'food-versus-fuel' debate. Land use competition can be partly overcome by using residues from crops for bioenergy (such as straw or rice husks) and the arisings from plantation forests (that remain on the land after log harvesting). At least, in theory, reducing food losses and wastes would enable more land and water to become available for energy cropping.

Typically, a large-scale wind or solar farm plant developer would lease land from a farmer who would receive a share (usually around 1–2%) of the revenue from the wholesale price obtained for the electricity generated. This revenue is often higher than what could have been earned from farming the land, although it is also often possible to graze animals beneath the solar panels or between the wind turbines or even to grow crops (Figs. 2 and 3). In addition, many governments provide some form of subsidy or incentive to encourage commercial investment into large-scale renewable generation systems.

As solar PV and wind power technologies have become more efficient over time through improved design and materials, the area of land needed to generate, say 100 GWh per annum of electricity, has declined. The land area needed depends on the solar and wind resources that vary for a given location.

Mean annual wind speeds of commercial wind farm sites vary from around 7 m/s (as in Denmark) to 10 m/s (as in New Zealand) giving capacity factors of around 30–50%, respectively. The power output of a wind turbine is proportional to the cube of the wind speed; hence, a wind turbine installed on a 10 m/s site will generate around three times as much electricity in a year as the same turbine located on a 7 m/s site. Hence, for example, a site with high mean wind speeds, to generate around 100 GWh/yr, would require 25 MW of installed wind capacity (say 10 turbines each of 2.5 MW) on around 300 ha of land. Each turbine will have a concrete foundation of around 70 t, access roads will be needed for maintenance and power lines could be overhead or underground. Thus, the wind farm would typically have around 270 ha available for grazing or cropping (Fig. 3). Larger wind turbines up to around 5 MW are being installed on land, and up to 8 MW for off shore installations.

A solar farm will generate more power in a location with high levels of annual solar radiation (such as California) than if installed somewhere with lower levels (such as Washington). Typically, a 1 MW installation of solar crystalline panels of around 18% efficiency will require 3000 panels and need around 1.5 ha of land, whereas for less-efficient thin-film technology, more panels would be required so perhaps 2 ha of land would be needed for the same power output.

Box 1 Case study – Solar farms

(Photo courtesy of Solisinverters.com)

A solar PV power plant constructed on a tidal flat in China has 690 000 monocrystalline solar panels, each of 440 W peak capacity. These are mounted 5 m above the ground to accommodate the tides and the structure is spread over a water surface area of around 1800 ha. Solis 1500 V string inverters convert the DC solar power to AC before it is exported from the on-site 220 kV sub-station to the grid. In this harsh environment, the structure has been designed to be sand- and waterproof as well as corrosion-resistant. It is claimed it will generate around 400 GWh of electricity per year, though this implies very high solar radiation levels (Shenvekar, 2021).

Solar farms have become popular in the UK due to government subsidies given that solar radiation levels tend to be relatively low. For example, a 150 ha dairy farm in central England has leased a 10 ha field to a solar farm developer.

The farmer has reduced the cow numbers due to the loss of land area, but this is more than compensated for by the revenue received from the developer. Given that pasture can still grow beneath the panels even if partially shaded, it would be possible to also graze sheep on the same land.

3.2 Distributed energy systems

Energy and water demands in the agri-food sector can be met with off-grid and decentralised energy systems. Renewable energy and energy-efficient technologies are mature and the cost of installation is declining. Many examples exist, from small-scale solar power systems for off-grid family farms to large-scale combined heat and power plants for supplying mini-grids and heat for food processing plants and district heating schemes. Technically viable and socially beneficial low-carbon technologies include wind- and solar-powered water pumps, solar water heaters, straw-fired crop drying heaters, biomass-fired heat and power plants, mini-hydropower turbines, insulated cool stores,

Figure 2 Animals can graze pasture growing under solar PV panels, and it is also possible to grow and harvest cereal crops beneath them, although greater capital investment for the support structures is required (photos courtesy of UN FAO).

efficient greenhouse lighting systems, precision irrigation systems, biogas for heat or transport fuel and solar photovoltaic milk coolers (UNFCCC, 2019).

Box 2 Mini-grids in rural communities

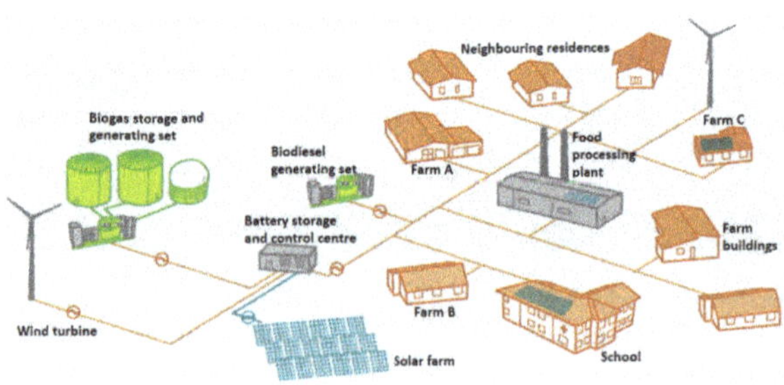

(Based on a figure from BINE Information Services, 2011)

A mini-grid (or slightly smaller micro-grid) is a small-scale local electricity distribution network that delivers power from a range of distributed energy generation sources, including renewable energy systems, combined heat and power plants and diesel- or gas-powered generating sets, to multiple users.

Stand-alone mini-grids - often based on solar and wind power and possibly with battery storage or back-up diesel plants - have the potential to electrify rural areas (Meier, 2017). In remote areas of developing countries, establishing a mini-grid can be more efficient, more economically feasible and more effective than extending a national grid to a remote location.

By the establishment of co-operatives, a mini-grid can enable social organisations to enter the energy market and benefit members by selling cheaper electricity for local residents and businesses than conventional electricity generation and distribution companies. A possible perception by rural communities is that mini-grids do not provide electricity supply reliably and have frequent outages and unstable voltage fluctuations. More demonstrations of mini-grids may need to be developed, monitored and widely promoted to build greater confidence in the concept.

Where the mini-grid is linked with automatically controlled appliances, demand-side load shedding at peak times (such as cool stores that can be turned off for a few hours without produce deteriorating), energy storage systems and so on, the term 'smart-grid' is sometimes used, although there is no firm definition as to what it implies or whether it can be large or small scale (MBIE, 2018).

Distributed renewable energy resources are playing an increasing role in providing energy services in developing countries, especially in rural areas that are remote and not easy to become grid-connected. Modern bioenergy and solar thermal systems for cooking and heating are displacing the inefficient use of traditional biomass in the form of scavenged fuelwood and dung. Local electricity generation, mainly from solar PV, small wind turbines, and micro-hydro, can provide lighting, motor drives, refrigeration and water pumping including for irrigation.

Cooling is a key component of modern energy services since in addition to preserving fresh agricultural products in cool stores and along the cool chain,

Figure 3 Te Apiti wind farm in New Zealand with 3 MW turbines spaced around 400 m apart and sheep and cattle shown grazing underneath.

it can improve labour productivity during hot weather. In developing countries, it can also improve health outputs as a result of refrigerated vaccine storage. Distributed energy systems can help provide cooling through conventional refrigeration using compressors, though independent, non-electric, solar-powered vapour absorption coolers are close to commercialisation. However, the variability of solar energy inputs both seasonally and diurnally can be a constraint to the deployment of sustainable solar cooling systems in some locations (Sharma et al., 2019). Innovative solutions for solar chilling, ice making powered by the gasification of farm residues and off-grid refrigeration are under development. In May 2020, Engineers Without Borders awarded grants between US$30 000 and US$ 50 000 to seven projects under its 'Chill Challenge' (EWB, 2020).

Given the rapid rate of cost reductions for solar, wind and battery storage technologies, distributed renewables can provide cost-effective energy access for remote farms as well as for small islands, rural communities, health carers, schools, rural businesses and holiday resorts. They can be scaled up over time. Distributed energy systems involve a wide range of technologies that generate, store and distribute energy in locations close to the end users. Hence, they can both improve energy access for those without reliable electricity and contribute to the restructuring of an existing power grid. Mini-grids are being widely developed (e.g. using mini-hydro power plants where a suitable watercourse exists) with the future potential to turn them into smart grids. Grid flexibility, including demand-side response to automatically shut down some pre-agreed loads when supply is constrained, together with battery storage systems, enables higher shares of variable wind and solar generation to be integrated without loss of reliability (Sims et al., 2011).

Taking a circular economy approach to the management of ecosystem services under the water–energy–food nexus approach, together with water and energy management, can help avoid the growing concerns of environmental impacts. To improve freshwater conservation and encourage the circular economy, government policies, measures and incentives are essential. Policies that reduce the on-farm demand for energy and water inputs as a result of nature-based solutions and ecosystem services to the agri-food chain involve many stakeholders who often have conflicting interests so need to be holistic. Coherent policy development requires dialogue and close collaboration among government ministries as well as between national and local authorities.

Many examples of the successful deployment of cost-effective energy-smart and climate-smart agri-food systems exist, but they have not yet been widely promoted or deployed. Therefore, off-grid and decentralised energy is yet to become mainstream. To overcome the constraints to uptake, including the relatively high capital costs, will require capacity-building, consultation with the private sector, innovative business models, implementation of national standards, supporting government policies and education of potential users.

Mini-grids will have a specific role to play to bring electricity to farms in remote rural areas wherever they are a more viable proposition than extending the main electricity grid. Solar PV and battery systems have started to displace diesel generating sets, and mini-hydro systems are also prevalent (Mini-Grids Partnership, 2020).

4 Policies and measures to encourage the deployment of renewable energy on farms

A number of government policies have been introduced to encourage clean energy and distributed energy systems for farms. For example, Canada under its climate plan has invested CA$165.7 M to support the agricultural sector develop 'clean technologies' (Environment and Climate Change Canada, 2021). It has also announced a CA$1.5 billion low-carbon and zero-emissions fuel fund to support biofuels and hydrogen and hence also support agriculture.

Jamaica's Ministry of Agriculture and Fisheries announced that all irrigation schemes operated by the government-owned National Irrigation Commission would be powered solely by solar PV by 2022 (Bellini, 2020). An investment of US$200 000 will link with the water treatment plant and have an impact on 700 small farmers. At present, the country has an installed renewable electricity capacity of 93 MW of which 43 MW were installed in 2019 with a further 500 MW planned for 2021.

Many African countries are also increasing their renewable energy capacity. In Togo for example, the government, in association with the private sector, aims to cover 50% of the capital cost for solar-powered irrigation pumps to make them more affordable for 5000 farmers (Renewable Energy World, 2020). As is common for many domestic solar PV installations, the cost will be under a 'pay-as-you-go' system.

Grid-based, large-scale electricity generation plants are often owned and operated by state-owned utilities. Distributed renewable energy generation is more reliant on the private sector. Hence, a range of innovative business models have evolved in the past decade with 'pay-as-you-go' systems successfully enabling energy access for many off-grid consumers, including farmers especially for solar-powered water pumping. Typical payment models are 'lease-to-own' or 'usage-based' (IRENA, 2020).

5 Farm-scale renewable energy technologies and systems

A range of renewable energy technologies are commercially available at a small scale suitable for farms and mini-grids. As for the large commercial-scale installations, they each have different characteristics, especially the capacity

Table 3 Characteristics of renewable energy technologies suitable for electricity generation and integration into rural mini-grids

Technology type	Generation plant size range[a] (MW)	Variability of generation[b] (timescale)	Degree of plant dispatchability[c]	Geographic diversity potential[d]	Predictability[e]	Capacity factor range[f] (%)
Wind power	0.1–200	Minutes to years	X	XX	X	20–50
Solar PV	0.003–100	Minutes to months	X	X	XX	12–27
Hydropower						
Run-of-river	0.01–15 000	Hours to years	XX	X	XX	20–95
Reservoir	1–20 000	Days to years	XXX	X	XX	30–80
Bioenergy						
Solid biomass	0.1–200	Seasonal, depending on biomass type and when available	XXX	X	XXX	50–90
Biogas	0.01–40		XXX	X	XX	30–70

Source: Adapted from IPCC (2011).
[a] Range of typical rated plant capacity.
[b] Timescales for which variability is significant for power system integration and reliability.
[c] Degree of plant dispatchability: X = low; XX = partial; XXX = easily.
[d] Degree to which specific siting of the technology may reduce variability and improve predictability: X = moderate; XX = high.
[e] Accuracy of power output prediction at relevant timescales to assist system operation: X = moderate; XX = high; XXX = very high.
[f] Varies with site location (e.g., wind resource level, solar radiation input, water flows), as well as reliability of the technology and maintenance requirements.

factors and dispatchability of the power generated into a grid when needed to match the varying demand (Table 3).

5.1 Mini- and micro-wind turbines

A large range of commercially available mini- and micro-wind turbines are suitable for farm and rural mini-grid applications. The most common mini-turbine used is three-bladed on a horizontal axis with a diameter between 8 m and 15 m and generating up to 30–300kW rated power at maximum wind speeds. A micro-turbine is smaller with a capacity of around 1–25 kW. Two-bladed and one-bladed designs exist, and vertical axis designs are gaining in popularity. Generators are usually three-phase alternating currents (AC) but some have direct current (DC) output for charging batteries (including those in electric vehicles) with power inverters needed to convert the DC to AC for powering motors and appliances. For designs with a fixed frequency, grid connection is also possible.

The power in the wind increases with the wind speed (Fig. 4). Typically, the cut-in wind speed when the rotor starts to rotate is around 3–4 m/s. At higher wind speeds (maybe 12–14 m/s), some form of the device (such as dynamic braking) is built into the design to maintain the rated speed and hence a constant rated power

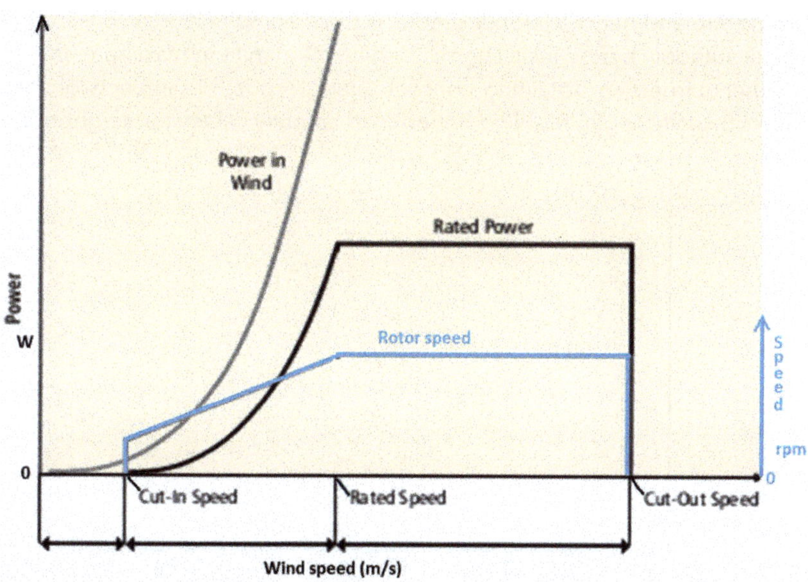

Figure 4 Typical characteristics of a wind turbine with the power output related to the variable wind speed. It increases as the wind speed increases until the rated speed is reached when the maximum power output is achieved and maintained until the wind speed becomes too high and a safety cut-out system activates.

output over a range of higher wind speeds. A safety cut-out system is common to protect the turbine at very high wind speeds (usually around 20–25 m/s). On smaller wind turbine designs, a tail fin keeps the turbine pointed into the wind, though some designs such as the Proven (manufactured in Scotland) operate downwind with flexible blades that bend inwards at higher wind speeds. Larger turbines are actively pointed into the wind by a motor as the direction fluctuates.

The average power output of a wind turbine is equivalent to the square of the rotor diameter and the cube of the mean annual wind speed. The annual wind resource for a site affects the capacity factor, which in turn affects the levelised cost of electricity (Fig. 5).

The capacity factor indicates the annual power output for a turbine on a given site. It is the ratio of the actual power output (kWh/yr generated) to its nominal or theoretical power output if it was producing its rated power output for 24 h every day, that is, 8760 h a year. The capacity factor is also reduced by any downtime for maintenance or repair of the turbine.

Therefore, when proposing to install a wind turbine on a farm where there are several potential sites, it is recommended to initially monitor the wind speeds on each site by erecting an anemometer to record wind data over several months in order for the best site selection. Shelter from prevailing winds by buildings or trees (that might continue to grow larger over the life of the turbine) should also be taken into account. A taller mast will often give higher wind speeds at the rotor hub and hence a higher output, but obtaining resource consent from the local authority before installing the turbine may limit the maximum height.

Other barriers to obtaining consent relate to possible impacts on birds and bats (usually negligible) and on reduced amenity values for neighbouring

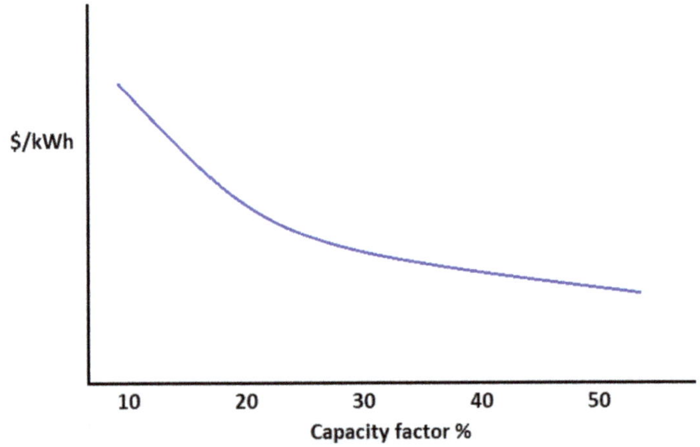

Figure 5 The greater the capacity factor due to a higher wind resource for a given location, the lower the levelised cost per kWh of electricity generated by a wind turbine installed there.

properties. Noise can also be a factor to consider as in order to optimise the rotor tip speed, a smaller diameter turbine has to rotate at higher speeds and this can result in higher noise levels. Typically, a 2.5-kW turbine would have a rotor diameter of 3 m and a rated speed around 300 rpm giving a tip speed around 160 km/h, whereas for a 15-kW turbine, a 9-m diameter rotor would have a lower rated speed of around 100 rpm to give a similar tip speed. For a three-blade turbine, the optimum tip speed ratio (tip speed/wind speed) to maximise turbine efficiency is around 5.

Noise can be measured at specific distances downwind from the turbine for different wind speeds using a decibel meter. As a guide, a car driving past 20 m away at 65 km/h will generate a noise level around 70–75 dB, and small turbines are often less noisy than this. (Note that decibels are measured on a logarithmic scale such that 60 dB is twice as loud as 50 dB and 70 dB is four times as loud.) Mitigating noise levels may be possible by planting trees and fences, though care is needed not to shelter the turbine from prevailing winds.

Given that wind speeds are very variable (Fig. 6), to meet the fluctuating power demand of a typical rural community, battery storage and/or integration with solar PV and a back-up diesel generator to give a reliable supply will be required. If the farm is grid-connected, then electricity can be purchased when needed to meet demand when the wind is not blowing, and power when generated that is surplus to demand at that time can be exported and sold to the utility.

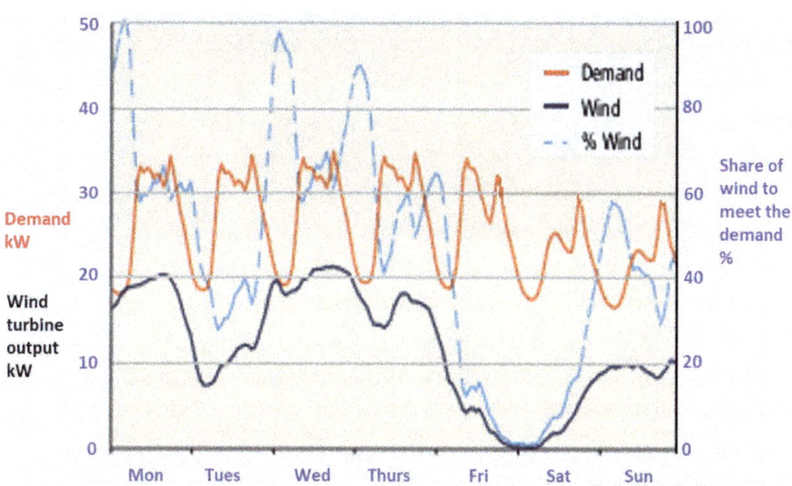

Figure 6 Energy demand variations for a small rural community over a week, here (showing lower demand at nights and at the weekend), are not matched by the wind turbine output that varies with wind speed fluctuations (with low available wind resource on Friday/Saturday). This leads to the share of wind power at times exceeding the demand but often being in deficit and needing back-up supply.

There is future potential for hydrogen to be a store and carrier of energy when integrated with a wind turbine system. As a demonstration, in the rural community of Totara Valley in New Zealand, a mini-wind turbine was installed on a hilltop around 2 km from the buildings, but the projected cost of installing a copper cable to carry the power was costly. Therefore, as a demonstration, the turbine was set up to power an electrolyser located next to it and the hydrogen produced was brought down to the community in a 16-mm medium-density polyethylene pipe that also acted as a storage system with the gas pressure increasing when the wind turbine was operating. Leakage was tested to be only around 1% through the pipe walls (Sudol, 2009) and a fuel cell located in a farm building generated the electricity when needed to meet the demand. This 'hy-link system' (Fig. 7) helps to overcome the problem that the wind does not always blow when there is demand for energy services. At the time of operation some 15 years ago, it was not an economically viable proposition but since then the costs of electrolysers and fuel cells have declined.

Box 3 Wind turbine integration

A 5-kW ANE turbine constructed on a small New Zealand farm has been integrated with solar PV and batteries to provide a reliable and economically viable power supply. Initially, twelve 270-W solar panels with Enphase micro-inverters were installed that generated around 4.5 MWh/yr. Export power was sold for US$0.05/kWh and when there was a deficit in supply, power was imported for US$0.20 /kWh. Various energy efficiency measures were introduced to reduce demand before 18 more 305 W panels were installed (that together generate around 6.5 MWh/yr on this site). An Alpha ESS Storion modular battery system with three modules was also installed together with a 5-kWh string inverter.

This system enabled an uninterruptible power supply for the water pump, lighting and refrigeration loads; purchase of lower-priced, off-peak power when imported from the grid; continued battery charging and use to meet priority demands during a grid power outage; and the export of around 220 kWh/month of electricity surplus to demand.

The ANE wind turbine was then purchased, consented, installed and integrated into the system. Even though the site has a relatively low mean annual wind speed, its addition has enabled less grid electricity to be imported as well as an electric vehicle to be charged on site. By also changing from the previous generator/retailer to a different retail power company (chosen for its greater flexibility from having three separate tariffs for peak, shoulder and off-peak rates and a willingness to buy exported power at the wholesale market spot price at the time of generation), the payback period for this well-balanced system is less than 8 years.

5.2 Solar energy

Besides large-scale solar farms and rooftop installations of several hundred MW that have become common, solar PV system installations have been widely deployed on farms and in rural communities in most countries. A single monocrystalline solar cell of around 0.3-mm thick and 100–200 cm² in area utilises the photo-electric effect of semi-conductors to directly convert solar radiation into direct current electricity. Typically, around 20–50 cells are fitted into a panel with a glass cover and aluminium frame. The brighter the sun, then with radiation at close to 1000 W/m² area, more electricity is generated by a panel but some generation does occur on cloudy days when the radiation can drop to less than 200 W/m² and some generation may be possible even under the bright moonlight.

The conversion efficiency (electricity energy output as a share of the solar energy input) of commercial PV panels is around 15–18% although specialist and expensive panels for space-craft applications can be double this. Amorphous poly-crystalline cells are less efficient (12–14%) but because

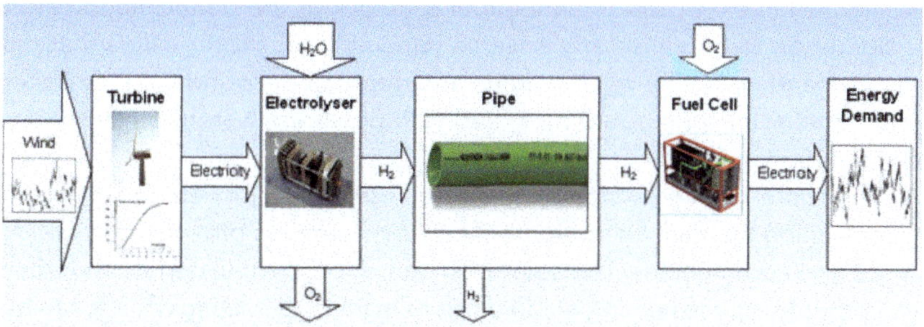

Figure 7 The 'hy-link' system uses electricity from a wind turbine to power an electrolyser with the green hydrogen produced, transported and stored in a gas pipe ready for a fuel cell that functions automatically when there is demand for electricity.

the higher efficiency has a greater cost/kW of peak capacity, they tend to be cheaper than monocrystalline cells. Thin-film cells are flexible so are useful for specialist applications but have a conversion efficiency of only around 8–10%.

Solar-powered water pumps are widely used on farms for irrigation applications. Solar electricity powers an electric motor that operates the pump. Often the water is pumped steadily into a raised tank when solar radiation levels are sufficiently high to drive the system, and the stored water is then used for irrigation when needed, delivered by gravity feed pipelines through sprinklers or drip irrigation nozzles.

Panels should be oriented correctly to maximise solar input and also placed at the correct angle. So in Spain, for example, a panel would ideally be facing due south and at an inclined angle of around 40° (roughly equivalent to the latitude of the site). If, for example, more electricity is required on a winter evening for some reason, then when installed, the panel can be adjusted accordingly to suit.

Tracking systems can be used to continually orientate the panels towards the sun as the day progresses to absorb the most radiation. This can increase generation by up to 30% compared with fixed panels. To provide a short return on the additional investment and maintenance costs, tracking systems are favoured where investment capital is available, government support is high and solar radiation yields are relatively high.

In the past decade, structures that float on lakes, water treatment ponds, shallow estuaries and so on are also being installed worldwide. The claim by one Dutch manufacturer (Texel4Trading, 2021) is that the floating, linked structures they make from recycled materials to which the solar panels are attached and are connected to the lake bed by concrete structures, will resist waves, can be walked on, have a 30-year warranty and the design allows around 25% of daylight to reach the water surface in order to keep the water body healthy along with any aquatic plant and fish life (Fig. 8). For new installations, the orientation and height of the panels need careful consideration to minimise any adverse impacts from strong winds that may arise. Depending on the water volume and depth of the pond, and the system design, the installation could be linked with powering water pumps to meet irrigation demands.

Solar water heating systems can also be usefully employed on farms (such as for cleaning the milking equipment on dairy farms) to provide hot water for houses and communities. The solar heat can also be circulated in underfloor water pipes for space heating of buildings or in thermal heating of air for animal buildings (Fig. 9).

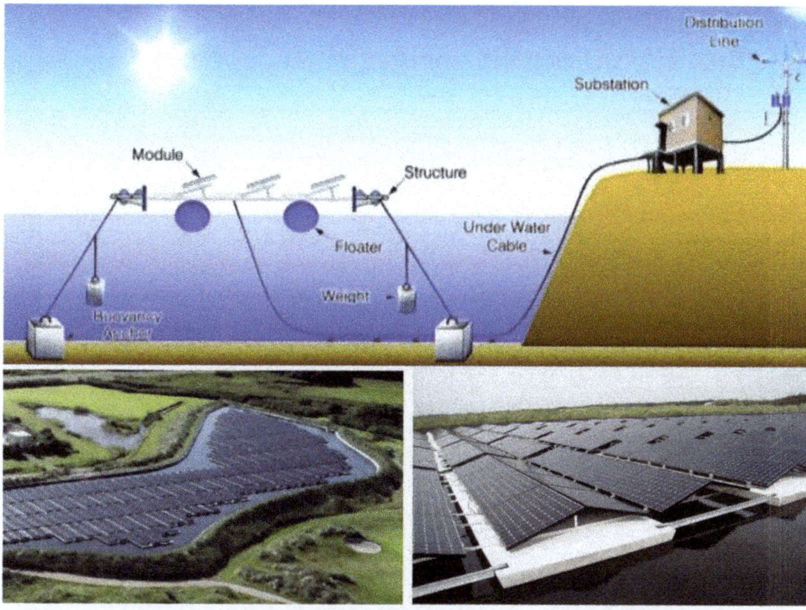

Figure 8 Floating solar panels can turn a body of water into a power generation system that does not compete with land use and allows fish and aquatic plants to live below (photos courtesy of Solar Float; Texel4Trading, 2021).

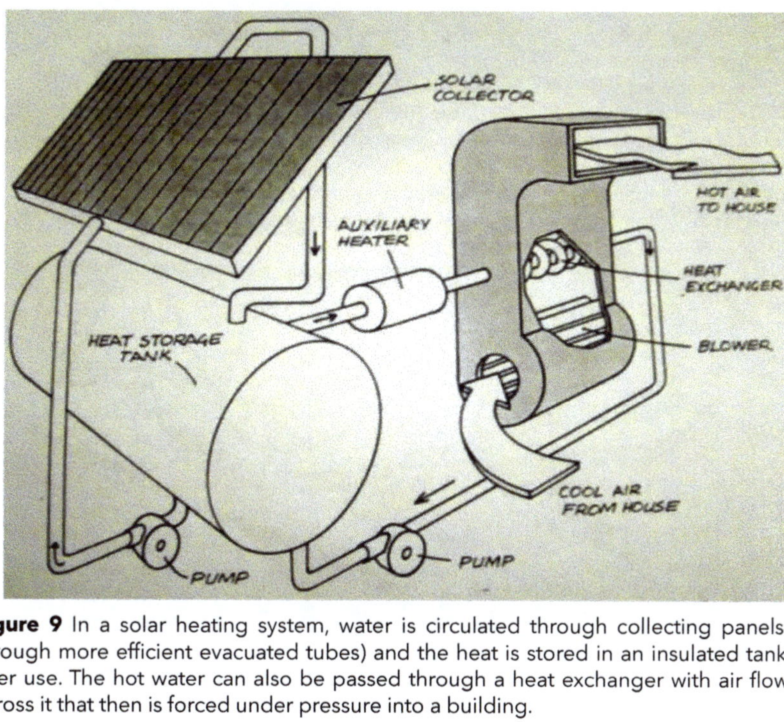

Figure 9 In a solar heating system, water is circulated through collecting panels (or through more efficient evacuated tubes) and the heat is stored in an insulated tank for later use. The hot water can also be passed through a heat exchanger with air flowing across it that then is forced under pressure into a building.

Box 4 Assessing the energy demand for a rural community

The electricity demand of a small New Zealand sheep and beef farming community consisting of three farms, each of several hundred hectares, a sheep shearing shed, five houses, a pumped water supply, and several workshops and sheds was measured over a 12-month period using many data collectors located close to the various loads (Murray, 2005). An analysis of the total electricity demand over 24 months for every day of the year showed there was a peak in the evenings (18:00–19:30 h), especially in winter (June–August).

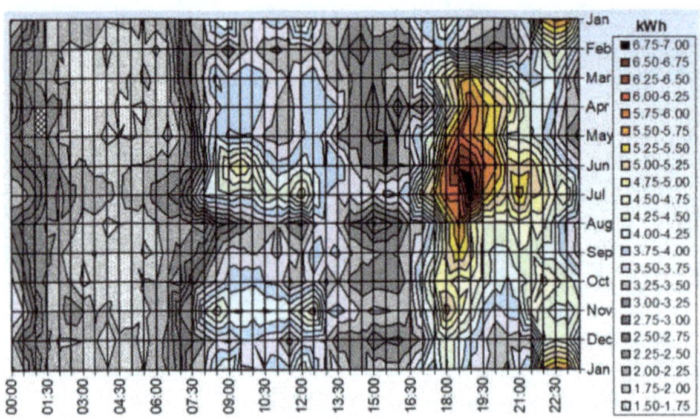

The wind resource was measured (m/s) over the same period by an anemometer on a 10-m mast on the preferred site on a hill above the community where a wind turbine could be located. There tended to be stronger wind in the afternoons, especially during the equinox of spring (September–November) and autumn (March–June), but with strong evening winds in the hotter month of February. Thus, there was a reasonable match between the seasonal and diurnal community energy demand and the available wind resource.

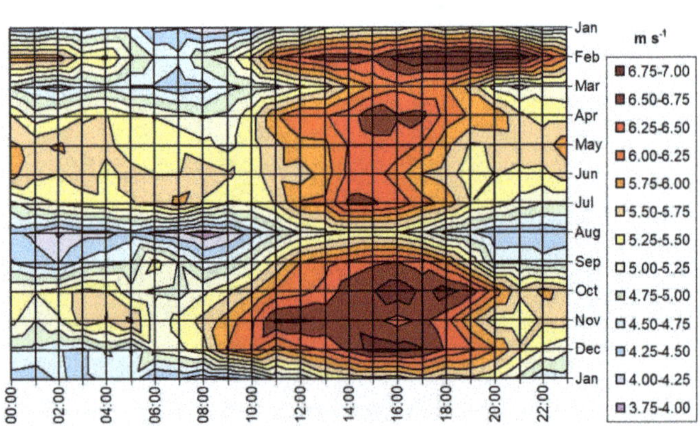

As might be expected, solar radiation levels (W/m^2 of land area) gave a different pattern with lower radiation levels in winter and zero during night times.

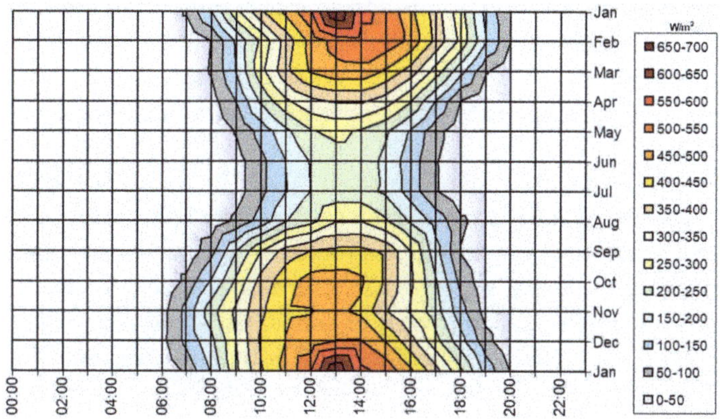

This analysis confirms the need to carefully measure the local renewable energy resources available prior to investing in heat and power generation technologies. How best to add some form of energy storage and to integrate a range of technologies to give a reliable, sustainable and cost-effective energy demand of the community or farm can then be assessed prior to investing.

A wide range of government incentives exists in many countries to encourage the deployment of smaller-scale solar PV systems. These include grants and feed-in tariffs, though these have evolved as the capital investment costs for PV have declined in recent years. Solar retail and installation businesses are also offering leasing arrangements whereby PV panels and battery storage are installed for no up-front costs. A fixed monthly fee is charged (with no increase promised for say 20 years), and some companies also offer price caps on imported power.

5.3 Mini- and micro-hydro

Streams and rivers suitable for hydro-power applications, often in hilly regions and if close to an electricity load, grid distribution network, or a mini-grid, can be used to generate electricity. In a run-of-river scheme, a portion of the water flow is diverted and then passed through a penstock (pipeline) to turbine propellers that spin a power generator to provide a resource-dependent, continuous power supply. Usually, only a portion of the water flow is approved to be diverted at the inlet and passed through the turbine after which the water is then returned to the waterway downstream. Resource consents are usually required.

In other circumstances, a reservoir behind a dam may need to be built to hold water reserves for dry periods and to increase the pressure (or head) of the water in the pipeline, which after passing through the turbine is returned to the waterway via the tailrace (Fig. 10). When the reservoir behind the dam is full, excess water overflows the dam and returns back to the waterway by way of a spillway.

Hydro-power plants can range from 1 kW to thousands of MW. The technology is mature and many examples exist at scales suitable for use on-farms, in the community or in the agri-food processing chain. Since generation will usually be for 24 h a day, exporting any surplus electricity or dumping it in shunt-loads if autonomous and not grid-connected is necessary during times of low power demand on site. Alternatively, if grid-connected, then the volume of water flowing through the turbine can be controlled automatically so that the generator output continually matches the fluctuating demand. A cost-efficient solution for electricity system balancing can often be employed to provide load control (Fig 10).

The choice of site is determined by proximity to the load demand or local distribution network, but the waterway needs to have sufficient head and flow. For example, a turbine required to generate 4 kW power output could be installed in a waterway with 10 L/s flow and 70 m head, or 80 L/s flow and 8 m head, or 200 L/s flow and 4 m head. Locations with 60 L/s available flow and 80 m head, 200 L/s and 30 m head, or 300 L/s and 16 m head would have enough potential energy in the water captured and passed down the pipeline (penstock) to drive a 30-kW generator.

As for wind and solar systems, careful monitoring of the resource, including when water flow is at low levels after dry periods or is in flood after heavy rainfall, should therefore be undertaken prior to designing and building an installation. Once the potential flow and head have been measured for a given site, there are

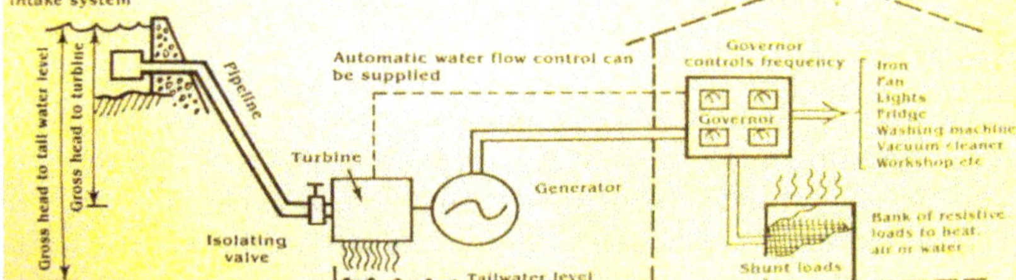

Figure 10 A typical micro-hydro project design with controls to maintain power quality as the load in a building varies – here shown as a dwelling but it could be a farm with a range of appliances, motors and pumps.

calculators available to assess the potential power output from the appropriate capacity of a turbine. In addition, the choice between turbines (including Pelton, turgo and low-head types) can then be made, based on the available pressure as determined by the available head (see e.g., PowerSpout, 2021).

Designers and installers of small-, mini-, or micro-hydro systems should take into account installing a filter or screen at the intake to prevent any debris from entering the turbine and damaging it. They should also account for any water supply fluctuations especially during dry periods, potential impacts on neighbouring and downstream ecosystems and fish life from extracting some of the water and any risk of damage to the installation during periods of flooding.

An international study reviewed about 20 mini-grids with micro-hydro technologies in developing countries and after returning 20 years later to the same sites, they found those with the original aim of using local electricity generation for agricultural development had a clear positive impact and had been maintained and survived. Others had been abandoned, often once a grid connection had become available (Stephens and Gallagher, 2015).

It can often be possible to install a micro- or mini-hydro turbine in an existing water supply system used for irrigation or stock water where a dam and pipe are already in place. Pumped storage systems may also have a place in mini-grids (or at a larger scale in main electricity grids to better manage large shares of variable generation from wind and solar). Surplus power from wind, solar or hydro-generation at times of low energy demand (e.g. during the night) can be used to pump water uphill to behind a storage dam so that additional hydro-generation can be undertaken at any time to meet peak power demands. There is an efficiency loss in pumping the water, but this is not an issue if it is only surplus power that is utilised and peak power prices can be obtained.

Box 5 Case study, micro-hydro

A small farming community evaluated options to generate hydroelectricity from several streams passing through their land. These waterways all had low gradients (and therefore low head) so would require building a dam and designing the system for major fluctuations in flow rates after heavy rain or dry periods. Therefore, the option finally chosen was to use a natural spring located up in the hills on one of the sheep farms. It had a low flow rate of only around 2 m/s but a very high head of 165 m above the farm buildings where the turbine has to be located.

Resource consents from the regional council were obtained to divert the surface water and discharge it to a stream. Due to the unusual nature of the consent, this was a costly process (around US$2000) that also took a lot of effort by the landowner to provide the requested documentation. After 4 months, the council finally granted a 20-year consent with 5 yearly reviews and with additional charges imposed to be paid annually for monitoring and administration. Such a consent process is necessary for a hydro-scheme as other downstream users of the water or the ecology could be impacted. However, the process could be simplified so as not to be seen as a barrier to the developers of other similar schemes.

On receiving the consent, a water intake grid was constructed at the spring outfall with a catchment connection to the pipeline installed below. The length of the medium-density polyethylene pipe needed to carry the water down to the turbine was 1.3 km, but in spite of the pipe friction losses, a high-pressure source was still available.

A Pelton wheel turbine was selected for this high head site (that gave a working pressure of 1260 kPa). The chosen manufacturer, Eco-Innovation, had designed a 1-kW product based around readily available and cheap washing machine motors running in reverse. A grid-interactive inverter was also selected, installed and successfully connected to the grid (Donnelly, 2008).

The technology continues to operate successfully some 15 years after installation.

5.4 Biomass, bioenergy, biogas and biofuels

Biomass resources come from a wide range of plant and animal products, originally derived from photosynthesis and which provide a chemical store of solar energy. Arising from energy crops, animal wastes and organic residues, these resources can be processed using a range of mature technologies, including combustion, gasification, gas turbines, pyrolysis, inter-esterification, anaerobic digestion, to produce heat, electricity and transport biofuels. A recent interest in algae and seaweed production could increase the available resource. The 'food-versus-fuel' conflict over competition for using limited areas of fertile arable land could be partly resolved by avoiding food wastes and losses (that account for around one-third of all the food produced). This would free up more land for producing biomass. There is also debate about the need to move away from the consumption of animal proteins per capita since the equivalent amount of plant protein to sustain human health needs less land, energy and water to produce it, thereby resulting in lower greenhouse gas emissions/kg protein. Again, existing farmland for growing pasture and animal feed crops could then be freed up for energy crop production.

Modern biomass and bioenergy have languished in the shadows of more 'sexy' renewable energy technologies. The concept gets confused with 'traditional biomass' where fuelwood, charcoal and dung are used for cooking and heating by millions of subsistence farmers and villagers in developing countries. Biomass production is often blamed for land-use change and deforestation. It commonly portrays an image of smoking chimneys, rubbish tips, and industrial wastes. Its carbon neutrality potential is complex and remains a mystery to many policy makers.

There is certainly 'good biomass' and 'bad biomass' but when produced and used sustainably it can provide a useful source of stored energy that can be efficiently and cleanly converted to useful forms of heat, electricity and liquid and gaseous fuels. Given that fossil fuels were originally formed from biomass millions of years ago, modern biomass has the technical potential to substitute for fossil fuels in all of its applications. It can also be used as a feedstock for manufacturing plastics and chemicals.

Modern bioenergy is mainstream in many countries with significant international trade in liquid biofuels and wood pellets in place. For example, the 4000-MW Drax power station in the United Kingdom has largely converted from coal to biomass and now imports over 7 Mt annually of sustainable biomass from 10 or more countries. It has also established its own wood pellet company subsidiary in the United States (Drax, 2021).

For smaller scale, on-farm bioenergy projects, the principles are the same in that the biomass fuel supply has to be reliable, economically competitive with fossil fuels and of suitable quality when delivered to the energy plant gate in terms of moisture content, particle size for ease of transport and handling and energy content.

Carbon dioxide is often produced during the production and processing of biomass and the utilisation of bioenergy. However, since the carbon is taken up during the photosynthesis of future growing crops (including pasture, feed crops and plantation forests), the atmospheric concentration of CO_2 is not increased over the long term. Biomass is therefore accepted internationally as being carbon neutral. Should there be deforestation undertaken in order to provide land to grow energy crops, then this would not be carbon neutral, and governments should certainly regulate robustly to prevent such activities.

Crops containing sugars (e.g. sugar cane and sugar beet), carbohydrates (e.g. wheat, corn and cassava) or triglycerides (e.g. soy bean, oilseed rape *Jatropha*) can be used to produce liquid fuels by farmers or village communities using a range of simple conversion processes. Biomass sources including municipal solid wastes (excluding the plastic components), crop residues, animal wastes and so on can also be used to supply distributed electricity generation using modern micro-turbines, Stirling engines, gasifiers, anaerobic digesters and so on and can also be a source of green hydrogen for fuel cells.

The anaerobic digestion of organic wastes (such as animal manure or crop residues) by bacteria to produce methane and carbon dioxide is a mature technology with many examples existing worldwide on farms, sewage treatment plants and at the community scale to convert a range of organic wastes including food. In addition, millions of micro-scale systems in developing countries are used to produce biogas for cooking. At the farm-scale, biogas can be used on farms for heating demands such as grain drying; as an engine fuel for tractors and other equipment, with spark-ignition engines (gasoline) being easier to convert than compression ignition engines (diesel); or for power generation through internal combustion engines powering generating sets. The effluent remaining after digestion is nutrient-rich, odourless and acts as a good soil conditioner.

The CO_2 should also be removed as it can contribute around 40% or more of the total raw gas volume. It is harmless but removal avoids it taking up space in the storage tank or when the methane is compressed into steel bottle containers. Ideally, the raw biogas should also be scrubbed of corrosive hydrogen sulphide prior to storage and use (especially if it is to be used as an engine fuel). This can be simply done at a small scale by passing the raw gas through drums filled with iron filings or compacted steel wool. The drums should be emptied occasionally and the filings (now partly converted to ferrous sulphide) then spread out and left to oxidise to drive off the sulphur before reusing them. The corrosive nature of the raw biogas can also necessitate regular maintenance of the biogas production plant by painting of steel components and the regular cleaning of pumps and motors. At a large scale, the biomethane component of biogas can be added to natural gas methane in gas distribution pipelines, but it must first be brought up to 90% purity.

Liquid biofuels for transport contain relatively low carbon compared with petroleum fuels, so have received much attention from policy makers in recent decades. Many countries have blending mandates for 'drop-in' biofuels with gasoline or diesel and in the future marine and aviation fuels. Others have established financial incentives and support for the development of the production, fuelling and blending infrastructure. Brazil, for example, now has its diesel fuel blended with around 13% of biodiesel (fatty acid methyl esters from soybean oil mixed with diesel) (B13), and bioethanol from sugarcane is also available at the pump as a 20% blend with gasoline (E20) or as a neat 100% fuel (E100) for suitably designed vehicle engines. Sweden has had an E85 blend for some years and also widely uses solid biomass from forest residues, municipal wastes and energy crops for providing heat and power (including in district heating and cogeneration plants). Overall, 32% of the nation's total primary energy comes from biomass. In both these exemplar countries, the farming community benefits from providing the biomass.

The global biofuels industry has had many successes and failures over the past few decades. The challenge to produce bioethanol from woody biomass residues and ligno-cellulosic crops such as miscanthus has yet to become commercially viable in spite of considerable investment in R, D & D, over several decades. By way of contrast, after first becoming commercial only a decade or so ago, hydrotreated vegetable oils that are synthesised from oils and fats using hydrogen to produce straight-chain hydrocarbons similar to petroleum fuels now supply around 6 billion litres of biofuels annually. Technical and financial developments within the biofuels industry are occurring frequently and can best be found in the monthly publication, *Biofuels International* (https://biofuels-news.com/).

Liquid biofuels of both ethanol and biodiesel as well as biogas can be produced at a small scale on the farm or in a community using simple and well-understood techniques with 'recipes' freely available on the internet. Many examples of small-scale biofuel production exist, for example, biodiesel from used cooking oils collected from take-away shops and restaurants. The chemical production process to manufacture biofuels at a small scale however involves considerable time and labour. There is also a risk of engine failure if the quality of the biofuels produced is not maintained and fails to meet certain characteristics (such as viscosity, flash point and contaminant levels). Engine manufacturers' warranties can be made null and void by improper biofuel use, so careful assessment is recommended.

The barriers to producing and using biomass for energy applications can include land use competition for food and fibre production; competition for water use; the mining of soil nutrients; monocultures reducing biodiversity, human health concerns from smoke, local air pollution and dioxins; exploitation of small landowners by large corporate companies; greenhouse gas emissions including black carbon (Sims et al., 2017); complex supply chain logistics;

and possible relatively high costs in terms of \$/t of CO_2 avoided. With careful management of the on-farm production system using regenerative farming methods based on the circular economy wherever possible (STAP, 2018) and the latest designs of conversion technologies, many of these barriers can be overcome.

The concept of linking bioenergy with carbon dioxide capture and storage (BECCS) has received considerable attention as a means of removing CO_2 from the atmosphere. Plantation-grown crops are harvested, and during their combustion for heat and power, the CO_2 emitted is captured and stored. CCS technology has been successfully implemented for natural gas extraction (e.g. the Sleipner Field off the coast of Norway since 1996) and also demonstrated for coal-fired power plants (though less successfully with, e.g. the Petra Nova plant in Texas recently closed down after 3 years). Nevertheless, many climate models show that staying below 2°C will be very challenging without the wide deployment of BECCS. At the farm scale, however, locking carbon up in the soil by increasing the organic matter content is a more practical means of helping to reduce atmospheric CO_2 concentrations.

6 Conclusion

Renewable energy systems at a scale suitable to provide heat, electricity and transport fuels for farms and rural communities are already well proven. Having evolved rapidly over the past two to three decades in response to the need to provide modern energy services for all as well as urgently reduce greenhouse gas emissions, many examples exist of successful and economically competitive technologies that have been implemented and deployed to some extent throughout most countries of the world. However, fossil fuels still dominate when meeting the energy demands for food production and processing activities, transport to the markets, as well as for retail, preparation and cooking. So innovative R, D & D is still required to enable the global food industry to become independent of this on-going demand for coal, oil and gas.

Analyses of the energy return on investment (i.e. the ratio of the energy input needed in say the manufacture, installation and disposal of a wind turbine to how much useful energy output is produced during the life of the turbine) are essential to fully understand the true benefits from renewable energy technologies. Similarly, life cycle assessment of all the material inputs (including rare earth metals, lithium, steel, etc.) and assessment of their sources and reserves, together with the recycling opportunities for the components of a given technology, is a useful tool but the full impacts on the UN's sustainable development goals are not yet fully understood.

The potential for innovative and sustainable technologies to be used on farms to improve productivity, reduce energy and water inputs, protect soil fertility, reduce post-harvest losses and so on needs further evaluation. Already smartphones now enable farmers in Africa to quickly gain assistance to diagnose and control a crop disease; drones are being used to check on livestock, pasture growth and water supplies; remote sensing and GPS systems enable soil moisture variations to be accurately measured; water can be applied precisely by smart irrigators that change the flow from individual nozzles as they pass over the ground to suit the varying moisture contents; animal welfare and health is being monitored through embedded chips; cows can enter a milking shed when they feel the need to be milked where they are electronically recognised, fed according to their stage of lactation and the milking machine then applied automatically and removed on completion; and autonomous electric-powered tractors (with no operator on board) can accurately cultivate weed between rows of crops and apply fertilisers and agri-chemicals precisely where needed. For such future technical developments to become mainstream on farms will take time but can lead to reduced fossil fuel demands and hence lower the carbon footprint per unit of food produced.

Farms will always need energy inputs. Even synthetic meat and milk production in urban factories will depend on reliable energy supplies. This energy will have to come from low-carbon systems, many of which are already well understood and commercially available. What is less certain is how society will accept changes to conventional farming and fishing systems in order to produce food more sustainably and reduce greenhouse gas emissions.

Behavioural change is already becoming evident, in that most restaurants now offer one or more vegetarian choices that simply was not the case 10 years or so ago. Healthier lifestyles in terms of regular exercise, choice of nutritious food options and less animal protein intake are becoming more common than in previous generations (though there are still many people becoming obese through excessive consumption of sugar drinks, take-away foods, etc.). Animal welfare issues are also changing farming practices such as moving away from intensively housed chickens and pigs to free-range.

Whether agriculture can move quickly enough to reduce its dependence on fossil fuels whilst aiming to feed a growing and more affluent population (that will need around 70% more protein per year by around mid-century than at present) is unclear. Constraining global temperature rise to well under 2°C whilst pursuing efforts to limit it to 1.5°C, as well as maintaining food supply security and eliminating hunger, as outlined in the Paris Climate Agreement, will not be possible unless the agricultural sector plays a major role. The rapid transition from fossil fuels to renewable energy will be imperative to meeting that goal.

7 Where to look for further information

- Opportunities for Agri-Food Chains to become Energy-Smart, UN Food Agricultural Organisation, Roma, http://www.fao.org/3/i5125e/i5125e.pdf.
- Sims, R. E. H. 2002. *The Brilliance of Bioenergy – In Business and in Practice*, James and James, London. https://www.taylorfrancis.com/books/brilliance-bioenergy-ralph-sims/10.4324/9781315067155.
- IPCC 6th Assessment Report, Mitigation (Working Group 3) 2022. Chapter 7 on Agriculture, Forestry and other Land Uses. AR6 Climate Change 2021: Mitigation of Climate Change – IPCC.
- Global developments in renewable energy technologies and analysis of new systems, some of which relate to the farm scale, are assessed and presented by the International Renewable Energy Agency, IRENA https://www.irena.org/.

8 References

Bellini, E. 2020. Jamaica turns to solar for irrigation and water treatment, *PV Magazine*, December. Available at: https://www.pv-magazine.com/2020/12/15/jamaica-turns-to-solar-for-irrigation-water-treatment/.

Crippa, M., Solazzo, E., Guizzardi, D., Monforti-Ferrario, F., Tubiello, F. N. and Leip, A. 2021. Food systems are responsible for a third of global anthropogenic GHG emissions, *Nature Food* 2(3), 198–209. https://doi.org/10.1038/s43016-021-00225-9. Available at: https://edgar.jrc.ec.europa.eu/edgar_food.

Donnelly, D. 2008. Totara valley micro-hydro development, Massey University Master of Applied Science thesis. Available at: https://mro.massey.ac.nz/handle/10179/695.

Drax. 2021. Available at: https://www.drax.com/sustainability/sourcing-sustainable-biomass/ and https://www.draxbiomass.com/.

Dunphy, S. 2018. *Could Bioenergy from Willow Enable Sweden to Become Completely Fossil Fuel Free?* European Scientist. Available at: https://www.europeanscientist.com/en/energy/could-bioenergy-from-willow-enable-sweden-to-become-completely-fossil-fuel-free/, 24 August.

Environment and Climate Change Canada 2021. Climate-smart agriculture. Available at: https://www.canada.ca/en/services/environment/weather/climatechange/climate-plan/climate-plan-overview/healthy-environment-healthy-economy/annex-climate-smart-agriculture.html.

EWB 2020. Chill challenge – affordable off-grid refrigeration. Available at: https://www.ewb-usa.org/chill-challenge/, Engineers Without Borders.

FAO 2011. Energy-smart food for people and climate. Issue paper. Available at: http://www.fao.org/3/i2454e/i2454e00.pdf, UN Food and Agricultural Organisation, Rome.

Flammini, A., Bracco, S. and Sims, R. 2019. Measuring impacts and enabling investments in energy-smart Agri-food chains. In: *Findings from Four Country Studies*, UN Food and Agricultural Organisation, Rome. Available at: http://www.fao.org/3/ca4064en/ca4064en.pdf.

Giovannelli, A. 2015. State of the art on small-scale concentrated power plants, *Energy Procedia* 82, 607–614, Elsevier. Available at: https://www.sciencedirect.com/science/article/pii/S1876610215026570?via%3Dihub.

IEA 2019. International Energy Agency, Share of total primary energy demand by fuel, 2010–2019 – Charts – Data & Statistics - IEA.

IEA 2020. Renewables. Available at: https://www.iea.org/fuels-and-technologies/renewables, International Energy Agency.

IPCC 2011. Integration of renewable energy into present and future energy systems. Special Report on Renewable Energy Sources and Climate Change Mitigation, Intergovernmental Panel on Climate Change, Working Group 3. Available at: https://www.ipcc.ch/report/renewable-energy-sources-andclimate-change-mitigation/, Cambridge University Press, New York.

IRENA 2016. Renewable energy benefits: decentralised solutions in the agri-food chain. Available at: https://www.irena.org/publications/2016/Sep/RenewableEnergy-Benefits-Decentralised-solutions-in-agri-food-chain, International Renewable Energy Agency, Abu Dhabi.

IRENA 2020. *Pay-as-you-go Models*, International Renewable Energy Agency, Abu Dhabi. Available at: https://www.irena.org/-/media/Files/IRENA/Agency/Publication/2020/Jul/IRENA_Pay-as-you-go_models_2020.pdf?la=en&hash=7A2E7A7FF8B5BAB7748670876667628A39DE40D5.

MBIE 2018. New Zealand smart-grid forum, ministry of business, innovation and enterprise community partners. Available at: https://www.mbie.govt.nz/building-and-energy/energy-and-natural-resources/energy-generation-and-markets/electricity-market/new-zealand-smart-grid-forum/.

Meier, T. 2017. Mini-grids. Available at: https://www.weadapt.org/knowledge-base/sdc-climate-change-environment-network/climate-change-environment-nexus-brief-mini-grids, Climate Change & Environment Nexus Brief, Swiss Agency for Development and Cooperation, Bern, Switzerland.

Mini-Grids Partnership 2020. State of the global mini-grids market report. Available at: https://minigrids.org/wp-content/uploads/2020/06/Mini-grids_Market_Report-20.pdf.

Murray, P. E. 2005. Designing sustainable distributed generation systems for rural communities, PhD thesis, Massey University. Available at: https://mro.massey.ac.nz/handle/10179/1566.

OECD 2017. Improving energy efficiency in the agri-food chain, *OECD Green Growth Studies*. Available at: http://www.oecd.org/publications/improving-energy-efficiency-in-the-agro-food-chain9789264278530-en.htm.

Piterou, A., Shackley, S. and Upham, P. 2008. Project ARBRE: lessons for bio-energy developers and policy-makers, *Energy Policy* 36(6). Available at: https://www.sciencedirect.com/science/article/abs/pii/S0301421508001109, 2044–2050.

Powerspout 2021. Available at: https://www.powerspout.com/.

REN21 2020. Renewables 2020 global status report, *Renewable Energy for the 21st*. Available at: https://www.ren21.net/gsr-2020/, Century.

Renewable Energy World 2020. Renewable companies team up with to go to expand solar-powered farming. Available at: https://www.renewableenergyworld.com/solar/renewable-companies-team-up-with-togo-to-expand-solar-powered-farming/December.

Sharma, D. K., Sharma, D. and Ali, A. H. H. 2019. A state of the art on solar-powered vapor absorption cooling systems integrated with thermal energy storage, *Environmental Science and Pollution Research International* 27(1), 158–189. Available at: https://link.springer.com/article/10.1007/s11356-019-06941-x.

Shenvekar, S. R. 2021. Solis supplies string Inverters for world's largest tidal-flat 300MW utility solar PV plant, *Solar Quarter*. Available at: https://solarquarter.com/2021/02/12/solis-supplies-215-255kw-string-inverters-for-worlds-largest-tidal-flat-300mw-utility-solar-pv-plant/.

Sims, R. E. H., Gorsevski, V. and Anenberg, S. 2017. Black carbon mitigation and the role of the global environment facility, GEF, Washington, DC. Available at: http://www.stapgef.org/node/1607.

Sims, R. E. H., Flammini, A. and Puri, M. 2015. Opportunities for Agri-food chains to become energy-smart, UN Food and Agricultural Organisation, Rome and USAID. Available at: http://www.fao.org/3/a-i5125e.pdf.

Sims, R. E. H., Mercado, P., Krewitt, W., Bhuyan, G., Flynn, D., Holttinen, H., Jannuzzi, G., Khennas, S., Liu, Y., O'Malley, M., Nilsson, L., Ogden, J., Ogimoto, K., Outhred, K., Ulleberg, Ø. and van Hulle, F. 2011. Chapter 8. Integration of renewable energy into present and future energy systems. In: *IPCC Special Report on Renewable Energy Sources and Climate Change*, Cambridge University Press, Cambridge, United Kingdom and New York, NY. Available at: https://archive.ipcc.ch/report/srren/.

SolarPACES 2021. *Member of the Energy Technology Network of the International Energy Agency (IEA)*. Available at: https://www.solarpaces.org/.

STAP 2018. Future food system for healthy humans beings and a healthy planet. Available at: https://www.thegef.org/council-meeting-documents/future-food-system-healthy-human-beings-and-healthy-planet.

Stephens, L. and Gallagher, M. 2015. The Energy-water-Food Nexus at Decentralized Scales - Synergies, trade-offs, and how to manage them, *Practical Action*. Available at: http://www.idaea.csic.es/medspring/sites/default/files/PPEBP3%2BONLINE.pdf.

Sudol, P. 2009. Modelling and analysis of hydrogen-based wind energy transmission and storage system, Master of Technology thesis, Massey University. HyLink system at totara Valley. Available at: massey.ac.nz.

Texel4Trading. 2021. Powering change: green energy solutions on water, Solar Float - Floating PV Systems.

Tubiello, F. N., Rosenzweig, C., Conchedda, G., Karl, K., Gütschow, J., Xueyao, P., Obli-Laryea, G., Wanner, N., Qiu, S. Y., Barros, J. D., Flammini, A., Mencos-Contreras, E., Souza, L., Quadrelli, R., Heiðarsdóttir, H. H., Benoit, P., Hayek, M. and Sandalow, D. 2020. Greenhouse gas emissions from food systems: building the evidence base, *Environmental Research Letters* 16(6). Available at: https://pubs.giss.nasa.gov/abs/tu03300h.html.

UNFCCC 2019. Off-grid and decentralized energy solutions for smart energy and water use in the agrifood chain, Technical Paper, United Nations Framework Convention on Climate Change. Available at: https://unfccc.int/sites/default/files/resource/tp2019_02_adv.pdf.

Chapter 7

On-farm biomass technologies for heat and power

Ralph E. H. Sims, Massey University, New Zealand

1 Introduction

Heat requirements on farms include activities such as drying cereal grains and other crops (such as tobacco), heating greenhouses or heating animal housing. There is also a relatively small demand for cooking and heating farm homes. These demands can all be met using biomass in its various forms, converted to useful heat using a range of thermo-chemical processes at both small (~5 kW_{th}) and larger (~5 MW_{th}) scales. Many practical and cost-effective examples of bioenergy technology applications exist on both large industrial farms (and other enterprises) and smallholdings. Biomass resources can also provide heat for on-farm electricity generation. In addition, some farmers earn revenue from the sale of biomass produced on-farm for use in off-farm industrial bioenergy applications. Biogas produced from biomass has been widely used on farms for supplying heat and also as a fuel for internal combustion engines driving a generator to produce electricity.

http://dx.doi.org/10.19103/AS.2022.0100.13

Biomass arises from organic plant material that, in effect, is a store of carbon and solar energy arising from photosynthesis. Unlike wind and direct solar energy, biomass can be easily stored and stockpiled, with the available energy content ready to release and use on demand. Moisture content varies with the nature of the biomass and may need natural drying prior to combustion for heat. Cereal straw is naturally dry at around 10–20% moisture content wet basis (MC_{wb}, which calculates water as a proportion of total mass) whereas woody biomass can be up to 60% MC_{wb} when first harvested, drying down naturally to 15–20% over a period of several weeks when it becomes more suitable for combustion or pelleting. Drying artificially is possible but is a relatively costly option compared with keeping the biomass sheltered from rain and allowing it to dry naturally.

'Wet biomass' in the form of animal manure,[1] vegetable processing residues, or purpose-grown green crops can be used as feedstock in anaerobic digestion plants. The biogas produced is mainly methane and carbon dioxide with some hydrogen sulphide. The H_2S is corrosive (as it can form sulphuric acid) and the CO_2 component is not combustible. Both are usually removed to provide 'scrubbed biogas,' though this may not be necessary for direct combustion for heat. After digestion of the organic matter by the anaerobic bacteria has been completed, the remaining effluent is odourless and can be applied to the land as a beneficial soil conditioner.

The potential for using green hydrogen as an energy carrier is gaining interest. It is not yet commercially viable so not widely used to supply heat directly but has been demonstrated at the farm scale using a fuel cell to generate electricity (Sudol, 2009) (see discussion below). The utilisation of biomass-derived 'producer gas,' a mix of carbon monoxide and hydrogen used widely in World War II as a vehicle fuel, has particular potential at the farm scale. Fuel cells, Stirling engines and micro-turbines can all be fuelled by either producer gas or biogas, making them promising technologies for distributed energy systems used for small-scale electricity generation on farms (UNFCCC, 2019).

Biomass is a potentially low-carbon source of renewable energy. This assumes that the equivalent number of trees or crops used to produce biomass is replanted. These should absorb over time a similar volume of atmospheric CO_2 to the volume released during biomass combustion. When using crop or forest residues, it is necessary to account for the CO_2 that would also be produced as if they were left to decay naturally (and also potentially methane if conditions are anaerobic).

1 The direct combustion of dried animal dung, as commonly practised in rural areas of many developing countries such as Nepal, is not included here.

Defining sustainable use of biomass is difficult. Where biomass production involves deforestation or the growing of energy crops on arable land that displace food and fibre crops, it cannot be classified as 'sustainable' and its use should be discouraged. It is recommended that, prior to collection and combustion, a careful assessment should be made of a specific biomass resource and its potential impacts on the atmosphere, soils and local air quality. To be more sustainable, after the combustion of the biomass, many of the mineral components contained in the ash can be returned to the soil.

There is usually some diesel fuel consumed during the collection, processing, storage and transport of crop or forest residues, but carbon emissions from these activities are relatively low. As an example, carting 30 t of wood chips 300 km in a diesel truck would emit less than 400 kg (0.4 t) $CO_{2.}$ However, if the chips are used as boiler fuel to displace coal, they would avoid around 30 t CO_2 of emissions; if displacing natural gas, they would avoid around 18 t CO_2. If the delivery truck is fuelled by biodiesel, renewable electricity or green hydrogen, the resulting direct emissions of CO_2 will be even less when assessed on a life cycle assessment basis.

Liquid biofuels for transport, such as bioethanol from corn in the United States and from sugarcane in Brazil and biodiesel from oilseed rape (canola) in Europe, have been criticised due to competition with food and fibre production on limited arable land.[2] Oil palm plantations in Indonesia and elsewhere for producing oil used to produce biofuels and other products are not sustainable where the land has first been deforested. The debate around producing and using liquid biofuels sustainably for transport over the past few years is not covered in this chapter but can be found elsewhere (e.g. IEA, 2008, 2017).

Bioenergy systems with installed capacities from 10 kW to 10 MW have the potential to displace fossil fuels by providing heat at the farm, village or community scale,[3] or for small-scale industrial uses. Electricity can provide similar heating services but can be more expensive than a bioenergy system where a reliable supply of biomass is readily available. In addition, as grid electricity demand increases in a region, this places extra load carrying capacity on the distribution lines which may then need costly upgrading. Bioenergy applications to generate electricity and reduce demand from the grid can offset such an investment. Larger-scale cogeneration (heat and power) plants up to several 100 MW capacities, using sustainable woody biomass fuels, are commonly used to supply both electricity and district heat for rural towns and communities in countries such as Sweden, Finland, Austria, Denmark but are outside the scope of this chapter.

2 If food wastage in the world, being around 30% of the total produced, could be reduced, this would save on water, labour and energy inputs as well as release some arable land that could then be used for producing biofuel crops.
3 Woody biomass and animal dung is also commonly used to provide heat for cooking or space heating in on-farm dwellings but this relatively small demand is not included in this chapter.

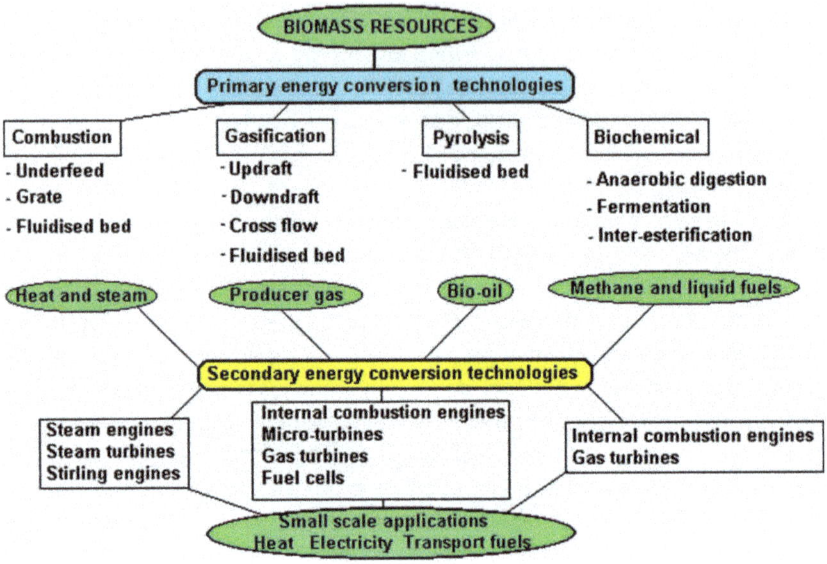

Figure 1 An overview of primary and secondary biomass conversion technologies suited to small-scale biomass projects as might be used on farms (Sims, 2015).

Primary conversion technologies convert biomass into heat as well as into gaseous or liquid products. Secondary conversion technologies then convert these energy carrier products into useful forms of energy (Fig. 1). The principles underlying thermo-chemical conversion technologies for combustion, gasification and pyrolysis, together with biochemical conversion technologies, are all well understood (Sims, 2015). The wide range of technologies that exist to convert the energy stored in the biomass to more useful forms of bioenergy can be classified according to the principal energy carrier produced in the conversion process. Production of energy carriers in the form of gas, liquid or solid fuel products depends on the extent to which oxygen, usually as air, is admitted to the conversion process. The three principal methods of thermo-chemical conversion corresponding to each of these energy carriers are as follows: combustion in excess air, gasification in reduced air, and pyrolysis in the relative absence of air. These are discussed in the next section.

2 Primary energy conversion technologies: combustion, gasification and pyrolysis

2.1 Combustion

Biomass can be combusted in a range of burner and boiler designs to produce heat. The method of feeding the fuel into the furnace characterises the technology.

Underfeed stokers feed the biomass into the combustion zone from underneath the firing grate. They are only suitable for small-scale systems up to a nominal boiler capacity of 6 MW_{th}. They are better suited for biomass fuels with a low ash content such as wood chips and sawdust since high ash fuels such as bark and cereal straw need more efficient ash removal systems. Sintered (melted) ash particles covering the upper surface of the fuel bed can cause problems in underfeed stokers due to unstable combustion conditions that can occur if the fuel and the air breakthrough the ash-covered surface.

Grate stokers are available as fixed grates for small-scale combustion systems less than 1 MW_{th}. Larger-scale systems use reciprocating grates and designs that enable a horizontal and vertical movement of the grate. When the supply of primary air is controllable, it is possible to operate grate firings efficiently, even at partial loads down to 25% of the nominal maximum furnace load. Water-cooled grate designs avoid slagging and extend the lifetime of the materials.

Grates are well proven and reliable and can tolerate wide variations in fuel quality including in moisture content, particle size and ash content. Advantages of grate furnaces for plants smaller than 10 MW_{th} include low investment costs, low operating costs and good operation at partial load. Disadvantages include the production of nitrous oxide (NO_x), excess oxygen decreasing thermal efficiency and combustion conditions being less stable than in other types of furnaces. More recent design developments have been driven by the desire to reduce emissions which has led to the development of the main alternative to grate-based systems, the fluidised bed combustor.

Fluidised bed combustors burn the biomass fuel on a bed of inert material (usually sand particles) suspended by blowing primary air through it. For plants with a nominal boiler capacity greater than 10 MW_{th}, bubbling fluidised bed (BFB) combustors start to become viable. If larger than 30 MW_{th}, circulating fluidised bed (CFB) combustors are more appropriate. The plant size where BFB and CFB technologies are economically viable is normally larger than 5 MW_e though several cogeneration plants are operating successfully in Scandinavia with electricity generation capacity typically around 3–4 MW_e linked with useful heat available of around 16–17 MW_{th}.

Conventional technology for generating electricity from biomass is to use a combustion system to raise steam and then to expand this steam in an engine or turbine used to produce mechanical power or drive a generator set. Whilst the production of steam is efficient, the conversion of steam to electricity is less so. Biomass combustion systems of all designs can be very efficient at producing hot gases, hot air, hot water and steam, typically producing 65–90% of the energy contained in the fuel as useful heat. Lower efficiencies are associated with burning wetter fuels (50–60% moisture content on a wet basis). Where the production of electricity is the main goal, the steam engine or turbine will exhaust into a vacuum condenser. Conversion efficiencies are then likely to be in the range 5–10% for plants smaller than 1 MW_e, 10–20% for plants at 1–5 MW_e and 15–30% for plants at 5–25 MW_e. Heat available from the condenser will be at low temperatures, usually below 50°C. This is insufficient even for low-grade heating applications, so it is normally wasted.

There is little scope for achieving significant efficiency improvements with steam technology because it is very mature. However, where there is demand for process heat as well as electricity, the plant can be arranged to:

- take some high-temperature steam directly from the boiler;
- partially extract expanded steam from a turbine designed for the purpose; and
- or provide the exhaust steam at the required temperature.

All three options will reduce the amount of electricity available, although the overall energy efficiency may be much higher, with 50–80% being common for such a cogeneration system.

Steam technology from biomass can be regarded as robust and well proven but relatively expensive compared with fossil fuel systems unless waste biomass fuels and residues are available at low or zero cost. Where the biomass has to be purchased at market prices, small-scale electricity production will probably not be competitive. In these circumstances, biomass-to-electricity schemes need other reasons for their existence such as carbon emission reduction. The main problems with bioenergy combustion technologies are the chemical reactions that can take place which cause corrosion and fouling, especially when potassium- and chlorine-rich biomass fuels such as straw, cereals and grass are used. Ash melting behaviour and its influencing impacts need further evaluation.

2.2 Gasification

The gasification of biomass is based on partial combustion in a restricted supply of air at elevated temperatures up to 1200°C. The variability of biomass fuels with respect to moisture content and particle size affects gas composition. This makes gasifier operation significantly more demanding than the operation of combustion systems.

Gasifiers that use air as a gasification medium (directly heated gasifiers) use the exothermic reaction between oxygen and the organics in the fuel to provide the heat necessary to devolatilise biomass and to convert residual carbon-rich chars. The heat to drive the process is generated internally within the gasifier. The high temperatures used during the gasification process make it unsuitable for fuels with low ash softening and melting temperatures, such as are typically found in many annual crop residues such as cereal straw.

The final fuel gas consists principally of carbon monoxide (CO), hydrogen (H) and methane (CH_4) with small amounts of higher hydrocarbons such as ethane (C_2H_5) and ethylene (C_2H_6). Because the combustible gases are diluted with carbon dioxide and nitrogen which have no heat value, the calorific value of the final fuel gas mixture is low (4–6 MJ/Nm^3). This is only around 10–14% of the heating value of natural gas methane, for which most commercial gas engines and gas turbines have been designed. The relatively low heating value (LHV) also makes the gas unsuitable for transport fuel because high on-board storage costs would be involved, such as in the form of compressed gas cylinders that also have a high payload weight.

The calorific value of the gas can be increased to 10–15 MJ/Nm^3 if pure oxygen is used for the gasification process instead of air. This would allow unmodified engines and turbines to be used to generate electricity. However, the cost of production of oxygen and potential hazards associated with its use have made the use of oxygen-blown gasifiers unattractive especially at the small scale suitable for on-farm applications.

A pressurised gasifier would produce gas under pressure, so it is therefore more suitable for direct gas engine or turbine applications in order to provide the highest overall process efficiency. To take full advantage of operating under pressure, however, a number of ancillary systems must be developed such as reliable feed systems and hot gas clean-up systems. Alternatively, gasifiers can be operated at atmospheric or low pressure and the product gas can then be cleaned and compressed to obtain the optimum pressure required for gas-turbine applications. The pressurised gasifiers are physically smaller than atmospheric gasifiers for the same energy output, but they are also more expensive to manufacture, so are not common. The pressurised gasifier would possibly be more efficient overall due to the more complete conversion of tars, the retention of sensible heat and the absence of the need for gas compression. However, the fuel feeding system would be much more complicated as it would involve pressure seals and would need purging with inert gas. These factors increase both the capital and operating costs of such a system and make it less suitable at a smaller scale. Evaluation of small, pressurised downdraft gasifiers has shown the comparative cost/kW installed is not competitive.

For heating applications only, it is possible to burn the hot fuel gas (as produced by the gasifier) directly in a close-coupled boiler or furnace. This was used by the Waterwide gasifier that was installed in several large-scale heated greenhouse complexes in the 1980s. Maintaining high temperatures prevents the condensation of tars which are cracked and burned at high temperatures in the combustor.

If the biomass fuel, or the gasification process, leads to the production of dust or ash, it may well be possible to remove it with a hot gas cyclone. If it is to be used as a fuel in internal combustion engines or small turbines, the gas may have to be cooled to intermediate or low temperatures due to temperature limitations in the fuel control system of the engine or turbine. Reducing the gas temperature will increase the volumetric heat value of the gas, but it will also increase the condensation of tars, making the gas even less suitable. Removing the tar residues from producer gas is difficult, particularly on the small scale. In these circumstances, a gas cleaning system is essential, possibly comprising cyclones, filters and wet scrubbers. Wet scrubbers are particularly effective as they capture the tars (which are water soluble), collect the inert ash and mineral contaminants and reduce the gas temperature in a single operation. However, they produce a contaminated effluent stream with potential toxic and carcinogenic properties. This undermines the clean image of biomass fuels and the concept of sustainability, especially at a smaller scale.

Gasifiers can be classified according to how fuel and air are fed into the conversion plant in relation to each other.

Fixed bed updraft (or counter-current flow) gasifiers have the air needed for combustion blown into the reaction chamber from below while the fuel is fed in from above. The great advantage of such updraft gasifiers is their suitability for a large range of fuel moisture contents and particle sizes and for fuels with a low slag melting point such as cereal straw. For heat applications up to 10 MW_{th}, updraft gasifiers are most popular and because the gas leaves at relatively low temperatures, the thermal efficiency is relatively high. This type of gasifier is not likely to be used for power generation due to the high tar production which would require extensive gas cleaning.

Fixed bed downdraft (or co-current flow) gasifiers have the fuel fed in from the top before it undergoes various processes as it moves downwards under gravity. Air is injected either into the middle section of the gasifier or from the top so that it flows in the same direction as the fuel (Fig. 2). In their traditional design, downdraft gasifiers are not suitable for fuels with a low ash

Figure 2 Starting a batch-fed downdraft gasifier from cold takes just 2-3 min before an internal combustion engine can be started on the producer gas to drive a generator.

melting point, such as straw. They are most popular for small-scale power generation. However, in order to operate properly, fuel moisture content and particle size have to be maintained within narrow limits. Many gasifiers of this design have been installed in developing countries where the problem of needing continuous monitoring is less of an issue due to cheap and abundant labour. Up-scaling of this design is difficult, and the maximum size is limited to about 3–4 MW_{th}. Several projects have been executed in Europe to improve downdraft gasifier operation leading to the development of small-scale, fully automatic units fuelled by one well-defined biomass type in order to reduce the risk of problems arising due to fuel particle size and moisture content variations.

Fixed bed cross flow gasifiers are suitable for very small-scale applications (smaller than or equal to 10 kW_e), using low-moisture charcoal as fuel. They operate at high temperatures up to 1500°C. This offers the possibility of thermal tar cracking to a level satisfactory for subsequent direct use of the fuel gas in an internal combustion engine without the need for extensive and costly cleaning. Opportunities for up-scaling are limited. The technical feasibility of this type of gasifier has been demonstrated using a small laboratory gasifier and a 100 kW_{th} pilot plant, mainly for use in developing countries.

Fluidised bed gasification of biomass was first implemented in the 1980s to meet the need for more efficient technologies for the utilisation of low-grade fuels such as biomass. Both bubbling and CFB gasification reactors operate under similar principles to comparable combustors but are more suitable at the larger commercial scale.

Two major reasons responsible for small-scale biomass gasifiers not being very reliable are as follows:

- The large variation in the key parameters determining the quality of the biomass fuel (moisture content, ash content, particle size, particle size distribution, chemical composition, bulk density, volatile matter content, etc.) causes extreme engine wear due to tar contamination and unstable operation due to slagging and variations in pressure drop.
- Small-scale gasification systems are characterised by minimal automatic measuring and control systems. As a result, systems often show variable performance and need continuous monitoring by experienced personnel (which adds to cost).

To make a gasifier more reliable, a standardised fuel could be produced using a mixture of different biomass fuels by combining pre-process technologies and advanced briquetting or pelleting technologies. Cheap and reliable hardware and software could then be used to control gasifier and gas engine operation automatically.

2.3 Pyrolysis

During the pyrolysis process, biomass is heated either in the absence of air (i.e. indirectly) or by the partial combustion of some of the biomass using a restricted air or oxygen supply. A wide range of reactor configurations have been devised. These include fluidised bed, ablative, entrained flow, rotating cone, transported bed, molten salt and vacuum reactors at scales from 20 kg/h to 650 kg/h of biomass fuel. Large-scale commercial operation is currently only being achieved using a CFB system.

The mix of products is extremely varied, consisting of gases, vapours, liquids, oils, solid char and ash. The composition and proportions of these products depend on temperature, input composition, pre-treatment and air composition conditions. At temperatures of around 500°C and short reaction times of 0.5–2 s, pyrolysis oils are produced. At higher temperatures of 700–800°C, pyrolysis reactions produce a much higher proportion of gas, with correspondingly less liquid and solid products. The gas typically has a heating value of 15–20 MJ/Nm^3, which is sufficient for many types of internal combustion engines and turbines without modification, though tars must be removed from the gas.

Pyrolysis liquid bio-oil has a high heating value of 22–23 MJ/kg measured on a dry basis or about 17 MJ/kg when produced with a moisture content of around 25% (wet basis). It is composed of a very complex mixture of highly oxygenated hydrocarbons which are chemically and physically unstable and require special treatment during handling, storage and use. It will not mix with hydrocarbon liquids. Density, viscosity, surface tension and heating value are key properties for combustion applications in boilers, furnaces and internal combustion engines. Other characteristics such as char level, particle size and ash content also have a major effect. It is theoretically possible to upgrade crude bio-oil into more familiar petroleum-like products although the technology for this is yet to be used commercially. Pyrolysis is the favoured technology for biomass from annual crops and their residues because the moderate process temperatures are less likely to cause ash melting or softening, and alkali metals are retained in the char. At larger than farm scale, pyrolysis is also being contemplated for the treatment of municipal solid waste as an alternative to disposal in landfills (Chen et al., 2014). The pyrolysis process can also be used to produce solid biochar, with the aim of integrating it into the soil as a means of locking up the carbon that had been extracted from the atmosphere by the growing biomass plants (Hedley et al., 2020).

A key advantage of producing bio-liquids rather than bio-gases is that fuel production can be de-coupled from heat and power demand. It is easier to meet peak demand by using stored bio-oil when needed. Higher heat and power plant availability using intermediate fuel storage may also result. Several pilot and near-commercial projects running on different biomass fuels have been developed in Europe, the United States and Canada. Bio-oil has been successfully combusted in diesel test engines where it behaves similarly to diesel fuel in terms of engine parameters, performance and emissions of NO_x, SO_x, etc. Continuous runs of 10 h have been achieved on raw bio-oil without any dilution or processing being first undertaken. The diesel engine may have to be modified to include a pilot injector ignition system if not already present since pyrolysis oil will not self-ignite. Diesel fuel is therefore needed for start-up, involving typically 5% of total fuel consumption in larger engines. Pyrolysis oil has also been successfully used as a boiler fuel, in a modified dual fuel diesel engine and in gas turbines, but these are more suitable for larger-scale applications than on-farm applications.

3 Secondary energy conversion technologies: internal combustion engines, steam turbines, steam engines and Stirling engines

The gaseous, solid or liquid energy carriers produced during the primary biomass conversion step (Fig. 1) are converted into a more useful form of energy during the secondary conversion step. The resulting energy products can be used for heating animal buildings and greenhouses, heating water in dairy sheds, drying crops and potentially generating process steam and electricity.

Small-scale heat systems and electricity-generating technologies can include:

- internal combustion engines;
- steam turbines;
- steam engines;
- Stirling engines;
- indirect-fired gas turbines;
- direct-fired pressurised gas turbines; and
- micro-turbines.

Gas turbines and micro-turbines are covered in Section 4 together with the potential for producing hydrogen from biomass for use in fuel cells.

A summary of indicative electric efficiencies and investment costs of these technologies (including the primary energy conversion technology) is given in Table 1. The information compiled was based on data from manufacturers and from feasibility studies.

3.1 Internal combustion engines

Internal combustion engines (ICEs) are among the most widely used prime movers for biofuels. Electricity conversion efficiencies (fuel energy to electricity)

Table 1 Summary of the electrical efficiencies and investment costs for energy conversion technologies including primary and secondary

Technology	Efficiency range (% based on LHV[a])	Investment cost range (US$/kW$_e$)
Internal combustion engine	25-30	1600-2400
Steam turbine	15-35	3400-9600
Steam engine	10-25	1400-4000[b]
Stirling engine	20-30	2000-9600
Indirectly-fired gas turbine	15-24	6000-12 200
Directly-fired pressurised GT	25-30	2400-3200
Micro-turbine	20-30	2600-3600

[a] Lower heating value.
[b] Investment cost for steam engine only.

of 25-30% (LHV) and low costs make ICEs an economic option. Several types are commercially available but the most relevant for stationary-power applications are four cycle-spark-ignition (Otto cycle) and compression-ignition (diesel cycle) engines. Diesel engines develop higher part load efficiencies than Otto cycle engines because of leaner fuel-air ratios at reduced load.

Advantages of ICEs include:

- being readily available over a wide range of power outputs;
- fast start-up;
- good part load efficiency;
- reliability; and
- long life.

Cleaning of biomass-derived gas fuel to remove tar and solid particles is the key challenge. None of the current gas cleaning systems currently meet gas quality requirements for satisfactory ICE applications. Hence, ICE manufacturers loathe to offer warranties.

Several projects have been evaluated linking biomass gasifiers with ICEs:

- In the UK, a 30 kW$_e$/60kW$_{th}$ downdraft gasifier (Fig. 2) was connected to an internal combustion engine generator (30 kW$_e$) and fuelled by short rotation coppice willow chips. The basic idea of the system was to provide electricity and heat supply to a single farm or to a co-operative of two or three small adjacent farms. Surplus heat produced was either used for wood chip drying or dumped. The system was feasible in producing energy, but mechanical problems were encountered including bridging of the biomass fuel in the hopper, poor flow of the charcoal/ash through the bottom of the reduction zone and inefficient operation of the gas cleaning chain. The economic feasibility of the system was not evaluated but is probably high, limiting the mass production of such small-scale gasifiers.

- In Germany, a fixed bed gasifier fuelled by forest arisings (residues from logging) was coupled to a spark-ignition engine. The cogeneration system had a capacity of 470 kW_e/1 MW_{th}, but the operation was not economic due to the high investment and maintenance costs involved.
- In Northern Ireland, a downdraft gasifier with the capacity to fuel a dual fuel gas/diesel engine driving a 100 kW_e induction-type generator was tested. The engine was designed to operate on 90% gas, with the 10% diesel fuel providing the necessary ignition. It was installed in a cogeneration configuration supplying 120 kW_{th} for a central heating system and 80 kW_{th} for drying the incoming short-rotation willow coppice fuel. Problems identified included fuel variation characteristics and gas clean-up. Modifications to fuel feed systems, hearth design and methods to remove particulates and tars resulted in the more consistent production of high-quality gas. The system costs an estimated US$1600/$kW_e$ including connection to the national electricity grid. At this cost, it showed good commercial potential wherever a suitable match could be made between customer requirements for both heat and power.
- Under a US-government funded 'Small Modular Biopower Project', Bechtel National Inc. undertook feasibility studies on a commercial KC gasifier (800 kW_e and 1600 kW_{th}) linked with a V-16 engine with a turbo-compressor and electronic ignition. The gasifier was tested satisfactorily utilising a wide range of biomass fuels including rice husks, straw, corn cobs and switchgrass.
- The US company BG Technologies marketed a small fixed-bed downdraft gasifier coupled to an ICE/generator in a range of 40–400 kW_e for an investment cost of US$1500–2000/$kW_e$.

3.2 Steam turbines

Steam turbines have traditionally been used in large-scale power generation and cogeneration plants with a nominal capacity of over 10 MW_{th}. The basic steam cycle is based on the closed Rankine cycle whereby water is heated, evaporated and superheated in a boiler. The energy contained in the steam is converted into rotational motion by expanding it through a turbine that is connected to an electricity generator. The exhaust steam is passed through a condenser, where it is cooled and turned into liquid again to complete the cycle. System efficiencies vary between 15% and 35%. At smaller scales, steam turbine efficiencies decrease and become less efficient than a steam engine at the same output. When under partial load, the efficiency of a steam turbine decreases further. Turbines are a well-established technology able to withstand high pressures and temperatures.

Capital costs for a steam turbine unit range from around US$2000/$kW_e$ at the 250 kW_e scale to $1000/$kW_e$ for 2 MW_e. Even though it is a mature technology with no technical barriers, the capital cost is the main obstacle to uptake. Other disadvantages include low efficiency at small scales, poor part-load efficiency, and low suitability if the steam is of poor quality in terms of pressure and water content.

Several small-scale bioenergy steam turbine projects have been evaluated:

- In the United States, Agrilectric Power Inc. developed a power generation unit in which rice hulls were combusted and steam from the connected boiler was directed to a turbine and generator. The company developed three model sizes of 500 kW_e, 1500 kW_e and 5000 kW_e with unit costs of around US$5000, $4000 and $3000/$kW_e$, respectively. The main markets are in developing countries.
- Carbona Corporation conducted a feasibility study for a small modular cogeneration project, using woody biomass in an air-blown fixed-bed updraft 1 MW_e gasifier. It produced steam for use in a back-pressure steam turbine. Calculated electrical efficiency was 18.6% and the investment cost was around US$4000/$kW_e$.
- A system developed in the United States consisted of a fluidised bed boiler coupled to either a back-pressure or condensing steam turbine, depending on the relative need for process heat or for electricity. The primary markets for the system included forest and agricultural processing facilities and remote rural communities. Investment costs for the 500–5000 kW_e plant size range were US$2200–5400/$kW_e$ for condensing turbine systems and $3600–5000/$kW_e$ for back-pressure turbine systems.

3.3 Steam engines

Steam engines are a well-proven technology for constant speed operations in industrial environments. However, they are only produced in small numbers which makes them relatively expensive per unit. The efficiency of steam engines largely depends on the quality of the steam, requiring boilers with good steam pressures of at least 10 bar and temperatures of at least 180°C. They are reliable and have low maintenance costs. The principle of a steam engine with reciprocating pistons is similar to that of an ICE, the main difference being that steam, as opposed to heat from combustion of the fuel, is the main output.

Most steam engines are double acting in that steam expands during both the forward and backward stroke of the pistons. As a result, a steam engine is lighter and smaller than an internal combustion engine for the same output.

Steam engines can be divided into traditional, modified, and those designs which use diesel engine components to reduce production costs. The main market for steam engines using biomass fuels is as cogeneration units where available biomass fuel quantities are too small for the economic operation of a steam turbine.

Steam engines are commercially available in different sizes ranging from 8kW$_e$ to 1400kW$_e$. Investment costs range from around US$7000/kW$_e$ for an 8 kW$_e$ engine to around $1000/kW$_e$ for a larger 1270 kW$_e$ engine. Compared to internal combustion engines, capital costs are too high and efficiencies too low to be cost competitive. Advantages include being a proven, robust technology that can withstand constant operation in an industrial environment; no internal contamination compared with an ICE or gas engine; and reasonably good efficiencies for small engines when used in connection with standard industrial boilers. Disadvantages include low efficiencies when used with low-pressure steam and the continued use of old designs and technologies.

3.4 Stirling engines

Stirling engines are external combustion engines. Fuel is combusted outside the engine and the heat is transferred into the cylinder either by a heat exchanger or via the cylinder wall. Any source of heat can be used, provided that it is of sufficiently high temperature. This characteristic makes the Stirling engine suitable for use in rural and remote areas where biomass fuels or biogas is available, especially where there is no electricity grid connection and low solar radiation levels make photovoltaics inefficient.

Typical designs of Stirling engines are single acting with one common cylinder containing two reciprocating pistons (Fig. 3). The working principle involves a gas enclosed in a cylinder moving two pistons continuously back and forth between hot and cold regions, as a result of differences in gas volumes due to temperature variations. Since there is no contact between the gas, its contaminants and the moving parts of the engine, engine lifetime is relatively long with lower maintenance requirements. The power output depends on the mean cycle pressure, swept volume of the pistons, speed and temperature difference between hot and cold regions.

Using any heat source, the air or helium gas sealed inside the expansion space (under the pressure of around 10 bar) is heated and the expansion drives the piston. Cold gas is compressed in the compression space by the second piston and heat produced by compression is removed by the cooling radiator and becomes available for cogeneration. The lower the cooled temperature, the higher the efficiency. The regenerator absorbs heat from the gas and then returns it later on in the cycle.

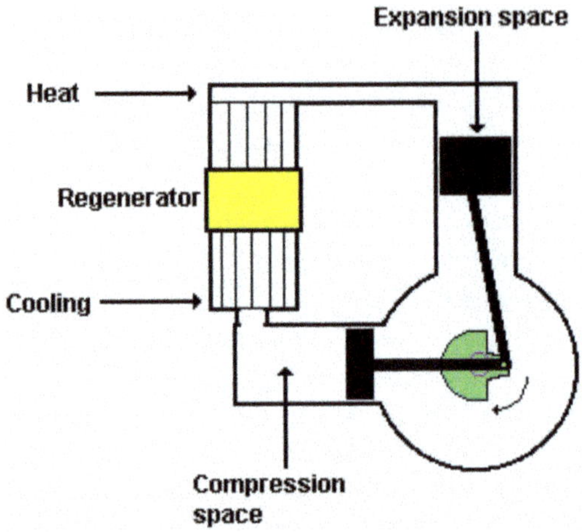

Figure 3 A Stirling engine uses a closed thermo-dynamic cycle that repeatedly expands and compresses a fixed mass of gas to obtain different temperatures (Sims, 2015).

During the late nineteenth and early twentieth centuries, several thousand Stirling engines were in operation in Europe and the United States with a maximum output of 4 kW. These were later replaced by more efficient and lighter internal combustion engines. Due to the low emissions and noise levels, the Stirling engine was revived by Philips Electrical Company in the 1930s. Challenges such as finding materials able to cope with the high pressures involved have been partly overcome in modern designs. Configurations include a double-acting design with four cylinders (containing one piston in each). Even simpler designs involve springs controlling the piston movements without a kinematic crank mechanism.

Stirling engines are commercially available in the 0.5–150 kW_e range and a number of companies, such as Microgen Engine Corporation in the Netherlands (https://www.microgen-engine.com/), are working on further development. Only a few engines have been successfully introduced to the market, and commercial viability remains a problem. The cost, for example, of an SPS V160 10 kW_e Stirling engine (that can be coupled to a 1 MW_{th} biomass atmospheric pressure fluidised bed combustion unit fuelled by wood chips and specifically designed for stationary use) is around US$6000/$kW_e$. Mass production would reduce the unit cost.

For Stirling engines specifically developed for biomass firing, overall efficiencies (fuel to shaft output) are around 30%. Advantages include good partial load efficiency, low noise levels, safe operation, low expected maintenance costs, suitability for a wide range of fuels, and a long engine lifetime (around 25 000 h compared to the average life of a car engine of around

4000 hours). Only very small engines have been built to date with little data on reliability. High heat exchange temperatures may cause corrosion while sealing off the engine to withstand high pressures remains a technical problem. The prospects for Stirling engines are good but further development is required.

4 Secondary energy conversion technologies: indirect-fired gas turbines, direct-fired pressurised gas turbines, micro-turbines and hydrogen

4.1 Indirect-fired gas turbines

Indirect-fired gas turbines are conventional gas turbines but with the combustion chamber replaced by a heat exchanger. Gas from a gasifier is burned in a close-coupled combustor-heat exchanger; the gas turbine is operated with the resulting hot clean air in a hot air turbine cycle. Before the hot combustion gas enters the heat exchanger, some cleaning is necessary. In a Canadian project, investment costs ranged between US\$7500/kW_e at the 530 kW_e scale down to \$6000/kW_e at 972 kW_e.

Advantages include using mass-produced components that reduce capital costs and high efficiency if the inlet temperature to the gas turbine can be increased to that used in directly fired designs. Disadvantages include low efficiencies of standard metal heat exchangers, the large size of the heat exchanger, the need for the particle cleaning system to work reliably at high temperatures and challenges in turbine regulation to match load. Whilst there are no technical barriers, extensive testing has yet to be done to prove that the system works well.

In Belgium, a fluidised bed gasifier designed for sawdust was coupled to an indirect-fired gas turbine through a high-temperature metallic air heater. The targeted electrical efficiency was 24%. To enhance power output and to allow flexible power-to-heat ratios, water is injected into the air heater. For a cogeneration capacity of 300 kW_e/1200 kW_{th}, investment costs are around US\$7000/kW_e, with the potential to lower to US\$4000/kW_e for a 1.8 MW_e design.

In the Netherlands, a small-scale stationary power generation plant of 500kW_e has been demonstrated for cogeneration using an indirect-fired gas turbine with clean air as the working fluid. The combination of low temperatures and pressures enabled the turbine construction materials to be made from stainless steel instead of ceramics. A two-stage combustor was chosen as the most promising system to couple to the gas turbine. The system had an electrical efficiency of 20% and capital costs were estimated to be US\$5200/kW_e, due to the relatively high costs of the heat exchanger.

4.2 Direct-fired pressurised gas turbines

Direct-fired pressurised gas turbines typically utilise hot combustion gas from a pressurised downdraft gasifier. As the hot gas temperature is usually higher than the maximum allowable inlet temperature of the gas turbine, it is first mixed with cooler air. The operation of a gas turbine with biomass-derived gaseous fuels appears promising in the range of 5–20 MW_e plant sizes, following several demonstrations in the United States, UK and Scandinavia.

Costs are around US$4600/$kW_e$ for a plant in the 5–6 MW_e range (of which around half is for the gas turbine). Costs are kept relatively low as:

- no steam ducting system is required; a standard gas turbine design can be used;
- there is relatively little erosion, corrosion or deposition occurring on the turbine blades due to effective hot gas clean-up; and
- residual ash collected in the cyclone can be used as a soil conditioner through the recycling of minerals from the biomass.

A small-scale, pressurised air-blown fluid bed gasification unit coupled to a gas turbine has been evaluated in North Carolina, USA. A 225 kW_e system was demonstrated with the gas being passed through a hot gas clean-up system prior to injection into the combustor chamber of the turbine. Barriers to wider commercialisation include the biomass feeding system, problems with hot-gas clean-up and relatively low conversion efficiency. The major engineering challenge is to design a turbine with a fuel and combustion system able to burn hot gas.

4.3 Micro-turbines

Micro-turbines are similar in principle to gas turbines. However, most designs incorporate a recuperator to recover part of the exhaust heat to preheat the incoming combustion air. This gives a higher efficiency of 20–30%. By simply changing the fuel injector, a range of liquid and gaseous fuels can be used including biogas, producer gas, biodiesel or bio-ethanol. The exhaust gas temperature is around 250°C so heat recovery is possible, with the recovered heat used either for heating or to drive an absorption chiller.

Air is drawn through a compressor, then mixed with a gaseous fuel in the combustion chamber and ignited to power the turbine section which drives the generator (Fig. 4). Most manufacturers, such as the Capstone Green Energy Corporation (https://www.capstonegreenenergy.com/), are pursuing a design where the only moving part is the shaft on which the compressor, turbine and generator are mounted. This aluminium and nickel alloy steel cast shaft

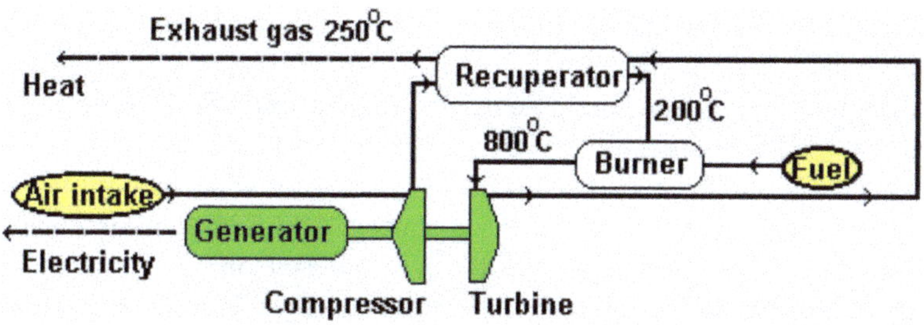

Figure 4 Components of a micro-turbine generator, showing the pre-heating of the intake air results from passing it through the compressor (Sims, 2015).

is suspended, lubrication free, on a wedge of high-pressure air compressed during the first few revolutions at start-up. This means wear and maintenance are minimal. Even at an operating speed of around 120 000 rpm, there is little noise or vibration. For models with power outputs of 25–30 kW_e electrical efficiency is 28–30%, biomass fuel consumption is around 430 MJ/h, total weight is less than 150 kg and the size is similar to a domestic refrigerator. The typical cost is around US$1000/$kW_e$ though, once mass-produced, this could be reduced.

Typically, micro-turbines operate at combustion temperatures below those where NO_x form, meaning these emissions are relatively low. Initially a design above 30 kW_e was limited by the shaft thrust bearing but several manufacturers have now developed engines in the 25–250 kW_e range. If greater electrical output is needed, multiple units can be integrated; this also provides additional reliability should one unit need maintenance. For single shaft machines, a standard induction or synchronous generator could be used to avoid any power-conditioning electronics. Net-metering and peak load matching are possible.

Advantages of micro-turbines include compact and light-weight design, a good size range availability and low noise levels of less than 65 dbA at 10 m distance. They can be fuelled efficiently by stored waste heat or biomass fuels as promoted by Flex Energy (https://www.flexenergy.com/heat-exchanger-applications-for-gas-turbines/). This design can run on fuel gases from the gasification of biomass, biogas from landfills and animal wastes, or on waste gases from petroleum and coal production operations. The investment cost of a Flex Microturbine™ coupled to a simple downdraft gasifier (including a small catalytic combustor) is around US$2600/$kW_e$ for capacities between 27 kW_e and 210 kW_e.

4.4 Hydrogen

Hydrogen is a promising energy carrier for providing heat from direct combustion as well as generating electricity from fuel cells. This is as a result of the flow of electrons when hydrogen combines with air (oxygen), with water as a by-product.

Hydrogen can be an energy-efficient, low-polluting fuel, produced from a wide range of processes, though producing and storing it ready for use can result in high-efficiency losses. For example, the efficiency of converting electricity to hydrogen via electrolysis is typically about 60% and the efficiency of converting hydrogen to electricity in a fuel cell process is about 70%. Therefore, the overall energy efficiency would be about 42%. As is also the case for biomass or fossil fuels, direct combustion to heat is reasonably efficient at around 70–80%.

Hydrogen is found in many diverse compounds such as water, fossil fuels and biomass. It is currently produced mainly from natural gas, naphtha and coal. In these circumstances, hydrogen is not renewable. CO_2 is also emitted during the hydrogen production stage, similar in volume to that released during the direct combustion of those same fuels. 'Brown' hydrogen is produced from the gasification of coal or 'grey' hydrogen from the steam reformation of natural gas. 'Green' hydrogen is produced from the electrolysis of water using renewable electricity and therefore has near-zero CO_2 emissions, much lower than brown or grey hydrogen. However, green hydrogen is presently much more costly to produce, though the capital cost of electrolysers is declining. 'Blue' hydrogen is the same as grey hydrogen but the CO_2 produced could be captured and sequestered, if and when CCS technology becomes commercially viable and mainstream.

The traditional use of coal gas (or 'town' gas), produced from gasifying coal, was replaced by natural gas in the UK in the 1970s. This change involved the minor conversion of millions of domestic and industrial appliances. The goal is to change from natural gas (that emits CO_2 when combusted) to green hydrogen in order to reduce greenhouse gas emissions. Demonstration houses have been established in Scotland using 100% hydrogen to fuel appliances such as hot water boilers, cookers, ovens and fireplaces (https://www.which.co.uk/news/2021/11/hydrogen-boilers-and-other-appliances-what-are-they-like-to-use/). Hydrogen could be used on farms for a similar range of applications with the hydrogen produced directly on the farm once this technology becomes economically viable. There has been a successful demonstration project producing on-farm hydrogen through an electrolyser driven by a small 1 kW wind turbine and storing the hydrogen in a gas pipe connected to a fuel cell (Sudol, 2009).

Hydrogen can also be produced from the thermochemical gasification of biomass feedstocks. Hydrogen atoms typically make up about 6% by weight of dry biomass. Hydrogen can be extracted by gasification to produce a gas containing around 10–25% hydrogen by volume. Using hydrogen from biomass compared to coal or natural gas is attractive from an environmental point of view. Although CO_2 is also emitted during the process, the carbon cycle would be closed by growing trees or crops to replace those that provided the biomass. The challenge is to overcome the economic barriers for converting biomass to hydrogen and using it as a clean fuel in efficient energy conversion devices.

There are several promising developments to produce hydrogen or methanol from biomass. Since producer gas coming from a biomass gasifier contains methane and carbon monoxide in addition to the hydrogen, a fuel cell's electrolytes cannot use it directly so it must receive appropriate pre-treatment. A Dutch modelling study compared the technical and economic performance of five hydrogen-producing systems at several scales. The most promising was ceramic membrane technology that enabled hydrogen to be separated from a mixed producer gas stream at high temperatures for direct use in a fuel cell. If the biomass feedstock is available for US$4-6/GJ, the hydrogen could be produced for US$8-10/GJ. This was cheaper than producing methanol as a fuel.

The sensitivity of a fuel cell design towards fuel impurities is mainly determined by the operating temperature, the electrolyte and the catalyst. Sulphur oxides, metal vapours, tars and high CO levels can also inhibit the operation of various types of fuel cells. If bark, wood wastes or other kind of biomass were to be used as a source of hydrogen, removal of any metals from the gas would be necessary as well as sulphur.

Purification steps for the producer gas depend on the type of fuel cell so the development of a small and efficient gas clean-up system would be a key step. Producer gas pre-treatment can involve adding steam at temperatures of around 600°C to convert the methane to carbon monoxide and hydrogen in a process known as steam reforming. At a lower temperature of 200–300°C, the steam reacts with the carbon monoxide to form hydrogen and carbon dioxide in the shift conversion reaction. Carbon dioxide and any sulphur compounds present can be removed. Methane reformers, shift reactors, CO_2 removal systems and hydrogen purifiers are well-established technologies in the chemical process industry.

Several other studies linking fuel cells with biomass fuels have been undertaken, including the following:

- The first biomass-based fuel cell power plant based on a small alkaline fuel cell module was demonstrated in 1999. The module consisted of 24 cells with a nominal power output of $350W_e$. The primary fuel input was producer gas obtained from gasification of wood char and agricultural residues. The power output using this feedstock showed potential as an efficient stand-alone power generator but the generating cost was high.
- High-temperature fuel cells and gas turbines have been integrated to boost electricity generation efficiencies. The hot exhaust from the solid oxide fuel cell was used to drive the gas turbine. Energy recovered from

the turbine exhaust was used in a recuperator to preheat the air from the turbine's compressor section before entering the fuel cell and the gas turbine. Any remaining heat energy from the turbine exhaust can be recovered for low-grade heating. for example of water.

- Using computer simulation techniques, evaluation of a pilot-scale air-blown fluidised bed gasifier in the United States was linked with a molten carbonate fuel cell when operated on gas derived from switchgrass feedstock. The system showed that a 1.75 MW_e power plant would operate at 46% efficiency. However, the effect of fuel gas quality on fuel cell performance and investment costs remain barriers to commercialisation.
- Several other systems have been compared using switchgrass or woody biomass in downdraft gasifiers in conjunction with a 200 kW_e fuel cell to generate electricity for on-farm installations:
  - A Batelle indirect gasifier coupled with a proton exchange membrane fuel cell (PEMFC).
  - A downdraft gasifier coupled with a molten carbonate fuel cell (MCFC).
  - A downdraft gasifier coupled with a phosphoric acid fuel cell (PAFC).

Including the cleaning of the gas, the investment cost of US$6400/$kW_e$ for the MCFC system was 25% cheaper than the PAFC system and around 30% less than for the Batelle PEMFC system. Together with these high capital costs, the primary barriers to development were relatively low grid electricity tariffs and high operating costs. Automation of the system to minimise operational costs is paramount for improving the return on investment for small systems.

- A technical and economic comparison has been made between a biomass gasification/gas turbine system and a biomass gasification/fuel cell system consisting of a Batelle gasifier, an MCFC and a steam turbine. Feeding the producer gas into the MCFC instead of the gas turbine allowed the system to cope with a higher level of contaminants. The fuel cell system also had a higher overall efficiency at 53% and a better economic performance at the larger scale of 1–2 MW_e.

An alternative approach to the production of hydrogen from biomass is by fast pyrolysis to generate bio-oil followed by catalytic steam reforming of the oil or its fractions. This approach has the potential to be cost competitive with other current commercial processes for hydrogen production.

5 Cogeneration system analysis at the small scale

There have been several studies comparing small-scale electricity and heat generation systems using different technologies. As an example, the technical and economic performance of a variety of small-scale systems has been assessed using the ECLIPSE process simulator programme and utilising short-rotation willow coppice as feedstock. This analysis gave the following results:

- At scales of around 5 MW_e, combustion plants have higher efficiencies than gasification/gas turbine systems in a simple cycle.
- Direct combustion for power generation was not feasible even up to 25 MW_e because of low efficiencies ranging between 16% and 26% and high capital costs ranging from around US$6000/$kW_e$ for a 350 kW_e system to US$3000/$kW_e$ at 4.75 MW_e. For cogeneration, sizes below 25 MW_e were considered more suitable.
- An atmospheric pressurised gasifier with ceramic filter hot gas clean-up coupled to a simple cycle gas turbine at power outputs of either 1.3 MW_e or 3.4 MW_e had low efficiencies of 13.9% and 16.0%, respectively, with investment costs of around US$7500/$kW_e$ and US$6000/$kW_e$.
- Gasification integrated with a spark-ignition engine/generator set resulted in moderate efficiencies ranging from 22.6% to 29.7% for plants with a capacity between 150 kW_e and 2.27 MW_e with investment costs of around US$3000 to $2000/$kW_e$.

Another study examined gasifiers of various sizes in combination with a PAFC and MCFC. The most appropriate gasifier was the Koppers-Totzek entrained-flow gasifier which was originally developed for coal gasification. It is able to produce gas with a low methane content from biomass so that reforming is not necessary. The technology has been commercially available and showed potential but is no longer in production (https://www.beyonddiscovery.org/gasification-technologies/532-the-kopperstotzek-atmospheric-process.html). The study concluded that, until fuel cell technologies are developed further and become commercially viable, gas-fuelled internal combustion engines are most appropriate for small-scale biomass systems.

Another study compared four different small-scale cogeneration technologies using biomass feedstocks and drew the following conclusions:

- The Stirling engine has good potential because of its relatively high efficiency and a wide spectrum of power ranges.
- To be efficient, the steam engine needs higher steam temperatures and pressures than those that can be achieved using standard small-scale industrial boilers.
- Steam turbines are commercially available and a well-proven technology but their efficiencies remain relatively low.
- Indirect-fired gas turbines have technical problems with high-temperature heat exchangers which are still to be resolved.

- Cogeneration is important for the viability of small-scale combustion applications, assuming there is a useful demand for heat. Cleaner and more robust technologies with high overall efficiencies need to be developed. The lower limit for cogeneration plants is considered to be at a boiler capacity of 5 MW_{th} due to the poor efficiencies of smaller-scale steam turbines and the higher investment costs of such systems.

To eliminate uncertainties associated with technologies that are not yet commercially available, estimates of the annual production expenses were calculated for a 3.12 MW_{th} capacity plant excluding capital costs. The operating costs were lowest for the Stirling engine, followed by the steam engine and steam turbine, and highest for the indirect-fired gas turbine.

Another study has evaluated the feasibility of a number of advanced small-scale power production systems based on biomass pyrolysis and gasification in the range of 5-60 MW_e capacity. Results are shown in Table 2. It was concluded that, for peak load generation at less than 2000 operating hours annually, a pyrolysis bio-oil fired diesel power plant concept appeared most promising in spite of having the lowest efficiency.

An overview study of gasification technologies for small-scale biomass cogeneration plant showed that, when integrated with steam turbines, efficiencies of 10–15% were obtained for capacities of 1–4 MW_e with investment costs of around US$10 000 to $5000/$kW_e$, respectively. For fixed bed gasifiers integrated with a gas engine in the range of 100 kW_e to 2 MW_e, the system efficiency was 32–40% and investment costs ranged between US$10 400 and $4600/$kW_e$.

Table 2 Comparison of system efficiencies and costs for small-scale biomass gasifier and pyrolysis projects

Primary system	Secondary system	System efficiency (%)	Indicative investment cost (US$)
Atmospheric pressure fluidised bed gasifier; dolomite tar cracker; gas clean-up	Integrated gas turbine combined cycle	37	$6600/$kW_e$
Pressurised fluidised bed gasifier; gas clean-up	Steam injected gas turbine engine	43	$7000/$kW_e$
Atmospheric pressure fluidised bed gasifier; dolmite tar cracker; gas clean-up	Dual-fuel diesel engine	34	$3200/$kW_e$
Pyrolysis plant	Gas turbine	33	3200kW_e$
Pyrolysis plant and bio-oil refinery	Diesel engine	30	$2400/$kW_e$

These comparisons between bioenergy cogeneration systems are only indicative and the broad ranges presented show the uncertainty in the data. The published information is conflicting and confusion arises between costs assessed from real projects or when estimated and between efficiency claims based on laboratory tests or the typically lower efficiencies achieved once the technology is scaled up and installed at a commercial scale. Some potential economies of scale are evident, with the exception of internal combustion engines, which are already mass produced.

In summary, gasification using a range of dry biomass feedstocks in conjunction with micro-turbines or fuel cells has the potential to generate electricity at the rural community scale. Automation is essential for any small system to realise a return on investment. Overall, the Flex Microturbine™ appeared to be the most cost-effective system when compared at the 200 kW_e level. Gasifiers linked with internal combustion engine systems were also found to be competitive.

One approach could be to gasify or pyrolyse solid biomass to produce a low calorific value gas. This could subsequently be combusted in a modified petrol or diesel engine to drive a generator; it could also provide heat (via a heat exchanger) for use to heat water or for other low-grade heat applications. Fast pyrolysis or atmospheric pressure gasification for power generation using a gas-fired internal combustion engine should remain competitive in the rural, off-grid small-scale electricity generation market.

Model simulations showed that fast pyrolysis has a market advantage in cases where biomass resource is available and located some distance from the electricity grid and where electricity is required at several remote sites (since the system can be de-coupled). This means that a pyrolysis plant could be located near the biomass resource and a generator near where electricity is needed, thus reducing transport costs of large volumes of biomass. Alternatively, a single pyrolysis plant could be constructed to supply fuel for all the generators located at several remote sites, saving investment costs through economies of scale.

6 Case study: conversion of coal boilers to biomass in greenhouse enterprises

A New Zealand horticultural greenhouse company, K&L Nurseries, that grows roses and gerbera, was consuming 30 t of coal a year to heat the greenhouse complex. In 2013, their old Vekos coal boiler was only 60% efficient and was ready for a major re-fit. In addition, under the emission trading scheme, the company was paying over US$1000 per year as a carbon charge which was likely to rise significantly as the carbon market price rose. After research, the company invested in a 90-kW biomass boiler.

The potential problems from using biomass instead of coal are fuel handling and high moisture content. This has been largely overcome by the Austrian manufacturer Polytechnik Biomass Energy Ltd (Fig. 5). Their boiler is engineered so the woody biomass is partially dried before it is fed into the furnace with a hydraulic ram. This allows the burner to handle fuels with moisture contents between 20% and 60% (wet basis). Using a wide range of biomass fuel types is also possible with good computer control of the large furnace and the use of a moving grate system in the boiler. Around 100 t per year of green waste produced from the nursery operation is now used for boiler fuel, avoiding previous disposal costs. The flexibility of the boiler operation means the system has one of the lowest emissions of any biomass boiler design and an overall efficiency above 90%.

The boiler has a smaller capacity heat output than the nursery needs at times, so the system was designed to also draw heat from a thermal storage unit when needed to meet the full heat demand. A thermal storage tank of around 400 m^3 was constructed to store the hot water prior to its circulation around the greenhouses. After looking at wood and steel tanks, a lined and insulated in-ground tank was constructed, with lower costs than for the other options. The company won a national sustainable energy award in 2014 (https://www.stuff.co.nz/business/industries/10095179/Energy-efficient-flower-grower-wins-awards).

Partly as a result of this successful investment, another greenhouse grower, Zealandia Horticulture Ltd., producing a wide range of nursery plant seedlings, recently replaced their 20-year-old 6 MW Vekos coal boiler (even though it was already being fuelled with dry biomass coming from old wood pallets, demolition waste, etc.). The company installed two modern, biomass-fuelled hot water boilers at its site in Christchurch. The smaller 1.5-MW boiler is used

Figure 5 Polytechnik boiler showing compressed air cleaners at the end of each of the boiler tubes (*https://biomass.polytechnik.com/en/products-and-components/*).

during periods when less heat is required and the second 2-MW boiler can be fired up to help meet higher heat demands, for example, in winter. With a total maximum heat output of 4100 kW_{th}, and a minimum of 350 kW_{th}, the system has been designed to utilise hogged and chipped wood, bar, forest residues and arboriculture wastes. This enables the company to purchase local biomass fuels a lot cheaper than coal. The atmospheric particulate emissions are significantly lower than from the old boiler (Zealandia Horticulture Ltd.|Bioenergy facilities).

The new boiler design includes a combustion air pre-heater, an oxygen control system, a fuel drying zone, compressed air cleaning of the boiler tubes and a Siemens programmable logic controller system that includes visualisation through furnace cameras to enable monitoring and optimisation of the plant, especially when running on fuels with different characteristics. Each boiler has three individual controlled combustion zones, a back-end flue gas temperature of 125–140°C and automatic cleaning and de-ashing systems (though only 1 m^3 of ash results from every 65 truckloads of biomass fuel). Each boiler is linked with a 10-m tall, 1000 m^3 thermal water storage tank to help meet the daily demand of the greenhouse complex and provide a 2–3 day backup for the smaller boiler without needing to fire up the larger boiler.

7 Conclusion

There is good potential for the use of sustainably produced biomass to provide heat at farm scale, together with electricity generation under specific circumstances. The overall capital and fuel costs for a bioenergy system can be a barrier. They are more affordable where grid electricity is less available.

The operation of steam or gas turbines for small-scale biomass plants is relatively uneconomic in terms of operation, maintenance and capital costs compared with using small gasifiers linked with internal combustion engines (Fig. 6). The cheap cost of such engines due to mass production is a benefit, although they are fairly inefficient.

There is a wide range of biomass feedstocks available on farms, coupled with a diverse choice of primary and secondary conversion technologies used to produce bioenergy in the form of heat, electricity and transport fuels. This makes it difficult to assess the most appropriate investment for any specific farm situation. Where heat is usually supplied from coal or natural gas, replacement of these fossil fuels by low-carbon biomass can be a cost-effective investment, especially where a high carbon price is added to the purchase price of the fossil fuels. There are many examples of commercial systems successfully using woody biomass, cereal straw or biogas at the on-farm scale to provide useful heat for a range of applications. Wider dissemination of such bioenergy technologies to displace fossil fuels would help reduce annual global greenhouse gas emissions.

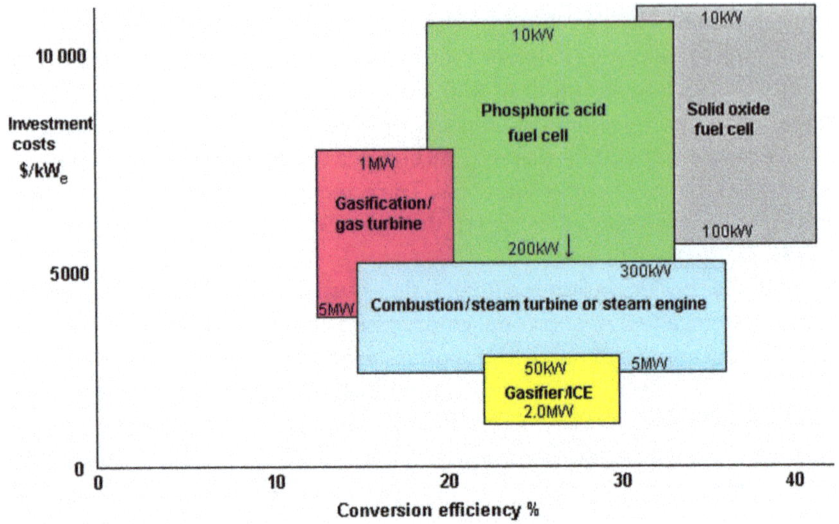

Figure 6 At the small scale suitable for on-farm applications, stand-alone power generation bioenergy and hydrogen fuel cell systems cannot yet compete with a gasifier and internal combustion engine either due to their poorer fuel-to-electricity conversion efficiencies or higher system costs or both (Sims, 2015). Note: It was assumed the fuel cells were linked with an oxygen-blown gasifer and wet gas scrubber system.

8 Where to look for further information

The IEA Bioenergy technology collaboration programme (TCP) was established over 40 years ago to enable member countries to share research, demonstration, deployment and commercial activities relating to biomass and bioenergy. It is structured into a number of specific tasks that have evolved over the years. The 11 present tasks include biomass supply, gasification, climate and sustainability, biogas, waste and circular economy, and energy systems and where relevant, link with other IEA TCPs. An IEA member country can opt to join one or more tasks and pay a joining fee. Details of the large number of publications together with details of forthcoming meetings and conferences can be found at Bioenergy | International Collaboration in Bioenergy (ieabioenergy.com).

9 References

Chen, D., Yin, L., Wang, H. and He, P. 2014. Pyrolysis technologies for municipal solid waste: A review. *Waste Management* 34(12), 2466–2486. DOI: 10.1016/j.wasman.2014.08.004.

Hedley, M., Camps-Arbestan, M., McLaren, S., Jones, J. and Chen, Q. 2020. A review of evidence for the potential role of biochar to reduce net GHG emissions from New

Zealand agriculture. Available at: http://www.nzagrc.org.nz/assets/Publications/Potential-Role-of-Biochar-in-NZ-2021.pdf.

IEA. 2008. From 1st to 2nd-generation biofuel technologies - An overview of current industry and *R*, D&D developments. International Energy Agency. Available at: https://iea.blob.core.windows.net/assets/594edd68-08d6-4258-95ca-180df4952b26/2nd_Biofuel_Gen.pdf.

IEA. 2017. Liquid biofuels. *IEA Bioenergy Task 39*. Available at: https://www.energy.gov/sites/prod/files/2017/10/f37/mcmillan_bioeconomy_2017.pdf.

Sims, R. E. H. 2015. *The Brilliance of Bioenergy - In Business and in Practice*. Routledge.

Sudol, P. 2009. Modelling and analysis of hydrogen-based wind energy transmission and storage systems: HyLink system at totara Valley. Palmerston North, New Zealand: Massey University. Available at: https://mro.massey.ac.nz/handle/10179/786.

UNFCCC. 2019. Off-grid and decentralized energy solutions for smart energy and water use in the agrifood chain. United Nations Framework Convention on Climate Change. Available at: https://unfccc.int/documents/200379.

Chapter 8

Developments in agrivoltaics: achieving synergies by combining plants with solar photovoltaic power systems

Stefano Amaducci, Eleonora Potenza and Michele Colauzzi, Università Cattolica del Sacro Cuore, Italy

1 Introduction

Agrivoltaic (AV) systems combine agricultural activities with the generation of electricity from solar photovoltaic (PV) panels constructed on the same area of land. To enable the access of tractors and machinery to carry out agricultural activities including crop cultivation and harvest, the PV panels should be mounted at a height above the soil to give adequate clearance. The alternative option - growing the crop in rows between the rows of PV panels - is less practical (Fig. 1).

Goetzberger and Zastrow (1982) introduced the concept of AV systems but only more recently have the international economic and political frameworks and increased environmental concerns stimulated a growing interest in this technology. Articles published on agrivoltaics in scientific journals since 2011 were assessed (Fig. 2). Using Scopus, the titles, abstracts and keywords were searched for the terms 'agrivoltaic' (with 95 results), 'agrivoltaic AND system' (85) and 'agrophotovoltaic' (24), while using Google Scholar, articles with those same terms in just their titles were 'agrivoltaic' (113), 'agrivoltaic system' (27) and 'agrophotovoltaic' (20), with citations not included.

http://dx.doi.org/10.19103/AS.2022.0100.22

- Increase water use efficiency (WUE)
- Decrease water inputs
- Reduce drought stress of crops
- Protect crops against adverse conditions
- Stabilise crop production
- Increase in specific leaf area (SLA)
- Improve land use efficiency
- Increase ecosystem services
- Minimise environmental impact
- Generate renewable electricity
- Mitigation of climate change
- Increase the economic value of crops
- Generate additional income for farmers
- Reduce dust and soil blown on to the panels (soiling)
- Increase PV cooling
- Crop albedo can increase electricity generation when using bi-facial panels

Figure 1 Synergies between crops and photovoltaic panels in agrivoltaic (AV) systems.

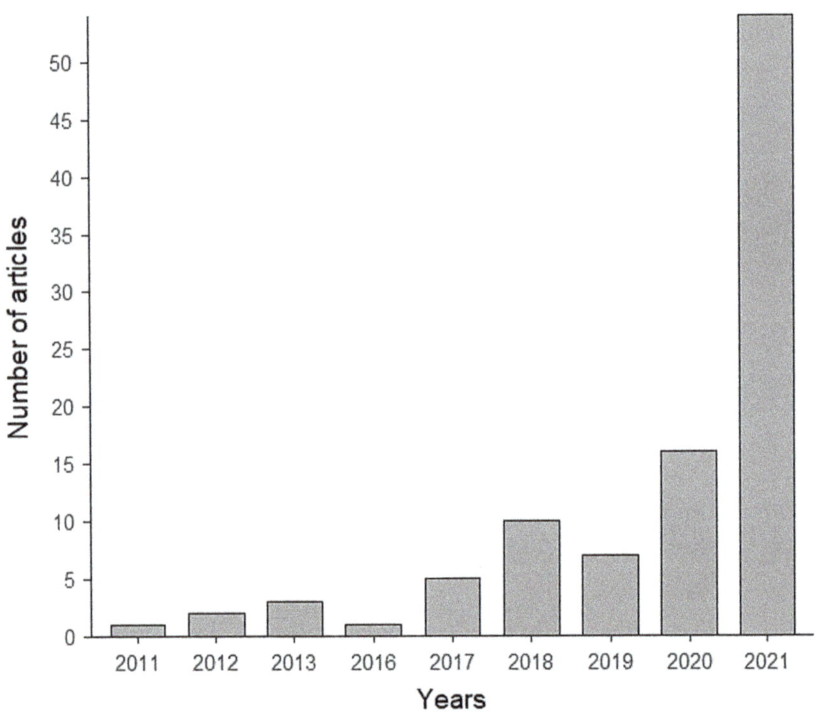

Figure 2 Articles related to agrivoltaics, published in scientific journals from 2011 until end of 2021.

The main driver of the recent interest in AV systems is the transition towards renewable energies. To reach the goal set by the Paris Climate Agreement to limit the temperature increase to below 2°C and aim for 1.5°C, the global power sector has to be fully decarbonised by 2050 (Jäger-Waldau, 2019).

Renewable energy sources, which include biomass energy, wind energy, hydroelectric power, geothermal energy and solar energy, are low-carbon but not totally free from adverse environmental effects. Being characterised by a relatively low power density, the deployment of renewable energies on a large scale can have a significant impact on land use. Biomass produced from energy cropping, in particular, has been considered a major driver of direct or indirect land-use change (Dale et al., 2011). (This is not the case for biomass sources arising from crop or forest residues or animal wastes.)

Solar photovoltaic (PV) systems will play a key role in achieving the Paris Climate Agreement goals and providing a sustainable electricity supply (Jäger-Waldau, 2019; IRENA, 2019). The installed global solar PV capacity was over 600 GW in 2019, after high growth rates, and is expected to reach 2840 GW by 2030 and 8519 GW by 2050 (IRENA, 2019). Integrated PV technologies on buildings are preferred as they would minimise land use impact, but to achieve these ambitious PV targets, the implementation of land-based PV systems also seems necessary (IEA, 2020). This could exacerbate the problem of land-use change, particularly when PV solar farms, with either fixed or tracking panels, occupy agricultural land currently dedicated to crop production (Van de Ven et al., 2021). To avoid land-use competition, floating PV systems on ponds, lakes and estuaries are also being constructed up to a capacity of over 40 MW (Vella, 2021).

In this chapter, the conditions under which AV systems could offer a sustainable solution to land-based PV deployments are discussed.

A fundamental issue related to the sustainable implementation of AV systems is the impact of PV panels on crop productivity. Photovoltaic support structures can affect micro-meteorological conditions in the field; the panels also cause shading, which reduces the solar radiation level available to the growing crops (Dupraz et al., 2011; Weselek et al., 2019). The radiation intercepted by plant canopies is the main driver of crop growth and development (Campillo et al., 2012). However, considering that only a fraction of the incoming radiation is actually absorbed by plants during photosynthesis, AV systems can be designed and managed as 'solar sharing' systems, where the surplus solar radiation, which is not contributing to crop production, is used for power generation (Sekiyama and Nagashima, 2019). In this way, AV systems can increase the overall productivity of the land.

Multiple studies have quantified the impact of AV systems in terms of land equivalent ratio (LER), an ecological index usually employed to evaluate the efficiency of intercropping and agro-forestry systems (see Section 2 for details) (Dupraz et al., 2011; Valle et al., 2018; Amaducci et al., 2018; Elamri et al., 2018a; Fraunhofer ISE, 2020; Trommsdorff et al., 2021). It is important to note that the positive impact of AV systems on land-use efficiency largely depend on the specific environmental conditions, the crop in question and the design features of the AV system.

On the other hand, the success and expansion of AV systems strongly depend on social acceptability and policy support. Emerging policies regulating the construction and management of AV systems support implementation where they preserve agricultural productivity but aim to prevent the speculative expansion of poorly designed ground-mounted PV structures that claim to be AV systems.

Preserving agricultural productivity in an AV system implies that conventional agricultural machinery is still able to operate in the field (e.g. when the PV panels are mounted at least 3 m above ground level) and that yield reductions do not exceed a specific threshold (Irie et al., 2019, Schindele et al., 2020). The design and implementation of sustainable AV systems should also be evaluated around the water–food–energy nexus (Barron-Gafford et al., 2019).

There is a broad consensus that cultivation under the shade of PV panels increases water-use efficiency (Marrou et al., 2013a,b; Hassanpour Adeh et al., 2018; Barron-Gafford et al., 2019). Reduction of crop evapo-transpiration rates can decrease irrigation needs, for example in lettuce by about 20% (Elamri et al., 2018a). An additional positive interaction of water management under an AV system is related to the possible integration of gutters in the infrastructure, with storage tanks (Hernandez et al., 2019). This would enable the collection of rainwater to be used for irrigation of cultivated crops (Weselek et al., 2019) or for washing any dust generated, inhibiting dust accumulation on the PV modules (Dinesh and Pearce, 2016).

The positive effect on crop evapo-transpiration rates, combined with the possibility of implementing rainwater harvesting systems, renders AV systems particularly suitable for hot and drought-prone environments (Barron-Gafford et al., 2019). They could increase the productivity of conventional crops grown in the region and/or enable the cultivation of other crops that normally could not withstand the heat or lack of water (Weselek et al., 2019; Willockx et al., 2020).

In AV systems, the interaction between electricity generation and crop production is synergistic: not only do crops benefit from the presence of the panels, but the PV system also benefits from the vegetative cover. In particular, transpiration from crops cultivated under PV panels lowers the air temperature, thereby significantly increasing the performance of the solar cells (Barron Gafford et al., 2019). Crops can also contribute towards reducing soil particles blowing onto the panels (Ravi et al., 2016) and the crop albedo can increase the electricity yield of the PV panels, particularly for bifacial modules (Fraunhofer ISE, 2020).

Agrivoltaic systems bring about most of the benefits of the so-called photovoltaic agriculture, which combines PV power generation with agricultural activities (Xue, 2017). On-farm production of solar electricity can boost the overall sustainability of farm activities by providing low-carbon energy used for wastewater purification, water pumping, controlled-environment buildings, and the cooling, drying and storage of agricultural products (Mekhilef et al., 2013; Xue, 2017; Weselek et al., 2019). In remote rural areas and in developing countries, AV systems can provide electricity to remote communities that have limited or no access to the electricity grid (Meah et al., 2008; Xue, 2017).

Before designing and developing an AV system, its integration into the landscape should be carefully considered. The design of a ground-mounted PV system is generally well-defined and solar farm installations are usually optimised for electricity production and land occupation following an energy-cost-oriented design approach. Consequently, in large ground-mounted PV arrays, the panels are arranged in rows, facing the optimal azimuth angle and have an optimal tilt angle based on the latitude of the site (Scognamiglio, 2016). This rigid structural design renders traditional PV plants aesthetically unattractive and poorly integrated into the landscape, and in general, they reduce the provision of ecosystem services from the land they occupy (Scognamiglio, 2016).

If properly designed and harmonised into the landscape, AV systems can respond better to the 'sustainable energy landscape approach' (Scognamiglio, 2016). This links energy and society in a design vision that aims to maximise food production, foster biodiversity, maintain landscape quality and favour ecosystem services. When considered from this point of view, AV systems can offer a more sustainable renewable energy solution than ground-mounted PV alternatives.

Having introduced AV systems and clarifying that 'photovoltaic agriculture' has many positive aspects, the following sections will focus on how system design, modelling and management activities can exploit the reciprocal benefits that can be reaped by combining agricultural activities with the production of solar electricity on the same land area.

2 Design of agrivoltaic systems to maximise the synergy between energy and agricultural production

The most crucial aspect in the development and expansion of sustainable AV systems is the optimisation of their design. Optimising PV systems has been dealt with extensively (Alsadi and Khatib, 2018), and their economic, energetic and environmental sustainability have also been well addressed (Tawalbeh et al., 2021).

The levelised cost of electricity (LCOE) is one of the most commonly used metrics when comparing different solutions for electricity generation (Branker et al., 2011). For PV systems, it is affected by various factors relative to their design and management. For ground-mounted PV systems, the optimisation process is mainly affected by economic evaluations that have supported the development of many PV plants with common design solutions (Scognamiglio, 2016).

A major factor in determining the construction costs of PV infrastructures and their environmental impact relative to resource depletion (Agostini et al., 2021) is the steel used for the supporting structure (IRENA, 2019). In a conventional ground-mounted PV system, the height of the panels from the ground is minimised to reduce the use of materials (Santra et al., 2017; Fraunhofer ISE, 2020). This renders the use of the land under the panels very difficult for most agricultural applications.

Moreover, the design of ground-mounted PV systems makes their integration into the landscape very difficult and significantly limits the provision of ecosystem services from the land they occupy (Scognamiglio, 2016). Recently, the possibility of growing plants that attract pollinators between the arrays of PV panels has been proposed as a measure to increase their environmental sustainability. A methodology to assess the impact of this management strategy in terms of increased yield of the crops cultivated in the land around the PV plants was proposed (Macknick et al., 2013; Ravi et al., 2016; Walston et al., 2018; Hernandez et al., 2019; Graham et al., 2021).

Agrivoltaic systems which facilitate the cultivation of agricultural crops underneath the PV panels, constitute a paradigm shift in the integration of PV infrastructures at a landscape level. Their design includes a set of parameters that go beyond those taken into account for ground-mounted PV systems. For example, the height at which the panels are mounted, the number of panels per array, the distance between arrays and the possibility of modulating these parameters are essential features for the growth and management of crops under the PV panels.

A minimum height from the ground is necessary to enable the access of machinery for agricultural operations (Dinesh and Pearce, 2016; Fraunhofer ISE, 2020). By ensuring freedom of movement for agricultural machinery, the choice of suitable crops is wider and crop rotation is similar to what is applicable in open-field conditions in the same region. Large cereal-harvesting machines require a height clearance of at least 5 m; therefore, a minimum height from the ground of 5–6 m is a requisite feature of an AV system that can then be integrated into conventional crop production systems.

The height of solar panels from the ground should also take into consideration the mature height of the crops that will be cultivated. A crop such as maize or sorghum should not exceed the height of the panels above the ground. Sekiyama and Nagashima (2019) showed that stilt-mounted AV systems (with panels placed well above the maximum height of traditional maize crops) can enable a wide range of commercial crops for which standard agricultural machinery is used. Perennial crops, grape vines and fruit trees can also be cultivated under AV systems, provided that their height can be easily managed, through pruning for example (Santra et al., 2017). In some circumstances, AV systems can be designed with PV panels installed at a relatively low height, for example when growing aloe vera (Ravi et al., 2016). Aloe vera has a crop cycle of around 5 years, short crop height, and a thick shallow root system that can maximise water-use efficiency when grown on drylands.

Existing AV prototypes and commercially available systems illustrate how crop choice and relative management practices can have an impact on design. The panel height above the ground can vary between 2.7 m (Sekiyama and Nagashima, 2019), 4 m (Dupraz et al., 2011; Marrou et al., 2013a), 5 m (Amaducci et al., 2018, Elamri et al., 2018a; Valle et al., 2018) and up to 8 m at the highest side of an inclined panel (Schindele et al., 2020; Fraunhofer ISE, 2020). Photovoltaic panels used may be monocrystalline, bifacial or thin-film (Dupraz et al., 2011; Valle et al., 2018, Cho et al., 2020, Fraunhofer ISE, 2020, Schindele et al., 2020; Trommsdorff et al., 2021). The panels can be fixed (Dupraz et al., 2011; Marrou et al., 2013a) or installed on single- or dual-axis sun-tracking systems (Elamri et al.; 2018a, Valle et al., 2018; Amaducci et al; 2018). The distance between the lower sides of two consecutive panels can vary between 0.7 m (Sekiyama and Nagashima, 2019) and 1.6 m (Dupraz et al., 2011; Valle et al., 2018; Sekiyama and Nagashima, 2019) and up to 3.2 m (Valle et al., 2018; Elamri et al., 2018a).

The optimisation of AV systems must consider a set of design solutions that affect the 'sustainability' of electricity production from PV panels in its broadest sense. At the same time, all the factors relevant to crop production should be considered. To date, very few studies have addressed the problem or proposed a methodology for the optimisation of AV systems. In a recent simulation study, Trommsdorff et al. (2021) analysed the effect of an AV system´s design on the availability of solar radiation to crops grown underneath, by calculating the land equivalent ratio (LER) as used by Dupraz et al. (2011) to calculate land productivity under an AV system.

$$\text{LER} = \frac{\text{crop yield in AV system}}{\text{crop yield in a monocrop system}} + \frac{\text{electricity output in AV system}}{\text{electricity output in PV system}}$$

LER values higher than 1 indicate a positive synergy between energy and food production on the same area for an AV system, as reported in several studies (Dupraz et al., 2011; Valle et al., 2018; Amaducci et al., 2018, Elamri et al., 2018a; Trommsdorff et al., 2021). This indicates that AV systems can significantly improve overall land use. The LER of an AV system further increases when solar-tracking PV systems are used instead of fixed panels (Valle et al., 2018; Amaducci et al., 2018; Elamri et al., 2018a).

Campana et al. (2021), in their seminal work, conducted a multi-object optimisation in order to explore the trade-off between competing AV key performance indicators. The design of vertically mounted AV systems with bifacial PV modules was optimised according to three objectives - electricity generation output, power fluctuations and LER - by varying the distance between PV arrays and their orientation.

This work highlighted the complexity of optimising AV systems. Besides electricity and land productivity, multi-objective optimisation should also include economic indicators. Studies have been conducted on a simple solar farm design with fixed, tilted bifacial PV panels (Trommsdorff et al., 2021) and on AV systems with vertically mounted bifacial modules (Campana et al., 2021). The design of appropriate AV systems should also include other relevant features that can affect crop yield and/or electricity generation including the PV panel area; the solar PV technology; transparency of the solar cells; the tilt angle; the type of system (whether fixed or tracking); the height of the PV panels above the ground; the pillars and the foundations.

In recent years, a reduction in the cost of solar PV panels has led to the construction of PV systems with larger area panels, reduced inter-row spacing, and consequently, increased ground coverage ratio (GCR, the ratio of total PV panel surface area to the ground area) (Sánchez-Carbajal et al., 2019; Fretzen et al., 2021). The GCR, indicated here as 'panel density,' is one of the most relevant parameters to be considered in the design of AV systems. It influences the level of shading and consequently the solar radiation levels available for crops to carry out photosynthesis. It is also the greatest determinant of annual electricity production per unit of land (Willockx et al., 2020).

The panel tilt angle also affects both electricity output and crop productivity, as does changing it dynamically with a solar tracking system. A single-axis tracker moves the solar modules horizontally, according to the elevation of the sun, or vertically, according to the azimuth, most commonly from east to west. A double-axis tracker orientates the modules both horizontally and vertically from both north to south and east to west, thus maximising generation (Valle et al., 2018), especially when compared with a fixed ground-mounted PV system (Seme et al., 2020).

The position of the panels in a tracking system can also be adjusted to enable easier access to large agricultural machinery (Fig. 3). Tracking systems can also enhance crop productivity by managing shading during specific growing stages when crop growth is favoured by either high or low levels of radiation.

Panel density alone is not sufficient to predict shading and crop productivity. Under the same panel density, crop and electricity production is significantly affected by PV technologies with contrasting transparency or energy conversion efficiency.

Transparency is a physical property that enables light to pass through the panels without interrupting the light transmission (Husain et al., 2018).

Figure 3 Panels mounted on double-axis trackers are kept horizontal to enable cereal harvesting with a conventional combine harvester (Courtesy RemTec https://www.remtec.energy/en/agrovoltaico/).

Semi-transparent panels have a transmission rate of 50% or more due to the distance between the silicon cells (if opaque cells are arranged widely apart, the panels become more transparent because of a larger surface of the glass) (Wong et al., 2008; Dinesh and Pearce, 2016). Semi-transparent panels reduce the shading of the plants beneath the panels, which can help to mitigate the negative effect of shading from AV systems (Dinesh and Pearce, 2016). Thin-film technology is among the most frequently employed for transparent PV panels (Husain et al., 2018). Light transmittance can be adjusted by the thickness of the film (Husain et al., 2018; Lee et al., 2020). In addition, the cost per unit area of thin-film modules is lower than monocrystalline and polycrystalline PV cells, which could help reduce the overall cost of an AV system (Fraunhofer ISE, 2020).

For a given panel density, the use of bifacial PV panels significantly affects the production of electricity per square metre and potentially that of crops too, compared to monofacial ones (Cho et al., 2020; Schindele et al., 2020). Bifacial panels increase the level of diffuse radiation available to the crops due to small gaps between single solar cells, which increases the transparency (Trommsdorff et al., 2021). In addition, the electricity output of bifacial panels is higher than in monofacial designs because the radiation reflected from the ground is intercepted by the rear side of the panel. As a result, bifacial panels enable AV system designs to have larger inter-row distances, which increases the levels of solar radiation available to crops while maintaining the same level of electricity generation per unit of land area (Schindele et al., 2020). Bifacial panels can be mounted either horizontally (facing the north/south direction) or vertically (facing the east/west direction), without affecting electricity production (Riaz et al., 2020). When installed vertically and spaced more than 6 m apart, regular agricultural practices can result in standard crop growth and development. Vertical mounting can also reduce dust and soil particles covering the surface of the panels (the tilt angle greatly affects this), which occurs as a result of tillage operations and machinery activity and reduces PV efficiency (Fraunhofer ISE, 2020; Riaz et al., 2020).

Tubular PV modules and organic photovoltaics (OPV) are promising innovative technologies for AV system applications (Meitzner et al., 2020). Tubular modules (©TubeSolar AG[1]) are flexible PV strips in a glass tube, which can be installed horizontally on supports and suspended over the cultivation area. This configuration renders the modules permeable to light and water which can then reach the crops, favouring uniform growth. Organic photovoltaics are still in the market launch phase and have low efficiency and durability (Fraunhofer ISE, 2020).

Other minor design features that can affect crop production and land accessibility are related to the supporting structures. The encumbrance by pillars that support PV panels should be minimised by using piled foundations or special anchoring with Spinnanker© (Fraunhofer ISE, 2020), or by using tensile structures such as Agrovoltaico® (as described in Agostini et al., 2021), instead of using large concrete blocks with environmental repercussions (Agostini et al., 2021). Harvesting operations are strongly influenced by the presence of pillars and other structures due to manoeuvring problems, leading to a loss of yield because approximately 8% of the arable land cannot be used (Fraunhofer ISE, 2020).

3 Physiological and agronomical aspects of crop cultivation under agrivoltaic systems

Agrivoltaic systems constitute a complex cultivation environment. The PV panels generate a maze of microclimatic conditions, and their supporting structures limit field access and the performance of agricultural operations. Full exploitation of the potential productivity of AV systems entails very specific choices and requires very intensive agronomic management. A critical issue is the selection and cultivation of species that can thrive in the micro-meteorological conditions generated by the AV system.

Beck et al. (2012) proposed a classification for the suitability of vegetable species for cultivation under AV systems based on the effect of shading on crop yield. Crops are divided into three groups:

1. crops tolerant to shade that do not show significant yield reduction or increase under shaded conditions;
2. neutral crops that are not significantly affected by light limitation, at least at a low level of shading; and
3. sensitive crops whose growth and productivity are significantly affected by shading.

1 © TubeSolar AG https://tubesolar.de/en/the-future-of-photovoltaics/

Information on the classification of crops according to these three groups, or in general based on their suitability for cultivation under shaded conditions, is scarce. Additional criteria to inform the selection of suitable species are based on physiological and morphological traits (Dupraz et al., 2011; Marrou et al., 2013a,b). The main traits and shade ranges that have been investigated on a selection of crops in scientific experiments carried out under actual or simulated AV conditions are summarised in Table 1.

Considering that data on the ability of crops to produce sustainable yield under the shaded conditions of AV are scarce (Dupraz et al., 2011; Sekiyama and Nagashima, 2019), it can be argued that the selection of suitable vegetable species to match AV conditions should be based on results from research carried out in agroforestry or on agricultural species cultivated under artificial shade (Weselek et al., 2019).

However, the radiative micro-environment generated by an AV system, usually characterised by an average reduction of incoming radiation, is actually dynamic, in both space and time (Amaducci et al., 2018; Slattery et al., 2018). Therefore, selecting a genotype for a given AV system based on its capacity to grow at a specific and constant level of shading might not be a suitable strategy. Another limitation in selecting a species for AV systems based on shade tolerance is the consideration that this is the ability of a species to survive at low light levels (Valladares and Niinemets, 2008), whereas under an AV system, the aim is to grow crops that not only survive but that can contribute to the overall sustainability by producing a marketable yield. It is therefore argued that crop selection should not be limited to shade-tolerant species but should involve agricultural crops that can provide a sustainable production under a specific set of AV conditions. Therefore, the choice could include crops other than shade-intolerant species, such as maize (Sekiyama and Nagashima, 2019), that have the capacity to produce a sustainable yield under the dynamic environment of AV systems. The complexity of the AV environment is determined by the combination of a specific set of design features (panel density, panel height, tracking system, etc.) as well as climatic factors and soil conditions. This renders the classification of crops for their suitability to AV conditions particularly complex and potentially misleading.

Collection of field data on crop responses under actual AV systems is essential, but the results cannot be generalized by applying them outside of the specific conditions in which they were obtained. It is therefore relevant to define a set of objective criteria that can be used for determining crop suitability to specific AV conditions, such as CO_2 light-response curves, phenotypic plasticity, specific leaf area (SLA) (ratio between leaf area and leaf dry mass), radiation interception efficiency and crop cover rate.

An evaluation of the suitability of vegetable species to be cultivated under AV conditions based on their light-response curves can be used to describe how the rate of photosynthesis varies as a function of light (Wang et al., 2017;

Table 1 List of crops and crop parameters that have been investigated in AV systems. The column 'shade range' indicated the level of shading that was experienced by the crop in the specific AV conditions reported by the authors

Crop	Crop parameters		Shade range	Reference
Lettuce	• Specific leaf area • Crop growth rate • Yield • Dry matter • Water-use efficiency • Fresh biomass • Leaf temperature • Stomatal conductance • Crop temperature • Crop cover rate • Crop cycle length	• Leaf number and emission rate • Evapotranspiration • Water productivity • Water consumption • Photosynthesis/transpiration ratio • Leaf length, thickness, width and dry mass • Radiation transmission efficiency • Radiation conversion efficiency • Radiation interception efficiency	20–50%	Marrou et al. (2013a,b,c); Valle et al. (2018); Elamri et al. (2018a)
Cucumber	• Leaf numbers and temperature • Leaf emission rate • Evapotranspiration	• Water-use efficiency • Soil cover rate • Dry matter accumulation • Crop cycle length	38–63%	Marrou et al. (2013a,b)
Potato	• Marketable yield		40%	Schindele et al. (2020)
Celeriac	• Diameter and weight of bulbs • Harvestable yield • Chemical composition	• Fresh matter • Dry matter • Leaf area index	30–40%	Schindele et al. (2020); Weselek et al. (2021)
Cherry tomato, jalapeño, chiltepin pepper	• Cumulative CO_2 uptake • Total fruit production • Water-use efficiency	• Net photosynthesis • Water loss through transpiration	N.D.	Barron-Gafford et al. (2019)
Tomato	• Yield • Water productivity	• Dry biomass • Plant water content • Evapotranspiration	N.D.	AL-agele et al. (2021)

Grass	• Yield • Dry biomass	• Water-use efficiency	40%	Hassanpour Adeh et al. (2018); Trommsdorff et al. (2021)
Maize	• Yield (grain and biomass) • Fresh weight	• Dry biomass • Crop revenues	5-50%	Amaducci et al. (2018); Sekiyama and Nagashima (2019)
Wheat	• Phenological stages • Crop temperature • Yield	• Dry matter content • Crop cycle length	29-57%	Dupraz et al. (2011); Marrou et al. (2013b); Schindele et al. (2020)
Grape	• Crop cycle length • Fruit quality: colouration • Sugar content	• Germination, bud -sprouting period and growth of flower	30%	Cho et al. (2020)
Flowers	• Bloom time	• Bloom abundance	25-95%	Graham et al. (2021)

Sekiyama and Nagashima, 2019; Willockx et al., 2020; Frauhnofer ISE, 2020;). In particular, CO_2 light-response curves are characterised by two meaningful parameters:

1 the point where photosynthetic activity of the plant equals respiration, defined as the light compensation point (LCP); and
2 the light saturation point (LSP) at which the photosynthetic rate reaches its maximum, above which further increments of light no longer increase assimilation.

Values of LCP and LSP vary among species, particularly depending on carbon assimilation pathways (C3, C4 or CAM), and within the same species at different phenological stages, in different environmental conditions (e.g. temperature), and in leaves present in different canopy layers (Kephart et al., 1992; Niinemets, 2007; Wang et al., 2017).

Plants with low LSP and LCP are light-demanding species that show low tolerance to shade conditions (Wang et al., 2017). Fraunhofer ISE (2020) hypothesised that crops with a low LSP are the most suitable for growing under an AV system. On this basis, the selection of species to be cultivated should be restricted to C3 plants that have a low LCP and LSP and are more shade-tolerant, but are less photosynthetically efficient than C4 plants (Kephart et al., 1992). Different values of LCP and LSP for the major crops were reported by Tazawa (1999). Generally, shade-sensitive crops (a third group as proposed by Beck et al., 2012) need an average daily photosynthesis activation radiation intensity (PAR) greater than LCP to achieve marketable yield (Wang et al., 2017). The relative yield reduction observed in various AV experimental conditions has been reported for several crops (Fig. 4). Data come from different conditions, which makes it difficult to compare results accurately, and only general conclusions on the relationship between crop production and shading can be drawn. In most cases, with a reduction of radiation intensity between 20% and 40%, yield reduction was between 20% and 25%. For grass, celeriac, potato and wheat crops, even with a radiation intensity reduction as high as 40%, a yield increment was observed. These data highlight that there is a strong variation in response to shading between crop species.

Maize is generally considered a shade-sensitive crop and is the only C4 species in the published literature that has been evaluated under AV systems in field experiments as well as in a simulation modelling study. The modelling study indicated that the yield of rainfed maize on average was higher under a set of AV conditions than under full light conditions, and the yield advantage under AV became highly significant under drought conditions (Amaducci et al., 2018). In the field trial (Sekiyama and Nagashima, 2019), maize was grown under two levels of shading at low and high PV panel densities and compared

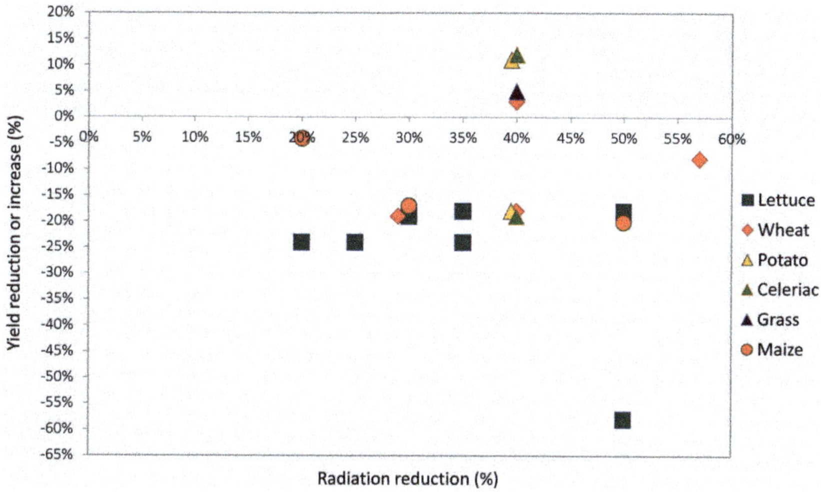

Figure 4 Relative yield reduction (%) compared to radiation reduction (%) for selected crops cultivated under AV conditions. (Data obtained from cited references.)

to plants grown under full light control. Surprisingly, the highest yield was obtained under the low panel density configuration, while at high density, the yield was slightly lower than under full light conditions. This confirmed the result of the modelling study and further validated the possibility of growing maize under AV conditions, especially considering that in the trial, the maize was irrigated and therefore not subjected to drought stress. These results indicate that a small fraction of the total incident radiation is required for plants to reach their maximum rate of photosynthesis (Sekiyama and Nagashima, 2019).

In a simulation similar to that conducted by Amaducci et al. (2018), maize grown in the Po Valley, Italy, showed yield reductions to be a function of the reduction in radiation obtained (Fig. 5). The radiation reduction level (RRL) was estimated from the relationship:

$$RRL = 100 * \frac{(rFl - rAV)}{rFL}$$

where rFl is the PAR intensity accumulated between crop emergence and harvest in full light conditions, and rAV is the mean PAR estimated in an AV system (https://www.remtec.energy/). The vertical bars represent the standard deviation of the relative yield reduction.

The simulation was conducted in the absence of water stress, to highlight only the effect of radiation. This simulated yield reduction for maize is in agreement with that reported by Sekiyama and Nagashima (2019) and confirms

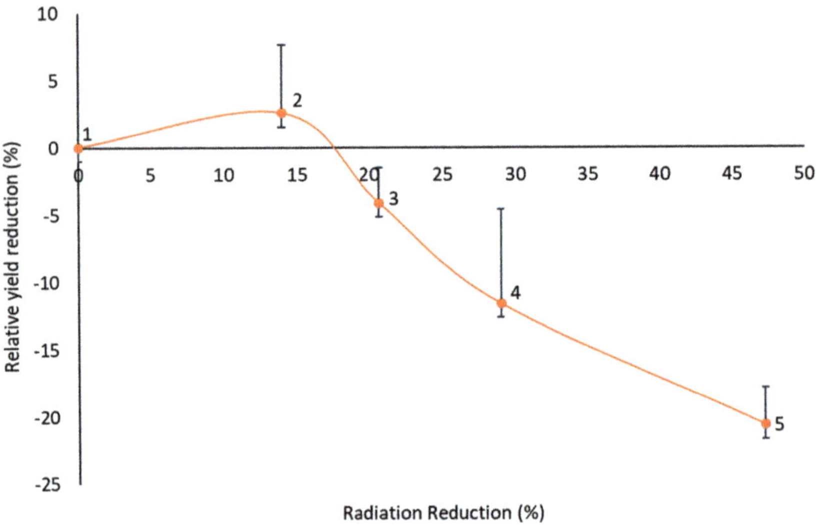

Figure 5 Relationship between radiation reduction (%) from shade and the simulated relative yield reduction (%) for a maize crop cultivated in the Po Valley, Italy (Amaducci et al., 2018), with data taken from four AV systems with increasing panel area/ground area ratios between 0.12 and 0.45.

that, in conditions of high irradiation such as around the Mediterranean, a significant reduction in radiation can result in a crop yield reduction of less than 20%. Therefore the shading conditions that are generated under AV systems should be examined in the context of the topology of full radiation and shaded areas. Diffuse radiation plays a fundamental role in photosynthesis because, in most situations, crops use diffuse light more efficiently than direct light (Brodersen et al., 2008; Li and Yang, 2008).

Phenotypic plasticity (PP) is the phenotypic adaptation of a genotype to the variation of environmental conditions (Gratani, 2014; Sage and McKown, 2005; Valladares and Niinemets, 2008; Nicotra et al., 2010). It is relevant when selecting species to grow under AV conditions (Marrou et al., 2013c). The PP is related to leaf traits including radiation conversion efficiency, net photosynthetic rate, SLA, leaf chlorophyll content, leaf lifespan and leaf insertion angle. These change due to the light gradient along the canopy layers and the decrease of the red/far red ratio (Gratani, 2014). Phenotypic plasticity in shade-tolerant species is generally lower than in shade-intolerant species, but it differs according to the leaf trait patterns (Portsmuth and Niinemets, 2007; Valladares and Niinemets, 2008).

Plasticity increases light capture by improving plant response in different environmental conditions such as adaptation to drought conditions or limited solar radiation (Valladares and Niinemets, 2008). A reduced potential in PP was reported for C4 species compared to C3 species. In fact, when the characteristics of acclimatisation to low light are considered, C4 plants seem to be lacking in one or more traits when compared to C3 species (Sage and McKown, 2005).

The adaptation of crops to shaded conditions consists of an increase in SLA (cm^2 g^{-2}), a decrease of the chlorophyll a/b ratio, and an increase in total chlorophyll content under low light conditions, which can increase the carbon gain in a shaded environment (Valladares and Niinemets, 2008). Leaves tend to show a high value of SLA in the shade to improve light interception (Gratani, 2014). The SLA reflects leaf thickness and the difference between sun leaves and shade leaves; it depends on leaf internal structure that performs an important role in light capture (Evans, 1999).

A relevant feature to consider under AV conditions is radiation interception, which depends on incident radiation, leaf area index and the extinction coefficient of canopy architecture (Sadras et al., 2016). In AV systems, the crop radiation interception efficiency (RIE) represents the major compensation mechanism to achieve high yield in shaded conditions (Marrou et al., 2013c). The RIE is the plant's ability to cover the soil under the PV panels, also indicated as 'crop cover rate' (Marrou et al., 2013c). Crops that show high plasticity of cover rate (proportion of soil area covered by the crop) are able to intercept more light in shaded conditions (Valladares and Niinemets, 2008; Marrou et al., 2013c). In AV systems, selecting crops with a high soil cover rate increases light capture, increases photosynthetic efficiency when there is a light reduction and reduces soil evaporation (Marrou et al., 2013a). The cover rate of crops contributes to an increased water-use efficiency at certain stages of the crop cycle. At the onset of vegetative growth, the plants cover the soil less efficiently because the canopy is not fully developed. During the vegetative stage, crops cover the soil efficiently and help reduce soil evaporation. The cover rate also depends on the plants acclimatising to shaded conditions and to the rapidity of their vegetative growth (Marrou et al., 2013a).

Lettuce is an example of a crop that can exploit a high cover rate or a high RIE by increasing the total plant leaf area, optimizing leaf canopy architecture to harvest light more efficiently, and reducing leaf angle (if the leaf angle is close to horizontal orientation, the crop would cover a significant proportion of the ground and intercept more light) (Marrou et al., 2013b,c). Conversely, cucumber and wheat crops grown under AV conditions did not show a high cover rate or significant morphological changes, indicating low plasticity to shaded conditions (Marrou et al., 2013b).

Despite this, the adaptation of cropping practices in AV systems can contribute to finding the optimum solution to cultivate crops underneath an array of panels. For example, an increase in the RIE can be obtained by selecting the optimum planting distance (by increasing inter- and intra-row spacing) in order to avoid the leaves' self-shading (Marrou et al., 2013c).

Additional synergies between electricity production and agronomic management in AV systems are related to the cooling of the PV panels, the modification of surface albedo (the fraction of the total solar radiation that is reflected by a surface) and the reduction of soiling, which can all have significant effects on electricity production.

Soiling, which is the accumulation of material on PV panels that reduces their power output and performance, is influenced by environmental conditions, and, in particular, by the presence of dust and its properties (i.e. the size, shape and weight of dust particles). Electricity output can technically be reduced from 2% to 50% by dust deposition on PV panels (Maghami et al., 2016). Agricultural management of AV systems can play a role in reducing the soiling effect by influencing the level of dust production and accumulation (Ravi et al., 2016). For instance, the implementation of conservation agriculture that promotes permanent soil cover and minimal or no soil disturbance (such as minimum tillage or no-tillage) can significantly reduce the production of dust and at the same time provide additional benefits such as increased water- and nutrient-use efficiencies and reduced land degradation (Baker et al., 2005)

The application of conservation agricultural practices could also contribute to the preservation or improvement of soil properties. Information on the effect of AV systems on soil conditions is limited. The impact of rain on the soil, which can be a major determinant of compaction in bare soils, can be influenced by the presence of overhead PV panels. In AV systems with one or two axes where the panel angles can be adjusted, Elamri et al. (2018b) indicated that keeping the panel horizontal during a rainfall event increases the spatial heterogeneity of rainfall distribution on the soil, with a potential increase in runoff and soil erosion.

Access to machinery can result in soil compaction (Soane et al., 1981). Under AV systems, this problem can be exacerbated by periodic maintenance operations that require heavy machinery. On the other hand, the presence of supporting infrastructures, which limit free movement in the field, confines all vehicles accessing the field to permanent traffic lanes. Implementation of controlled-traffic farming, which is a strategy to reduce the problem of soil compaction (Tim Chamen et al., 2015), could be particularly effective for AV systems.

Air temperature plays a key role in affecting PV module efficiency, and the implementation of cooling strategies can significantly increase electricity outputs (Dwivedi et al., 2020). In AV systems, crops generate a passive cooling effect on PV panels, thanks to plant transpiration (the release of water from aerial organs of the plant to increase heat dissipation). Transpiration is a natural cooling mechanism that can dissipate up to 32.9% of the total absorbed solar energy by the leaves (Othman et al., 2020). Barron-Gafford et al. (2019) demonstrated that during the cultivation of cherry tomato, chiltepin pepper and jalapeno under AV systems, the PV panel temperature was significantly cooler than in a traditional ground-mounted array. The temperature decrease documented by Barron-Gafford et al. (2019) in the growing months of May–July led to a 3% increase in electricity generation over those months and a 1% annual increase.

Another aspect that influences PV output, especially for bifacial modules, is the surface albedo. The performance of bifacial panels is positively affected by albedo and by the installation height of the panels.

Bifacial panels generate more electricity than similarly sized monofacial panels by producing electricity from both sides of the panels (Rodríguez-Gallegos et al., 2020; Guo et al., 2013). Crop albedo depends on leaf characteristics, such as colour, presence of surface waxes, orientation and trichome density (Doughty et al., 2010), plant height, agricultural practices, soil moisture and vegetation cover (Todd and Hoffer, 1998). It varies within the crop growth cycle among phenological stages (Oguntunde et al., 2007; Oguntunde et al. 2004). Mulching techniques can influence the surface albedo and thereby also the surface energy budget. Fan et al. (2014) showed that grassland albedo was increased by 23.5% and 33.9% on clear and cloudy days, respectively, when it was covered by agricultural white plastic film. Therefore, combining the use of a white mulch to avoid weed infestation and in conjunction having crops growing, and not bare ground, can be a strategy to increase albedo, and consequently electricity production. In turn, bifacial panels can improve the availability of sunlight for crops by multiplying the reflection of incoming light to the ground. Schindele et al. (2020) reported a gain in electricity generation of 8% for bifacial panels in an AV system with potato, wheat or celeriac crops in the first year of operation.

4 Crop modelling applications for the management of agrivoltaic systems

Since the 1970s, the application of modelling has complemented agronomic studies to support the understanding and interpretation of the physical and physiological phenomena (such as soil, nitrogen, water dynamics, radiation interception, photosynthesis) involved in crop production. However, the pioneering phase, in which crop models were mainly aimed at estimating crop production, has passed. The available modelling solutions enable the development of computing platforms. These are used for a multitude of scientific and technical applications, for example, to support decisions related to crop management (Thorp et al., 2008, Bonfante et al., 2019); to aid plant breeding (Cooper et al., 2014); and to perform scenario analysis, for example, when related to the impact of climate change.

Research on AV systems is in its infancy and the application of modelling to understand the response of crop growth and development under AV conditions has been extensive (Dupraz et al., 2011; Marrou et al., 2013a; Dinesh and Pearce, 2016; Valle et al., 2018; Elamri et al., 2018a; Amaducci et al., 2018). The complexity of AV systems and the trade-offs that exist between electricity generation and crop production suggest that the use of computing platforms should not be limited to the management of AV systems but should also include their design. This is particularly important when the construction and operation of AV systems is constrained by regulations or by specific agronomic requirements. In this context, modelling can be performed to simulate how various design parameters (such as PV panel dimension, height of the panels from the ground, panel density, sun-tracking configurations, plant layout and interaction with site topology) can affect the microclimate under the panels and particularly the level of radiation. Computing platforms using simulated micro-meteorological data will therefore be able to simulate the potential physiological and productive responses of crops cultivated underneath the panels. In principle, it is not necessary to develop a new crop model specifically designed for an AV system, given that numerous crop models with varying levels of sophistication are available (Di Paola et al., 2015). Crop models to be used for applications related to AV systems can implement either complex mechanistic functions or simple empirical functions. Mechanistic models implementing sophisticated algorithms that simulate the response of individual physiological processes to environmental variables have been successfully applied to AV systems to model crop response to the daily variation in radiation. The mechanistic model 'STICS' was used for the first time by Dupraz et al., (2011) to simulate crop productivity under the shade of PV panels. 'GECROS' was used by Amaducci et al. (2018) for a long-term simulation of maize cultivation under AV in Italy. For large-scale simulations that include a complex scenario analysis requiring high computational complexity, a simple model is appropriate. Campana et al. (2021) aim at developing a techno-economic optimisation model for AV systems with the 'EPIC' model, implemented for its low parameterisation requirements.

4.1 Simulation aspects in agrivoltaic platforms

What makes AV simulation particularly complex is the extreme dynamism of the environmental variables in the cultivation area under the panels. Shade conditions vary continuously throughout the day, with significant effects on the soil and microclimate. Marrou et al. (2013b) and Barron-Gafford et al. (2019) identified that the variability of micro-climatic conditions under the panels resulted with respect to full light, detectable both in hourly observations during a day as well as during the entire season. The reality is likely to be even more complex than reported by Marrou et al. (2013b) because the daily shading dynamics vary considerably with space and time at ground level. This appears in ground radiation maps (Dupraz et al., 2011, Amaducci et al., 2018, Elamri et al., 2018a) and has significant effects on the balance of ground radiation, temperature and evapotranspiration (Elamri et al., 2018a; Amaducci et al., 2018).

The capacity of a model to handle complex simulations should be evaluated considering the specific objectives of the simulations and the available resources (Liman Harou et al., 2021). Considering the modelling work with regard to the AV systems already performed in this context, for any modelling objective, good radiation estimations and an efficient system to map shading generated by the PV panels are essential prerequisites. Coupling radiation data (measured or simulated) with crop models enables an estimation of plant growth with a level of precision that is similar to that achievable under normal conditions in the open field.

Crop simulations under AV systems can be performed to verify whether certain construction choices, such as panel density or height from the ground, can guarantee the level of radiation necessary to achieve a certain production target. The simulation can be performed for specific crops or for entire cropping systems following defined agronomic management, while the target production level can be dictated by economic, social or political constraints.

Simulation models could also greatly facilitate management operations, as presented in the work of Elamri et al. (2018a) who developed a model to support irrigation in an AV environment.

Regardless of the specific objectives of the modelling activity, the fundamental steps necessary to launch a simulation are:

1 create a radiation dataset for the specific AV environment;
2 complete the dataset with all other microclimate variables;
3 select soil and crop parameters; and
4 run the simulation.

The realization of time-related radiation maps and dynamic shading models implies computational strategies for:

- ray-tracing and shading;
- definition of the areas cultivated on the ground with an estimate of all the shading trajectories in each single portion (plot); and
- estimation of direct and diffuse radiation.

The above-mentioned steps use rather well-established computational technologies that rely on the light-interception literature for the estimation of electricity production of solar panels (Quaschning and Hanitsch, 1995, Khan et al., 2017, Mousazadeh et al., 2009) and calculation of ground radiation (Dupraz et al., 2011; Diaz et al., 2015; Riaz et al., 2020).

In the creation of a radiation dataset, an important feature is the spatial resolution for the simulations. In this regard, two options are available:

1. Create input datasets (hourly or daily) that spatially integrate the different radiation conditions. The calculations are considerably simplified but the simulation loses the possibility to distinguish the effects of contrasting values of radiation that are present within the same plot.
2. Create radiation (direct and diffuse) datasets for each element of a mosaic of small pixel-plots (e.g. 0.12 m) as performed by a multi-year study on maize (Amaducci et al., 2018). This could be performed on all or just a few 'representative' plots of the cultivated area, as performed for a lettuce simulation (Marrou et al., 2013b; Dinesh and Pearce, 2016). Both methods have advantages and disadvantages.

Here, the calculations are demanding in terms of computational resources, and time consuming, especially for hourly (or shorter) time steps. This option can be used in scientific studies that aim at understanding the effect of variability under an AV system when managed as vegetable gardens, with multiple crops present at the same time. In this case, the position of each crop under the PV panels can be decided according to the spatial variation of meteorological parameters and the specific requirements of each species.

Besides radiation, for other climatic variables needed for completing climate datasets (including air temperature, relative humidity, wind speed and rain), two solutions were adopted in the literature:

- To use actual data measured under AV systems in the specific positions where the simulations were carried out;
- To use the data of a weather station positioned nearby the AV system but in the open air, assuming that the influence of AV infrastructures on microclimatic conditions, with the exception of radiation, is negligible.

'Adjustments' can be made to this simplification, as done, for example, by Elamri et al. (2018a) who implemented a model to simulate a precipitation pattern modified according to PV constructive details. So far, no study has been published for the simulation of the microclimatic conditions under AV systems in which all the variables, including changes in air humidity, wind speed and their interactions with the AV system, are integrated into a microclimate-model, though a model similar to this was proposed for AV greenhouses by Fatnassi et al. (2015).

Another important limitation relating to the measurement of microclimatic variables in the literature on AV systems is that measurements have been carried out in pilot AV plants, which are probably too small to capture the effect that full-scale AV systems can have on meteorological variables.

Despite this limitation, once an appropriate climatic dataset is available, the simulations of crop growth under AV environments can be carried out in the same way as for normal crop models. However, the effectiveness of the model will depend on the parameterisation and the collection of sufficient data for the necessary operations of calibration and validation.

4.2 Design and optimisation

As already pointed out, model exploitation goes beyond the potential of simulating the effect of shading and microclimate on crop production. Advanced use of the models can be of value in the design phase to create an AV system design suitable for a given location (annual solar irradiation and its seasonality), a specific cropping system and a set of specific economic, environmental and social constraints. Altogether, these aspects can only be addressed with a calculation platform that combines simulation models (for the crops and for the meteorological parameters, including radiation) with multi-objective optimisation systems.

- A paradigmatic example addressing the optimisation problem in AV systems was proposed by Campana et al. (2021), in which a set of variables including the LER and economic and constructive features were used as driving variables to identify optimal AV solutions. To analyse most of the trade-offs involved in the design and operation of AV systems, the optimisation algorithm maximises LER, minimises the fluctuation of electricity injected into the grid in terms of annual standard deviation (STD) and maximises the annual electricity production. Although this study was carried out for a northern European environment in Sweden, with vertical panels, interesting conclusions emerged that underline the broad potential of this approach.
- By studying the contributions of LER (i.e. the crop yield and the electricity production contributions), it became evident that the optimal row distance varies according to the crop. This leads to important consequences in terms of the long-term optimal design of AV systems.
- An optimal design of AV systems should consider multi-year and multi-crop simulations and optimisation based on conventional farm activities.
- LER cannot be used as the main variable for the optimal design of AV systems. More objective functions should be included for a better estimation of the synergies between crop and electricity production. Maximising the LER tends to drastically reduce electricity production and this could undermine the economic sustainability of the system.

5 Conclusion and future trends in research

Agrivoltaic systems are expected to be resilient energy/food systems that, by combining the production of food with that of renewable energy, could provide valuable solutions to major future challenges for humanity. They have the potential to support the multiple Sustainable Development Goals (SDGs) mainly 'Affordable and Clean Energy' (SDG 7), 'climate action' (SDG 13) and, when applied in developing countries or where food production is insufficient, 'Zero Hunger' (SDG 1). In addition, the possibility of developing AV systems in urban and peri-urban areas (Majumdar and Pasqualetti, 2018) renders them relevant for 'Sustainable Cities and Communities' (SDG 11), while growing species under the shade of the panels that attract pollinators or that increase biodiversity, making AV systems compatible to 'Life on Land' (SDG 15).

There is an urgent need to transition to renewable energy sources, and AV systems appear to be a valid contributor within the renewable sector. Therefore, it is essential that additional research needed to validate their contribution to the above-mentioned SDGs is facilitated and accelerated.

This chapter has outlined the promising results obtained from preliminary research on AV systems. It has highlighted the need to co-optimise electricity and food production and evaluated the effectiveness of AV systems to help meet global environmental goals.

The benefits of AV systems have been found across the food-energy-water nexus, particularly where water is a limiting factor (Barron-Gafford et al, 2019). Future research and development will help to identify the conditions under which AV systems can increase water-use efficiency and mitigate water stress, particularly in areas that will be negatively affected by climate change (McAusland et al., 2016).

The availability of AV installations is limited and experiments to validate their potential under various environmental conditions, and with a large number of crops and cultivation systems, are expensive and time consuming. Therefore, at least in the short term, simulation studies will be used extensively. Future research is needed to calibrate and validate available models, which can thereafter be used to optimise AV systems and, more specifically, simulate the conditions under which they can:

- lower the impact of heat and high radiation stress to increase the resilience of the agricultural sector against the threat of climate change;
- stabilise and increase crop production;
- turn barren and degraded land into viable areas for agricultural production and ecosystem service provision;
- maximise land-use efficiency and energy gain from high solar radiation; and
- reduce GHG emissions by avoiding fossil fuel combustion.

Significant improvements in optimising AV systems will be achieved when modelling solutions developed specifically for the AV environment become available. They will enable simulation of the physical environment, particularly the microclimate and soil moisture dynamics, and the capacity of a crop to acclimatise to the AV system by adjusting its morphology and physiology.

These research goals are mainly focussed on the productivity of AV systems. The impact that cultivation under AV systems can have on the quality of the crops can be evaluated. So far, only a few studies have addressed this issue (e.g. the chemical composition of celeriac by Weselek et al. (2021)). Cultivation of horticultural crops, fruit trees and even grapevines has high potential under AV systems, given the effect that AV conditions can have on quality parameters such as fruit colour and sugar content (Cho et al., 2020).

Most experiments to date have focussed on the effect of shading on crop production, on the water-use efficiency, and on the identification of shade-tolerant species. Little attention has been paid to the modulation of other production factors such as plant nutrition or on the incidence of pests and diseases (Weselek et al., 2021).

Thus, AV systems can play a pivotal role in the transition toward the deployment of renewable energy systems. Guidelines for their development will require multi-disciplinary research. Engineers and plant scientists should work together to optimise AV system designs; agronomists must elaborate appropriate crop rotation and cultivation techniques; and landscape designers should propose strategies to integrate AV systems into the landscape so that they can respond to the 'sustainable energy landscape approach' and help meet the sustainable development goals.

6 Where to look for further information

6.1 Key conferences, institutes and websites

- Agrivoltaics conference: https://www.agrivoltaics-conference.org/.
- Agrivoltaics - Fraunhofer Institute for Solar Energy Systems ISE: https://www.ise.fraunhofer.de/en/key-topics/integrated-photovoltaics/agrivoltaics.html.
- Sustainable agrivoltaics - ENEA: https://www.agrivoltaicosostenibile.com/en/.
- The Colorado Agrivoltaic Learning: https://www.coagrivoltaic.org/.
- SANDBOX SOLAR: https://sandboxsolar.com/agrivoltaics/.
- Wikipedia: https://en.wikipedia.org/wiki/Agrivoltaic.
- **RemTec** https://remtec.energy/en.
- **Sun'Agri** https://sunagri.fr/en/.
- **BayWa r.e.** https://www.baywa-re.com/en/solar-projects/agri-pv.
- **Next2Sun** https://www.next2sun.de/en/homepage/.

7 References

Agostini, A., Colauzzi, M. and Amaducci, S. (2021). Innovative agrivoltaic systems to produce sustainable energy: an economic and environmental assessment. *Applied Energy* 281, 116102.

Alsadi, S. and Khatib, T. (2018). Photovoltaic power systems optimization research status: a review of criteria, constrains, models, techniques, and software tools. *Applied Sciences* 8(10), 1761.

AL-agele, H. A., Proctor, K., Murthy, G., and Higgins, C. (2021). A case study of tomato (Solanum lycopersicon var. Legend) production and water productivity in agrivoltaic systems. *Sustainability* 13(5), 2850. https://doi.org/10.3390/su13052850

Amaducci, S., Yin, X. and Colauzzi, M. (2018). Agrivoltaic systems to optimise land use for electric energy production. *Applied Energy* 220, 545–561.

Baker, J. B., Southard, R. J. and Mitchell, J. P. (2005). Agricultural dust production in standard and conservation tillage systems in the San Joaquin valley. *Journal of Environmental Quality* 34(4), 1260–1269, 10.2134/jeq2003.0348.

Barron-Gafford, G. A., Pavao-Zuckerman, M. A., Minor, R. L., Sutter, L. F., Barnett-Moreno, I., Blackett, D. T., Thompson, M., Dimond, K., Gerlak, A. K., Nabhan, G. P. and Macknick, J. E. (2019). Agrivoltaics provide mutual benefits across the food-energy-water nexus in drylands. *Nature Sustainability* 2(9), 848–855.

Beck, M., Bopp, G., Goetzberger, A., Obergfell, T., Reise, C. and Schindele, S. (2012). Combining PV and food crops to agrophotovoltaic - optimization of orientation and harvest. *Solar Energy Conference and Exhibition*, 4096–4100.

Bonfante, A., Terribile, F. and Bouma, J. (2019). Refining physical aspects of soil quality and soil health when exploring the effects of soil degradation and climate change on biomass production: an Italian case study. *SOIL* 5(1), 1–14.

Branker, K., Pathak, M. J. M. and Pearce, J. M. (2011). A review of solar photovoltaic levelized cost of electricity. *Renewable and Sustainable Energy Reviews* 15(9), 4470–4482.

Brodersen, C. R., Vogelmann, T. C., Williams, W. E. and Gorton, H. L. (2008). A new paradigm in leaf-level photosynthesis: direct and diffuse light are not equal. *Plant, Cell and Environment* 31(1), 159–164.

Campana, P., Stridh, B., Amaducci, S. and Colauzzi, M. (2021). Optimization of vertically mounted agrivoltaic systems. (preprint). Available at: https://arxiv.org/abs/2104.02124.

Campillo, C., Fortes, R. and Henar Prieto, M. D. (2012). *Solar Radiation Effect on Crop Production. Solar Radiation*, 167–189.

Cho, J., Park, S. M., Park, A. R., Lee, O. C., Nam, G. and Ra, I. (2020). Application of photovoltaic systems for agriculture: a study on the relationship between power generation and farming for the improvement of photovoltaic applications in agriculture. *Energies* 13(18), 4815.

Cooper, M., Messina, C. D., Podlich, D., Totir, L. R., Baumgarten, A., Hausmann, N. J., Wright, D. and Graham, G. (2014). Predicting the future of plant breeding: complementing empirical evaluation with genetic prediction. *Crop and Pasture Science* 65(4), 311.

Dale, V. H., Kline, K. L., Wright, L. L., Perlack, R. D., Downing, M. and Graham, R. L. (2011). Interactions among bioenergy feedstock choices, landscape dynamics, and land use. *Ecological Applications* 21(4), 1039–1054.

Diaz, M., Wang, L., Li, D., Zhao, X., Conrad, B., Soeriyadi, A., Gerger, A., Lochtefeld, A., Ebert, C., Opila, R., Perez-Wurfl, I. and Barnett, A. (2015). Tandem GaAsP/SiGe on Si solar cells. *Solar Energy Materials and Solar Cells* 143, 113–119. 10.1016/j.solmat.2015.06.033.

Di Paola, A., Valentini, R. and Santini, M. (2015). An overview of available crop growth and yield models for studies and assessments in agriculture. *Journal of the Science of Food and Agriculture* 96(3), 709–714.

Dinesh, H. and Pearce, J. M. (2016). The potential of agrivoltaic systems. *Renewable and Sustainable Energy Reviews* 54, 299–308.

Doughty, C. E., Field, C. B. and McMillan, A. M. S. (2010). Can crop albedo be increased through the modification of leaf trichomes, and could this cool regional climate? *Climatic Change* 104(2), 379–387.

Dupraz, C., Marrou, H., Talbot, G., Dufour, L., Nogier, A. and Ferard, Y. (2011). Combining solar photovoltaic panels and food crops for optimising land use: towards new agrivoltaic schemes. *Renewable Energy* 36(10), 2725–2732.

Dwivedi, P., Sudhakar, K., Archana Soni, A., Solomin, E. and Kirpichnikova, I. (2020). Advanced cooling techniques of P.V. modules: a state of art. *Case Studies in Thermal Engineering* 21, 100674.

Elamri, Y., Cheviron, B., Lopez, J.-M., Dejean, C. and Belaud, G. (2018a). Water budget and crop modelling for agrivoltaic systems: application to irrigated lettuces. *Agricultural Water Management* 208, 440–453.

Elamri, Y., Cheviron, B., Mange, A., Dejean, C., Liron, F. and Belaud, G. (2018b). Rain concentration and sheltering effect of solar panels on cultivated plots. *Hydrology and Earth System Sciences* 22(2), 1285–1298.

Evans, J. R. (1999). Leaf anatomy enables more equal access to light and CO2 between chloroplasts. *New Phytologist* 143(1), 93–104.

Fan, X., Chen, H., Xia, X. and Yu, Y. (2014). Increase in surface albedo caused by agricultural plastic film. *Atmospheric Science Letters* 16(3), 291–296.

Fatnassi, H., Poncet, C., Bazzano, M. M., Brun, R. and Bertin, N. (2015). A numerical simulation of the photovoltaic greenhouse microclimate. *Solar Energy* 120, 575–584.

Fraunhofer Institute for Solar Energy Systems ISE (2020). Agrivoltaics: opportunities for agriculture and the energy transition. Available at: https://www.ise.fraunhofer.de/en/publications/studies/agrivoltaicsopportunities-for-agriculture-and-the-energy-transition.html.

Fretzen, U., Ansarin, M. and Brandt, T. (2021). Temporal city-scale matching of solar photovoltaic generation and electric vehicle charging. *Applied Energy* 282, 116160.

Goetzberger, A. and Zastrow, A. (1982). On the coexistence of solar-energy conversion and plant cultivation. *International Journal of Solar Energy* 1(1), 55–69.

Graham, M., Ates, S., Melathopoulos, A. P., Moldenke, A. R., DeBano, S. J., Best, L. R. and Higgins, C. W. (2021). Partial shading by solar panels delays bloom, increases floral abundance during the late-season for pollinators in a dryland, agrivoltaic ecosystem. *Scientific Reports* 11(1), 7452.

Gratani, L. (2014). Plant phenotypic plasticity in response to environmental factors. *Advances in Botany* 2014, 1–17.

Guo, S., Walsh, T. M. and Peters, M. (2013). Vertically mounted bi-facial photovoltaic modules: a global analysis. *Energy* 61, 447–454.

Hassanpour Adeh, E., Selker, J. S. and Higgins, C. W. (2018). Remarkable agrivoltaic influence on soil moisture, micrometeorology and water-use efficiency. *PLoS ONE* 13(11), e0203256.

Hernandez, R. R., Armstrong, A., Burney, J., Ryan, G., Moore-O'Leary, K., Diédhiou, I., Grodsky, S. M., Saul-Gershenz, L., Davis, R., Macknick, J., Mulvaney, D., Heath, G. A., Easter, S. B., Hoffacker, M. K., Allen, M. F. and Kammen, D. M. (2019). Techno-ecological synergies of solar energy for global sustainability. *Nature Sustainability* 2(7), 560–568.

Husain, A. A. F., Hasan, W. Z. W., Shafie, S., Hamidon, M. N. and Pandey, S. S. (2018). A review of transparent solar photovoltaic technologies. *Renewable and Sustainable Energy Reviews* 94, 779–791.

IEA (2020). Renewables 2020, IEA, Paris. https://www.iea.org/reports/renewables-2020.

IRENA, International Renewable Energy Agency (2019). *Future of Solar Photovoltaic: Deployment, Investment, Technology, Grid Integration and Socio-Economic Aspects (A Global Energy Transformation: Paper*, International Renewable Energy Agency, Abu Dhabi. ISBN 978-92-9260156-0.

Irie, N., Kawahara, N. and Esteves, A. M. (2019). Sector-wide social impact scoping of agrivoltaic systems: a case study in Japan. *Renewable Energy* 139, 1463–1476.

Jäger-Waldau, A. (2019). EUR 29938 EN, Publications Office of the European Union, Luxembourg, 2019, ISBN 978-92-76-12608-9, doi: 10.2760/326629, JRC118058.

Kephart, K. D., Buxton, D. R. and Taylor, S. E. (1992). Growth of C3 and C4 perennial grasses under reduced irradiance. *Crop Science* 32(4), 1033–1038.

Khan, M. R., Hanna, A., Sun, X. and Alam, M. A. (2017). Vertical bi-facial solar farms: physics, design, and global optimization. *Applied Energy* 206, 240–248.

Lee, K., Um, H. D., Choi, D., Park, J., Kim, N., Kim, H. and Seo, K. (2020). The development of transparent photovoltaics. *Cell Reports Physical Science* 1(8), 100143.

Li, T. and Yang, Q. (2008). Advantages of diffuse light for horticultural production and perspectives for further research. *Frontiers in Plant Science* 6, 704.

Liman Harou, I., Whitney, C., Kung'u, J. and Luedeling, E. (2021). Crop modelling in data-poor environments – A knowledge-informed probabilistic approach to appreciate risks and uncertainties in flood-based farming systems. *Agricultural Systems* 187, 103014.

Macknick, J., Beatty, B. and Hill, G. (2013). Overview of opportunities for co-location of solar energy technologies and vegetation. *NREL-TP-60240*, National Renewable Energy Laboratory, Golden, CO.

Maghami, M. R., Hizam, H., Gomes, C., Radzi, M. A., Rezadad, M. I. and Hajighorbani, S. (2016). Power loss due to soiling on solar panel: a review. *Renewable and Sustainable Energy Reviews* 59, 1307–1316.

Majumdar, D. and Pasqualetti, M. J. (2018). Dual use of agricultural land: introducing 'agrivoltaics' in Phoenix Metropolitan Statistical Area, USA. *Landscape and Urban Planning* 170, 150–168.

Marrou, H., Dufour, L. and Wery, J. (2013a).How does a shelter of solar panels influence water flows in a soil-crop system? *European Journal of Agronomy* 50, 38–51.

Marrou, H., Guilioni, L., Dufour, L., Dupraz, C. and Wery, J. (2013b). Microclimate under agrivoltaic systems: is crop growth rate affected in the partial shade of solar panels? *Agricultural and Forest Meteorology* 177, 117–132.

Marrou, H., Wery, J., Dufour, L. and Dupraz, C. (2013c). Productivity and radiation use efficiency of lettuces grown in the partial shade of photovoltaic panels.*European Journal of Agronomy* 44, 54–66.

McAusland, L., Vialet-Chabrand, S., Davey, P., Baker, N. R., Brendel, O. and Lawson, T. (2016). Effects of kinetics of light-induced stomatal responses on photosynthesis and water-use efficiency. *New Phytologist* 211(4), 1209–1220.

Meah, K., Fletcher, S. and Ula, S. (2008). Solar photovoltaic water pumping for remote locations. *Renewable and Sustainable Energy Reviews* 12(2), 472–487.

Meitzner, R., Schubert, U. S. and Hoppe, H. (2020). Agrivoltaics-the perfect fit for the future organic photovoltaics. *Advanced Energy Materials* 11(1), 2002551.

Mekhilef, S., Faramarzi, S. Z., Saidur, R. and Salam, Z. (2013). The application of solar technologies for sustainable development of agricultural sector. *Renewable and Sustainable Energy Reviews* 18, 583–594.

Mousazadeh, H., Keyhani, A., Javadi, A., Mobli, H., Abrinia, K. and Sharifi, A. (2009). A review of principle and suntracking methods for maximizing solar systems output. *Renewable and Sustainable Energy Reviews* 13(8), 1800–1818.

Nicotra, A. B., Atkin, O. K., Bonser, S. P., Davidson, A. M., Finnegan, E. J., Mathesius, U., Poot, P., Purugganan, M. D., Richards, C. L., Valladares, F. and van Kleunen, M. (2010). Plant phenotypic plasticity in a changing climate. *Trends in Plant Science* 15(12), 684–692.

Niinemets, U. (2007). Photosynthesis and resource distribution through plant canopies. *Plant, Cell and Environment* 30(9), 1052–1071.

Oguntunde, P. G., Olukunle, O. J., Ijatuyi, O. A. and Olufayo, A. A. (2007). A semi-empirical model for estimating surface albedo of wetland rice field. *Agricultural Engineering International: The CIGR Ejournal* IX, 1–12.

Oguntunde, P. G. and Van de Giesen, N. (2004). Crop growth and development effects on surface albedo for maize and cowpea fields in Ghana, West Africa. *International Journal of Biometeorology* 49(2), 106–112.

Othman, N. F., Yaacob, M. E., Mat Su, A. S., Jaafar, J. N., Hizam, H., Shahidan, M. F., Jamaluddin, A. H., Chen, G. and Jalaludin, A. (2020). Modeling of stochastic temperature and heat stress directly underneath agrivoltaic conditions with *Orthosiphon stamineus* crop cultivation. *Agronomy* 10(10), 1472.

Portsmuth, A. and Niinemets, Ü (2007). Structural and physiological plasticity in response to light and nutrients in five temperate deciduous woody species of contrasting shade tolerance. *Functional Ecology* 21(1), 61–77.

Quaschning, V. and Hanitsch, R. (1995). *Shade Calculations in Photovoltaics Systems.*

Ravi, S., Macknick, J., Lobell, D., Field, C., Ganesan, K., Jain, R., Elchinger, M. and Stoltenberg, B. (2016). Colocation opportunities for large solar infrastructures and agriculture in drylands. *Applied Energy* 165, 383-392.

Riaz, M. H., Imran, H. and Butt, N. Z. (2020). Optimization of PV Array density for fixed tilt bi-facial solar panels for efficient agrivoltaic systems. 2020 47th IEEE Photovoltaic Specialists Conference (PVSC).

Rodríguez-Gallegos, C. D., Liu, H., Gandhi, O., Singh, J. P., Krishnamurthy, V., Kumar, A., Stein, J. S., Wang, S., Li, L., Reindl, T. and Peters, I. M. (2020). Global techno-economic performance of bi-facial and tracking photovoltaic systems. *Joule* 4(7), 1514–1541.

Sadras, V. O., Villalobos, F. J. and Fereres, E. (2016). Radiation interception, radiation use efficiency and crop productivity. In: *Principles of Agronomy for Sustainable Agriculture* Villalobos, F. and Fereres, E. (Eds), Springer, Cham.

Sage, R. F. and McKown, A. D. (2005). Is C4 photosynthesis less phenotypically plastic than C3 photosynthesis? *Journal of Experimental Botany* 57(2), 303–317.

Sánchez-Carbajal, S. and Rodrigo, P. M. (2019). Optimum array spacing in grid-connected photovoltaic systems considering technical and economic factors. *International Journal of Photoenergy* 2019, 1–14.

Santra, P., Pande, P., Kumar, S., et al. (2017). Agri-Voltaics or solar farming: the concept of integrating solar PV based electricity generation and crop production in a single land use system. *International Journal of Renewable Energy Research* 7(2), 694–699.

Schindele, S., Trommsdorff, M., Schlaak, A., Obergfell, T., Bopp, G., Reise, C., Braun, C., Weselek, A., Bauerle, A., Högy, P., Goetzberger, A. and Weber, E. (2020). Implementation of agrophotovoltaics: techno-economic analysis of the price-performance ratio and its policy implications. *Applied Energy* 265, 114737.

Scognamiglio, A. (2016). 'Photovoltaic landscapes': design and assessment. A critical review for a new transdisciplinary design vision. *Renewable and Sustainable Energy Reviews* 55, 629–661.

Sekiyama, T. and Nagashima, A. (2019). Solar sharing for both food and clean energy production: performance of agrivoltaic systems for corn, A typical shade-intolerant crop. *Environments* 6(6), 65.

Seme, S., Štumberger, B., Hadžiselimović, M. and Sredenšek, K. (2020). Solar photovoltaic tracking systems for electricity generation: a review. *Energies* 13(16), 4224.

Slattery, R. A., Walker, B. J., Weber, A. P. M. and Ort, D. R. (2018). The impacts of fluctuating light on crop performance. *Plant Physiology* 176(2), 990–1003.

Soane, B. B., Blackwell, P. S., Dickson, J. W. and Painte, D. J. (1981). Compaction by agricultural vehicles: a review II. Compaction under tyres and other running gear. *Soil and Tillage Research* 1, 373–400.

Spinnanker. Available at: www.spinnanker.com.

Tawalbeh, M., Al-Othman, A., Kafiah, F., Abdelsalam, E., Almomani, F. and Alkasrawi, M. (2021). Environmental impacts of solar photovoltaic systems: a critical review of recent progress and future outlook. *Science of the Total Environment* 759, 143528.

Tazawa, S. (1999). Effects of various radiant sources on plant growth: part 1. *Japan Agricultural Research Quarterly*, 163–176.

Thorp, K. R., DeJonge, K. C., Kaleita, A. L., Batchelor, W. D. and Paz, J. O. (2008). Methodology for the use of DSSAT models for precision agriculture decision support. *Computers and Electronics in Agriculture* 64(2), 276–285.

Tim Chamen, W. C., Moxey, A. P., Towers, W., Balana, B. and Hallett, P. D. (2015). Mitigating arable soil compaction: a review and analysis of available cost and benefit data. *Soil and Tillage Research* 146, 10–25.

Todd, S. W. and Hoffer, R. M. (1998). Response of spectral indices to variations in vegetation cover and soil background. *Photogrammetric Engineering and Remote Sensing* 64(9), 915–921.

Trommsdorff, M., Kang, J., Reise, C., Schindele, S., Bopp, G., Ehmann, A., Weselek, A., Högy, P. and Obergfell, T. (2021). Combining food and energy production: design of an agrivoltaic system applied in arable and vegetable farming in Germany. *Renewable and Sustainable Energy Reviews* 140, 110694.

Valladares, F. and Niinemets, Ü. (2008). Shade tolerance, a key plant feature of complex nature and consequences. *Annual Review of Ecology, Evolution, and Systematics* 39(1), 237–257.

Valle, B., Simonneau, T., Sourd, F., Pechier, P., Hamard, P., Frisson, T., Ryckewaert, M. and Christophe, A. (2018). Increasing the total productivity of a land by combining mobile photovoltaic panels and food crops. *Applied Energy* 206, 1495–1507.

Van de Ven, D. J., Capellan-Peréz, I., Arto, I., Cazcarro, I., de Castro, C., Patel, P. and Gonzalez-Eguino, M. (2021). The potential land requirements and related land use change emissions of solar energy. *Scientific Reports* 11(1), 2907.

Vella, H. (2021). Power technology. Available at: https://www.power-technology.com/features/inside-the-worlds-largestdam-based-floating-solar-power-project/.

Walston, L. J., Mishra,S. K., Hartmann, H. M., Hlohowskyj, I., McCall, J. and Macknick, J. (2018). Examining the potential for agricultural benefits from pollinator habitat at solar facilities in the United States. *Environmental Science and Technology* 52(13), 7566–7576.

Wang, D., Sun, Y., Lin, Y. and Gao, Y. (2017). Analysis of light environment under solar panels and crop layout. 2017 IEEE 44th Photovoltaic Specialist Conference (PVSC).

Weselek, A., Ehmann, A., Zikeli, S. et al. (2019). Agrophotovoltaic systems: applications, challenges, and opportunities. *A review. Agron. Sustain. Dev.* 39, 35. https://doi.org/10.1007/s13593-019-0581-3.

Weselek, A., Bauerle, A., Hartung, J. et al. (2021). Agrivoltaic system impacts on microclimate and yield of different crops within an organic crop rotation in a temperate climate. *Agron. Sustain. Dev.* 41, 59. https://doi.org/10.1007/s13593-021-00714-y.

Willockx, B., Herteleer, B., Ronsijn, B., Uytterhaegen, B., and Cappelle, J. (2020). A standardized classification and performance indicators of agrivoltaic systems. *EU PVSEC Proceeding*. http://dx.doi.org/10.4229/EUPVSEC20202020-6CV.2.47.

Wong, P., Shimoda, Y., Nonaka, M., Inoue, M., and Mizuno, M. (2008). Semi-transparent PV: thermal performance, power generation, daylight modelling and energy saving potential in a residential application. *Renewable Energy* 33(5), 1024–1036. https://doi.org/10.1016/j.renene.2007.06.016

Xue, J. (2017). Photovoltaic agriculture - new opportunity for photovoltaic applications in China. *Renewable and Sustainable Energy Reviews* 73, 1-9. https://doi.org/10.1016/j.rser.2017.01.098

Part 3

Energy-smart farming: case studies

Chapter 9

Tools and technologies to reduce fossil energy use on dairy farms

Philip Shine and Michael D. Murphy, Munster Technological University, Ireland; and John Upton, Teagasc Moorepark, Ireland

1 Introduction

A 22% growth in global milk production is forecasted for the ten-year period between 2018 and 2027 (FAO 2018). This increased production is required to meet the projected 19% increase in the global consumption of milk and dairy produce by 2050, compared to the 2005–2007 levels (Bruinsma and Alexandratos, 2012). This increase in demand is largely fuelled by the projected economic growth across developing countries, where a 46% increase in the consumption of milk and dairy products is expected by 2050 compared to the 2005–2007 levels (Bruinsma and Alexandratos, 2012). From an environmental perspective, the increased dairy herd required to meet this increased demand comes with its own significant challenges regarding the consumption of on-farm energy resources and greenhouse gas (GHG) emissions.

On dairy farms, energy consumption is composed of both direct on-farm energy use, such as electricity and tractor fuel, and the indirect energy required to produce and deliver fertilisers, machinery, concentrate feed, and other necessary inputs. (Shine et al. 2020). In the absence of energy-efficient managerial strategies and/or farm infrastructure, increasing milk production will result in an increased consumption of energy resources, greater production costs and emission of GHGs. The magnitude and efficiency of energy input

http://dx.doi.org/10.19103/AS.2022.0100.16

per unit of milk produced (e.g. MJ/kg milk solids) will vary according to the level of production, the type of production system, the type of milking system employed, milking schedule, farm infrastructure as well as environmental conditions (O'Brien et al. 2012; Upton et al. 2013; Shortall et al. 2016; Shine et al. 2018a). The sustainable utilisation of energy in dairy farming has both environmental and economic consequences:

- Environmentally: the use of grid-sourced electricity, natural gas, liquefied petroleum gas (LPG) and liquid fossil fuels on dairy farms emit GHGs and negatively impact the local environment and air quality.
- Economically: increased future energy costs due to an overreliance on fossil fuels and grid-sourced electricity could potentially reduce future profits (Upton et al. 2013). In addition, due to the volatile nature of milk prices, dairy farmers must also prepare for potential periods of reduced revenue as a result of the global supply of, and demand for, milk and dairy produce (Shine et al. 2020).

Another challenge facing dairy farmers is the management of greater numbers of livestock in a herd while producing less GHG emissions and avoiding impacts on waterways and groundwater. Researchers are aiming to identify optimum technologies and strategies to help improve energy efficiency on-farm while ensuring that these technologies are financially beneficial to the farmer, in order to improve their likelihood of adoption (Smith et al. 2007; Lanigan et al. 2018).

This chapter summarises and assesses literature relating to the monitoring and prediction modelling of energy consumption and energy saving strategies, and outlines the development of the Dairy Energy Decision Support Tool (Murphy et al. 2018). This tool allows dairy farmers, researchers and policy-makers to assess the monetary and environmental feasibility and the applicability of various energy-efficient and renewable energy technologies on a farm-specific basis.

2 Energy monitoring

Research has largely focused on the physical monitoring and subsequent analysis of energy consumption on dairy farms to establish key performance indicators related to energy use. Monitoring energy usage is essential for benchmarking consumption metrics; comparing energy inputs across various farming systems; making international comparisons; and identifying the major contributors to overall energy use in order to target energy mitigation efforts. Often incorporated into wider lifecycle analyses (LCA), the monitoring of total energy consumption includes direct and indirect energy use (Barnett and Russell, 2010; von Keyserlingk et al. 2013; Upton et al. 2013). In addition, comparisons of energy consumption between conventional and organic farming practices have been made as have comparisons between confinement (where cows are housed indoors all year round) and pasture-based farming systems.

Internationally, a range of key performance indicators relating to milk analyses include litres of milk (l), kilograms of milk (kg), fat and protein-corrected milk (FPCM) (IDF 2010), energy-corrected milk (ECM) (Sjaunja et al. 1990), and kilograms of milk solids (kg MS). The most common functional unit that has been used in recent literature is ECM, and therefore it is used here, when possible, for comparative purposes.

2.1 Total energy

Shine et al. (2020) reviewed 36 studies that monitored energy consumption on dairy farms across 17 countries, while differentiating between energy demand in conventional, organic, confined and pasture-based systems. In confined systems, cows are housed indoors all year round, while on pasture-based farms, cows are predominantly grazing outdoors and feed on fresh grass. Although they found a disparity between reported key performance indicators, 23 of these studies either reported total energy use in terms of per kg energy-corrected milk (MJ/kg ECM), or reported fat and protein percentage values allowing for conversions between usage metrics (Table 1).

Total energy use was 54% greater on conventional farms compared to organic farms, primarily because conventional farms consume 75% more in indirect energy – from fertiliser and agrichemical inputs – while also consuming 13% less direct energy on average. In conventional dairy farms, confinement farms (Conv-c) required 68% greater energy resources compared to grazing systems (Conv-g). Conversely, organic confined farms (Org-c) required 28% less energy resources compared to grazing systems (Org-g).

Table 1 Mean energy consumption on dairy farms and characteristics from a number of studies (Shine et al. 2020)

	Conventional		Organic	
Characteristic	Conv-g*	Conv-c	Org-g	Org-c
Number of studies	5	11	5	2
Total energy (MJ /kg ECM)	2.8	4.7	2.9	2.1
Mean	4.1		2.7	

*Conv-g, conventional grazing farm; Conv-c, conventional confinement farm; Org-g, organic grazing farm; Org-c, organic confinement farm.

2.2 Direct energy

On conventional farms, direct energy accounted for 29% of the total energy consumption on average, compared with 50% on organic farms (Shine et al. 2020). The share of direct energy due to electrical energy consumption was similar, at 48% and 56% for conventional and organic farms respectively, with the remainder due to liquid and/or gaseous fuel use.

Electrical energy is consumed on-farm mainly for milk cooling (refrigeration), milk harvesting (vacuum pumps) and water heating, as well as for water pumping, lighting and other miscellaneous usage throughout the farm. On average, across conventional farming systems, 48.9 Wh of electrical energy is consumed per kg of milk (Wh /kg). Milk cooling is the greatest consumer of electricity (31%), followed by milk harvesting (29%) and water heating (19%), (Fig. 1) (Edens et al. 2003; Hörndahl 2008; Murgia et al. 2013; Upton et al. 2013; Rajaniemi et al. 2017; Shine et al. 2018a; Todde et al. 2018a). Five out of seven of these studies were carried out on farms with herringbone milking systems, while the remaining two studies did not mention the type of milking design employed.

On average, Conv-c farms used 45% more electricity per kg milk when compared to Conv-g farms (Fig. 1). This was a result of a 40% increased demand for milk cooling, 119% for milk harvesting and 14% for water heating.

The electricity consumption on farms employing automatic milking systems (AMS) was also assessed, which showed increased milk-harvesting electricity use (Hörndahl 2008; Calcante et al. 2016; Rajaniemi et al. 2017;

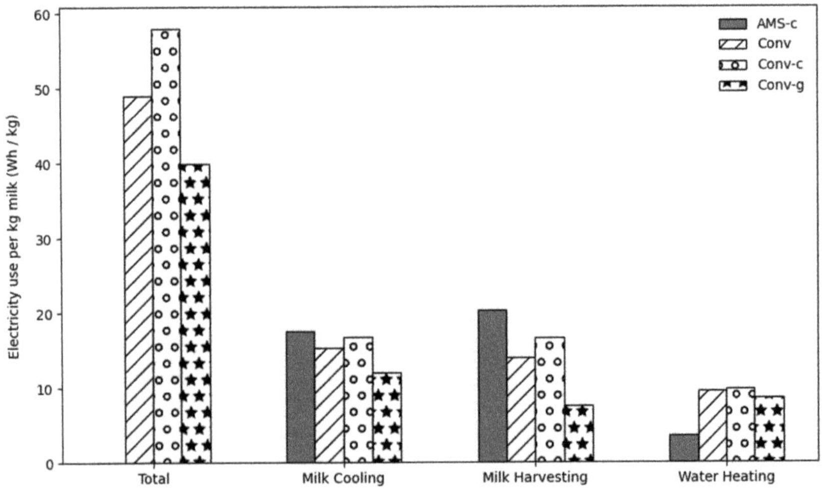

Figure 1 Electricity consumption for confined, grazed and average conventional farming systems and comparison with automatic milking machine removal systems

Shortall et al. 2018). The electricity used for water heating in AMS farms was 62% less than in conventional farms, possibly due to AMS having a fixed daily water heating procedure, where only the required volume of water is heated each day, thereby reducing excess energy demand. Shortall et al. (2018) found that 60.8 Wh/kg of electricity was consumed, which was 24% greater compared to farms without AMS.

Murgia et al. (2013) conducted a partial LCA to quantify the energy intensity and GHG emissions associated with electricity and diesel use on 20 Italian dairy farms in 2011. They found that diesel fuel contributed 70% to the total direct energy consumption, and electricity 30%. Todde et al. (2018a) analysed the direct energy use of 285 Italian dairy farms through a detailed survey of electricity, diesel and LPG consumption and the related on-farm activities and processes. Diesel fuel consumption was 40 kg/t FPCM, LPG demand was 0.4 kg/t FPCM, and electricity use was 73 Wh/kg FPCM. In agreement with Murgia et al. (2013), diesel fuel accounted for 72% of the total direct energy requirement while electricity accounted for 27%. On Irish dairy farms, 60% of direct energy consumption was attributed to electricity (Upton et al. 2013). However, the total electricity consumption on Italian farms was only 13% greater than the 39.1 Wh/kg FPCM, as calculated by Upton et al. (2013). This suggests that more diesel fuel is used on Italian dairy farms.

Liquid and gaseous fuels include diesel, kerosene, natural gas, LPG and lubricants. They are utilised on-farm for powering tractors and machinery, water heating and other miscellaneous use throughout the farm.

Only a small number of studies assessed petroleum fuel demands on dairy farms (Table 2), thus making it difficult to arrive at useful conclusions.

Regarding diesel fuel use, confined farms (five conventional and one AMS farm) consumed 1.0 MJ/kg ECM compared with grazed farms at 0.2 MJ/kg ECM. However, higher diesel fuel consumption was measured on pasture-based farms, due to the increased tractor requirement for pasture management (Cederberg and Mattson 2000).

Table 2 Consumption of petroleum-based fuels per kg ECM based on various studies

Fuel	Number of studies	MJ/kg ECM	References
Diesel	6	0.73	Hospido et al. 2003; Meul et al. 2007; Thomassen et al. 2008; Upton et al. 2013; Sefeedpari et al. 2014, 2015; Todde et al. 2018a
Kerosene	2	0.02	Upton et al. 2013; Sefeedpari et al. 2014, 2015
LPG	1	0.02	Todde et al. 2018a
Lubricants	2	0.03	Meul et al. 2007; Upton et al. 2013

Wells (2001) calculated liquid fuels to be responsible for 20% of the total primary energy consumption on the average New Zealand pastoral dairy farm, equating to 0.40 MJ/kg ECM, whereas in Ireland, O'Brien et al. (2012) calculated the liquid energy consumption on grazed dairy systems (Conv-g) to be similar, at 0.30 MJ/kg ECM, and higher than that for confined systems (Conv-c), at 0.18 MJ/kg ECM.

2.3 Indirect energy

Indirect energy consumption is categorised as either being ancillary or embodied energy (Table 3). The energy utilised for the production of agricultural products utilised on the farm (e.g. fertiliser and feed) fall under the ancillary energy category, while embodied energy refers to the energy consumed in the mining of raw materials for, and the production of capital inputs, averaged across their expected working life (Wells 2001). Such capital farm inputs include the construction of buildings and facilities, and the production of machinery and equipment.

Baldini et al. (2017) reported that embodied energy calculations are rarely included in LCA studies, due to the lack of available data. On average across conventional farms (Conv-c and Conv-g), the use of fertiliser required 0.7 MJ/kg ECM, while the production of feed required 2.3 MJ/kg ECM. Pagani et al.

Table 3 Indirect energy inputs for ancillary fertiliser manufacture and concentrated feed, and embodied energy values for buildings and machinery for conventional (Conv) and organic (Org) confined (c) and grazed (g) dairy farms

	Ancillary		
	System	No. of studies	MJ /kg ECM
Fertiliser manufacture and delivery	Org-c	1	0.00
	Org-g	1	0.00
	Conv-c	4	0.59
	Conv-g	4	0.85
Concentrated feed and delivery	Org-c	1	0.85
	Org-g	1	1.22
	Conv-c	5	2.85
	Conv-g	2	0.97
	Embodied		
Buildings & Facilities	Conv	3	0.21
Machinery & Equipment	Conv-g	1	0.27
	Org-c	1	0.39
	Org-g	1	0.62
	Conv-c	5	0.65

(2016) attributed no energy demand to fertiliser use across six organic farms (0.0 MJ kg^{-1} ECM), as the regulatory requirement for organic certification is through utilising cow manure produced on the farm or pig/poultry manure purchased elsewhere and brought in. For the production and delivery of concentrate feed, organic farming systems required 1.0 MJ/kg ECM, 55% less than conventional farming systems.

Fertiliser used on grazed farms (Conv-g) (Wells 2001; O'Brien et al. 2012; Upton et al. 2013; Pagani et al. 2016) required 44% more energy on average compared to confined farms (Conv-c) (Meul et al. 2007; O'Brien et al. 2012; Pagani et al. 2016; Todde et al. 2018b). The energy used for supplying feed on confined farms (Conv-c) (Meul et al. 2007; Sefeedpari et al. 2014; Sefeedpari et al. 2015; Pagani et al. 2016; Aguirre-Villegas et al. 2017) was almost three times more per kg ECM than for grazed pasture farms (Conv-g) (Upton et al. 2013; Pagani et al. 2016). Overall, Conv-c farms reported 90% higher ancillary energy demand than Conv-g farms.

The average embodied energy used on conventional dairy farms was 0.8 MJ/kg ECM. This encompasses 0.2 MJ/kg ECM for the construction of building and facilities (Wells 2001; Kraatz 2012; Todde et al. 2018b) plus 0.6 MJ/kg ECM for the manufacture of machinery and equipment (Meul et al. 2007; Kraatz 2012; Sefeedpari et al. 2014; Sefeedpari et al. 2015; Pagani et al. 2016; Todde et al. 2018b).

3 Energy modelling

When energy consumption predictions are required for a single farm, using average values per kg ECM may prove problematic because energy use can vary with country, farming system, milk-cooling equipment, water-heating technology, vacuum pump type, managerial strategies and other variables. LCAs often require the physical metering of dairy farm inputs, such as liquid fuel usage and electricity consumption, which requires the installation of expensive and time-consuming metering equipment. The development of prediction models for dairy farm energy consumption has enabled bypassing the requirement for physical metering equipment. The modelling methods evident across the dairy energy literature include mechanistic modelling, regression modelling and machine-learning.

3.1 Mechanistic modelling

In Ireland, a mechanistic model for electricity consumption on dairy farms (MECD), capable of predicting electricity consumption, cost and related GHG emissions, has been developed (Upton et al. 2014). It has a monthly prediction resolution, with the prediction variable being total electricity demand (kWh). It is a mathematical representation of the seven main infrastructural systems on a dairy farm (milk cooling, milk harvesting, water heating, lighting, water pumping, wash pumping and winter housing). The MECD was developed through Microsoft Excel, and calculates electricity consumption from first principals, which requires a large number of input variables. These include milk-to-water ratio, water temperatures, milking times, milk-collection interval, hot water temperature set point, and water pump motor sizes. Mathematical equations representing the seven infrastructural systems were constructed using a 12-month × 24-h matrix structure, and the MECD was validated on three farms sizes (small – 45 cows; medium – 88 cows; and large – 195 cows). It was found to predict annual electricity consumption with a relative prediction error (RPE) of less than 10%, thus providing excellent prediction capability (Fuentes-Pila et al. 1996).

3.2 Regression

Regression models have been widely applied in dairy energy literature in order to find a pattern for energy use in dairy farming systems. Edens et al. (2003) developed four multiple linear regression (MLR) models to predict the annual electricity consumption of the vacuum pumps, water heaters, refrigeration compressors, and air compressors used on dairy farms in the USA, as well as a fifth MLR model to predict their combined consumption. Using data from a single farm (herd size of 160 cows) throughout a 14-year period, the MLR models were developed using data related to the volume and quality of milk produced, the number of lactating cows and dry bulb air temperature values. In addition, five variable selection methods were utilised to help maximise the prediction capability of each MLR model in order to predict monthly the unseen electricity consumption for each individual component. Prediction capabilities varied from being able to explain 18% of the variation in energy consumption of the air compressors ($R^2 = 0.18$), to being able to explain 74% of the variation of milk cooling energy use ($R^2 = 0.74$) (Table 4).

Eden et al. (2003) noted that leaks in the air system could have been the primary cause of the poor predictive capability of the air compressor data, and concluded that the quantity of milk had a greater impact on dairy farm electricity consumption than the quality of milk produced, with the number of cows milked and the ambient temperature having weaker statistical impacts. The model validation method that was utilised involved only the known data, whereby the R^2 value between predictions and data used to develop the MLR coefficients was calculated. This method risks the effect of over-estimating the prediction capability of the model if it were used to estimate energy consumption on new data.

In South Africa, Mhundwa et al. (2017) utilised data from a single dairy farm (mean of 500 cows) monitored over three months to develop two MLR

Table 4 Comparison of regression models developed for predicting energy consumption on dairy farms

Study	Country	Res	Prediction variable	Validation method	R^2	RMSE	RPE (%)
Edens et al. 2003	USA	M	Milk harvesting (kWh)	n/a	0.44	n/a	n/a
Edens et al. 2003	USA	M	Milk cooling (kWh)	n/a	0.74	n/a	n/a
Edens et al. 2003	USA	M	Water heating (kWh)	n/a	0.34	n/a	n/a
Edens et al. 2003	USA	M	Air compressors (kWh)	n/a	0.18	n/a	n/a
Edens et al. 2003	USA	M	Combined (kWh)	n/a	0.62	n/a	n/a
Sefeedpari et al. 2014	Iran	A	Output energy of milk (MJ /cow)	Test set (20%)	0.11	0.2	n/a
Mhundwa et al. 2017	South Africa	D	Morning milk cooling (kWh)	Test set (30%)	0.92	n/a	n/a
Mhundwa et al. 2017	South Africa	D	Evening milk cooling (kWh)	Test set (30%)	0.9	n/a	n/a
Todde et al. 2017	Italy	A	Total electricity use (kWh)	LOOCV	n/a	n/a	11.4
Todde et al. 2017	Italy	A	Total diesel use (kg)	LOOCV	n/a	n/a	15
Shine et al. 2018c	Ireland	M	Total electricity use (kWh)	10-fold CV	0.72	543	26.1

Res, prediction resolution; M, monthly; A, annual; D, daily; LOOCV, leave-one-out cross-validation; 10-fold CV, 10-fold cross-validation; n/a, information not available.

models to predict the milk-cooling-related electricity consumption of a direct expansion (DX) bulk milk tank without pre-cooling. The first model predicted the electricity requirements of the morning milking while the second predicted the electricity requirements of the evening milking. The MLR models were developed using input data related to the volume of milk produced, the milking room and ambient temperatures, and relative humidity. The model was developed using 70% of the overall dataset, while the remaining 30% was utilised to test model accuracy. The R^2 values were 0.92 and 0.90 for the MLR models when estimating the electricity consumption for milk cooling at the morning and evening milking respectively,

Todde et al. (2017) utilised data from 285 Italian dairy farms to develop dairy energy prediction (DEP) models using polynomial regression to predict annual electricity and diesel fuel consumption. For both electricity and diesel model development, the authors assessed the herd size, the number of lactating cows, milk production and land area (hectares) as input variables. The methodology employed a univariate statistical method (correlation matrix) for variable selection, while also incorporating second and third-order polynomial terms for herd size and number of lactating cows. In addition, they employed 'Leave-one-out cross-validation' (LOOCV) to measure the prediction accuracy of the developed models. LOOCV is an adaptation of 'ten-fold cross-validation', where the number of folds equals the number of data points, therefore requiring far greater computational power. Through this LOOCV method, the authors were able to predict the annual electricity consumption of the Italian farms to within 11.4% (RPE), and annual diesel consumption to within 15.0% (RPE).

Shine et al. (2018c) employed MLR modelling to predict total monthly dairy electricity consumption using consumption data monitored on 56 Irish dairy farms throughout the January 2014–May 2016 period. MLR modelling was carried out utilising farm details related to milk production, cow numbers, infrastructural equipment, farm management processes and environmental conditions. Predicting consumption values at the monthly resolution allowed for consumption trends throughout the year to be modelled. In total, 15 farm variables were assessed, and the subset of variables that maximised the prediction accuracy of unseen electricity consumption was selected by applying a univariate variable selection technique and variance inflation factor (VIF) in conjunction with all subset regression and ten-fold cross validation. The final subset of variables was found to be herd size, the volume of milk produced, whether an ice bank DX or DX milk bulk tank was used, whether ground water was utilised for pre-cooling milk, the number of air compressors, the frequency of hot washing and the total water heater volume. This MLR model predicted monthly dairy electricity consumption to within 26% (RPE). A standardised regression analysis was carried out, whereby milk production and herd size had the largest impact on electricity consumption.

Similar to the varying functional units utilised for assessing energy consumption (as detailed in Section 2), there also exists a variety of model assessment criteria and methods used in the literature. The most common model assessment criteria were the coefficient of determination (R^2), root mean square error (RMSE), and RPE (Table 4). However, not all studies report all performance metrics and not all studies estimate prediction accuracy on unseen data using k-fold cross-validation or LOOCV, which makes comparisons between studies difficult.

3.3 Machine learning

Machine learning algorithms have also been employed for the prediction of energy consumption on dairy farms, as summarised in Table 5.

Energy demands on Iranian dairy farms were predicted through the development of an artificial neural network (ANN) (Sefeedpari et al. 2013) and adaptive neural-fuzzy inference system (ANFIS) models (Sefeedpari et al. 2014). Detailed surveys were carried out to quantify the amount of fossil fuels (diesel, gasoline, kerosene and natural gas) and electricity that were consumed in conjunction with the volume of milk produced, amount of cow manure, farm area and cow numbers. Energy intensity values were then utilised to calculate the total energy consumption values. The authors utilised 60% of the original dataset for model training, 20% for validation and 20% for testing. The ANN and ANFIS models predicted energy consumption with R^2 values of 0.88 and 0.79 respectively.

Shine et al. (2018b) tried to improve upon the prediction accuracy of dairy farm electricity consumption achieved by MLR modelling by employing various machine-learning algorithms. Using the same dataset employed by Shine et al. (2018a), the applicability of predicting monthly dairy farm electricity consumption was assessed using a support vector machine (SVM) algorithm, a CART decision tree algorithm, a random forest ensemble algorithm, and an ANN. The methodology employed backward sequential variable selection in conjunction with hyper-parameter tuning and nested cross-validation to calculate the predictive capability of each algorithm, while minimising the risk of over-fitting. The SVM algorithm provided a prediction accuracy of 12% (RPE), representing an improvement of 54% compared to the previously developed MLR model.

Machine-learning algorithms are capable of providing improved prediction accuracy when compared to MLR approaches, due to their ability to quantify non-linearities and interactions between input variables. Concurrently, these algorithms also offer an increased flexibility regarding data multi-collinearity, input data distributions and missing data points compared to standard statistical methods. Shine et al. (2019b) further assessed the ability of the SVM

Table 5 Comparison of machine learning models developed for predicting energy consumption on dairy farms

Study	Country	Model Type	Res	Prediction variable	Validation method	R^2	RMSE	RPE (%)
Sefeedpari et al. 2013	Iran	ANN	A	Output energy of milk (MJ /cow)	Test set (20%)	0.88	0.015	n/a
Sefeedpari et al. 2014	Iran	ANFIS	A	Output energy of milk (MJ /cow)	Test set (20%)	0.79	0.1	n/a
Shine et al. 2018b	Ireland	SVM	M	Total electricity use (kWh)	Nested CV	0.94	241	12
Shine et al. 2019b	Ireland	SVM	A	Total electricity use (kWh)	16 farms	0.94	2,040	10

ANN, artificial neural network; ANFIS, adaptive neural-fuzzy inference system; SVM, support vector machine; M, monthly; A, annual; Nested CV, nested cross-validation; n/a, information not available; Res, prediction resolution.

model to predict electricity consumption at an annual resolution, both at the farm-level and the catchment-level. A negative correlation between prediction resolution and RPE was found, whereby, as the prediction resolution increased, the RPE was reduced from 12% when predicting monthly consumption to 10% for annual farm-level consumption, and down to 5% for annual catchment-level consumption.

4 Smart energy strategies

Two smart energy approaches may be implemented on dairy farms, both of which are inherently related. An energy mitigation approach may be employed to maximise energy efficiency per litre of milk produced, thereby minimising energy consumption, the related costs and GHG emissions without hindering milk production volumes. In addition, the on-farm generation of renewable electricity and heat (such as solar water heating and biogas for grain drying) may be employed to reduce the dependence on fossil fuels and grid-sourced electricity use (where coal- and gas-fired power plants are dominant), thereby reducing the GHG emissions associated with the production of milk and minimising potential production cost increases due to any future increase in energy prices.

4.1 Energy mitigation

Studies focusing on energy-mitigation strategies on dairy farms have primarily focused on reducing electricity consumption by improving efficiency. Significant savings in electricity consumption can often be made by carrying out upgrades to milk cooling, milk harvesting and water heating infrastructure. Although electricity is a minor component of the overall production costs, relatively inexpensive methods of energy mitigation may result in a significant return on investment. In addition, the priority given to reducing on-farm energy use is likely to increase in the future as the cost of energy rises and greater incentives are given to reduce GHG emissions.

Energy efficiency strategies include pre-cooling of milk, installing variable speed drives (VSDs), heat recovery and solar thermal systems, as well as farm management strategies that also reduce direct and indirect energy use.

4.1.1 Milk pre-cooling

Pre-cooling warm milk with cold water through a plate heat exchanger (PHE) prior to entry into the milk bulk tank is an effective method to reduce refrigeration energy demands. Hartman and Sims (2006) highlighted that 90% of the energy required for milk cooling is used for initial cooling, while the remaining 10% is required to maintain the milk temperature below 4°C. Therefore, they calculated that a 5–7% reduction in cooling energy requirement may be achieved for each 1°C drop in milk temperature set point prior to entry to the bulk tank. A milk-cooling electricity reduction of 15% was reported by Ludington et al. (2003) due to the installation of a pre-cooling system by reducing the bulk tank set point and milk temperature differential. Similarly, Karlsson et al. (2012) calculated a 50% reduction in milk-cooling electricity consumption when pre-cooling milk through a PHE using a 1:1 milk-to-water flow rate, and measured a reduction in milk temperature from 37°C to 17°C. Employing cold ground water for milk pre-cooling was also found to reduce electricity consumption by 25% on average across 45 Irish dairy farms (Shine et al. 2018a). The optimum milk-to-water flow rate will depend upon the size of the PHE, the power of the milk pump and the coefficient of performance of the refrigeration unit. Murphy et al. (2013) developed two proportional integral derivative controllers (feedback and feedback-feedforward) to minimise milk-cooling electricity costs per litre of milk, in accordance with the specific coefficient of performance (COP) of the refrigeration units and varying milk flow during milking. This was achieved by varying the flows of ground water and ice-cold water through the dual-stage PHE, while considering a static cost of water. Murphy et al. (2013) identified the potential for a milk-cooling operational cost reduction of 34.5% by the selection and utilisation of the optimum water-to-milk ratio.

4.1.2 Variable speed drives

Conventional vacuum pumps operate at higher air flows than is required for milk harvesting as a higher vacuum reserve is required for plant washing, thus resulting in air admission to the system and energy wastage (Morison et al. 2007). By comparison, VSDs vary the speed of the vacuum pump motor to balance the flow rate of air that is removed and admitted to the system while still producing adequate vacuum stability (Morison et al. 2007; Upton et al. 2013; Rajaniemi et al. 2015). Hartman and Sims (2006) suggested that a 70% reduction in milk-harvesting energy (equating to 20% of total electricity demand) could be achieved by the installation of a VSD. Dunn and Butler (2010) reported milk-harvesting savings of between 40% and 50% with the installation of a VSD on UK dairy farms, while Ludington and Johnson (2003) calculated milk-harvesting savings of 33% across 32 dairy farms in the United States. Similarly, on Danish AMS farms, Brøgger Rasmussen and Pedersen (2004) highlighted farms that employed a VSD and had reduced their electricity consumption by 20 kWh daily compared to using a regular vacuum pump. On conventional Irish dairy farms, Upton et al. (2015) reported that the installation of a VSD for milk harvesting reduced milking costs by 60%, on average.

4.1.3 Heat recovery

Savings in water heating savings may be achieved through heat recovery systems that pre-heat the cold inlet water by extracting heat from the milk flow and utilising a heat exchanger on the chiller–condenser coils, or recycling residual heat from the used washing water (Hartman and Sims 2006). Breen et al. (2019a) developed a multi-objective optimisation method to identify farm infrastructure and managerial practices to balance maximising farm net profit and minimising farm electricity-related CO_2 emissions, over a ten-year time horizon. A heat recovery model was developed and validated using empirical data and combined with the MECD model developed by Upton et al. (2014). However, heat recovery systems were only selected in the optimal scenario, when the combined objective function was weighted heavily towards minimising CO_2 emissions.

4.1.4 Farm management

Strategic changes to farm management can have a significant impact on reducing direct and indirect energy use on a dairy farm. As shown in Table 1, organic farming practices use 34% less energy resources compared to conventional dairy farms, and conventional pasture-based systems use 40% less energy compared to confined systems, which may also result in financial savings. Finneran et al. (2010) identified a cost saving of 61% in switching a cow's diet from concentrate feed to grazed pasture due to reductions in indirect energy use associated with fertiliser (100%) and feed inputs (55%) (Table 3).

Regarding diesel fuel savings, Hartman and Sims (2006) identified the importance of the correct utilisation of the hydraulic systems, keeping the machinery well maintained, matching tractor size and ballast to the task in hand and checking tyre pressures. Barnett and Russel (2010) also mention the potential for biodiesel to reduce fossil fuel consumption and emission of GHGs, although they fail to mention any studies that directly focus on this aspect.

In their New Zealand-based study, Hartman and Sims (2006) highlighted that the majority of farms could reduce their electrical energy usage and costs by up to 20% through implementing various electrical energy reduction and load shifting strategies. Relatively inexpensive methods of load shifting may result in a greater return on investment compared to infrastructural upgrades, especially as time-of-use electricity pricing tariffs become more commonplace. Upton et al. (2013) reiterated the importance of load shifting with regard to dairy farms that milk twice a day with an inherent bi-modal electricity consumption load profile, highlighting evening milking times coinciding with peak electricity demand on electrical grids, and likely the highest rate of electricity cost.

4.2 Renewable energy technologies

Research focusing on the application of renewable energy technologies on dairy farms is becoming more prominent. Breen et al. (2015) added small wind turbine and solar PV models to the MECD (Upton et al. 2014), and assessed the applicability of installing them on a dairy farm across four technology scenarios using three electricity tariffs and three government-supported feed-in-tariffs to encourage deployment. For both wind turbine and solar PV systems, the optimum scenario for all three feed-in-tariffs was a farm with an ice bank milk-cooling system and night-time electric water-heating schedule using a day/night electricity tariff. The electricity output of the solar PV system matched the load profile of the direct DX scenario, resulting in similar monetary savings for a grid-connected solar PV system or a standalone system. To assess the applicability of a solar thermal system in conjunction with heat recovery for water heating on Irish dairy farms over a ten-year time horizon using multi-objective optimisation, a solar thermal model was developed and validated using empirical data and combined with the MECD model (Breen et al. 2019a). Solar thermal heating was never selected by the model in the optimal farm configuration, suggesting poor financial viability.

Multi-objective optimisation was then employed to further assess the applicability of solar PV on dairy farms by identifying the optimal combination of dairy farm equipment and management practices with respect to balancing farm net profit and renewable energy contributions (Breen et al. 2020). A trade-off parameter was incorporated to quantify the relative importance of net profit or renewable energy. When they are of equal importance, on a 195-cow dairy farm milking twice per day, the optimal configuration was for an 11 kWp PV system with a scheduled water heating load at 10:00am.

Barnett and Russell (2010) assessed the potential for renewable energy supplies, including anaerobic digestion, to produce biogas, small-scale wind power, small-scale hydro-electric power, solar thermal power and biodiesel. Economically, anaerobic digestion was more suited to confined dairy systems, where almost 100% of the manure can be collected for digestion, compared to between 10% and 20% collected in outdoor grazing systems during milking times. For biogas production, a minimum herd size of between 250 and 300 cows would be required for confined systems to be economically viable while a pastoral grazing system would require a minimum herd size of 1000 cows.

In the short to medium term, small-scale wind energy of around 10 kW capacity was not economically viable. Small-scale hydro-electric power generation would not be suitable for the vast majority of dairy farms as most would not have a waterway with sufficient head or flow rate required to generate electricity economically. There is potential for solar water heating to enable a 50% reduction in electricity consumption for water heating but the economic viability has not been discussed.

5 Case study: Dairy Energy Decision Support Tool

Some dairy farmers will purchase new energy efficient and renewable energy technologies to reduce their fossil fuel use, the associated costs and GHG emissions. The implementation of such technologies will reduce the demand for grid-sourced electricity and increase the penetration of renewable energy contributing to overall energy production. However, if not managed and sized correctly, such technologies may lead to greater long-term running costs, though capital costs are declining. Forecasted cost savings described by technology suppliers cannot be relied upon as these figures are typically calculated for an average dairy farm, and the solar, wind and hydro resources vary with location. For accurate calculations, the information related to herd size, specific dairy farm operations, any available government grant aid and current electricity tariffs are required, as the impact of installing a particular technology depends upon specific operating and financial conditions. Without this information, the cost savings of a particular technology may be greatly over-estimated, reducing the overall return on investment.

The Dairy Energy Decision Support Tool (Murphy et al. 2018; Shine et al. 2019a) bridges the gap between dairy farmers, researchers, policy makers and cutting-edge scientific outcomes related to the farm-gate economic, environmental and energy impact of dairy farming systems in Ireland. This tool represents the current state-of-the-art in agri-energy decision support systems and assesses the monetary, environmental and energy applicability of energy-efficient and renewable energy technologies on a farm-specific basis. The decision support tool incorporates the following.

1. The MECD model, used to estimate farm-level energy use, energy costs and CO_2 emission reductions on dairy farms operating a 300-day lactation cycle (Upton et al. 2014). The model requires input data such as (but not limited to) herd size, morning and evening milking times, number of milking units, type of milk cooling system, type of water heating system, hot wash frequency and milk-collection interval.
2. Six renewable and energy-efficient technology models, that were developed and validated using empirical data including plate cooler, solar PV, variable speed drive, solar thermal, heat recovery and wind turbine models (Breen et al. 2015; Breen et al. 2019a; Breen et al. 2020).
3. Optimisation functionality, to allow dairy farmers to autonomously identify their optimum milking times, water heating system, water heating start time and solar PV system size to either maximise return on investment or minimise energy-related CO_2 emissions (Breen et al. 2019b).
4. Added technology assessment functionality, including demand-side management, localised technology assessments, dynamic CO_2 intensity value and dynamic milk pre-cooling temperature set point levels.

5 Dairy energy usage statistics, highlighting graphical and numerical breakdowns of mean electricity consumption, and cost and carbon accounting according to each energy consuming process on a cohort of 58 dairy farms.

This functionality provides the tool's ability to provide cutting-edge research outputs to individual farmers on a large scale as well as to researchers looking for information related to dairy farm energy demands and to policy makers requiring tools to make informed decisions related to agricultural policy.

The Dairy Energy Decision Support Tool (Fig. 2) is available at: https://messo.cit.ie/agri-energy.

6 Conclusion and future trends

It is difficult to quantify the potential energy savings with the implementation of best practice methods. However, from the literature, some estimations can be made as to what areas are worth focusing sustainability measures upon. Conventional farming systems use 34% more energy than organic farming systems per unit of milk output. However, the analyses failed to incorporate the economic viability of switching to an organic farming system. Furthermore, conventional pasture-based systems use less energy than confined systems, where feed represents the largest energy input from an LCA perspective. Energy consumption may be 40% less on a conventional pasture-based system than a confined system. There are a host of complex interactive effects driving the dominant dairy system type within a region, such as land suitability and local environmental conditions, which are outside the scope of this review.

Strategies for reducing electricity demand, such as the installation of milk pre-cooling systems and/or variable speed drives on vacuum pumps can be relatively inexpensive, with the potential for a significant return on investment. Depending on local renewable energy resources, solar PV has been shown to be the preferable technology over wind turbines, solar water heating and small-scale hydroelectric systems. However, the effectiveness of a particular technology is directly related to the resource availability at a given location. Anaerobic digestion systems showed viability on large-scale confined dairy farms. The installation of cost-effective energy mitigation or renewable energy technologies can benefit both farmers and electricity system operators by reducing the demand load from the grid, increasing the penetration of renewable energy, reducing GHG emissions and reducing the impact of future energy price increases on dairy farm production costs.

The actual energy and cost savings associated with the installation of these technologies will vary on a farm-to-farm basis. Thus, research has also focused

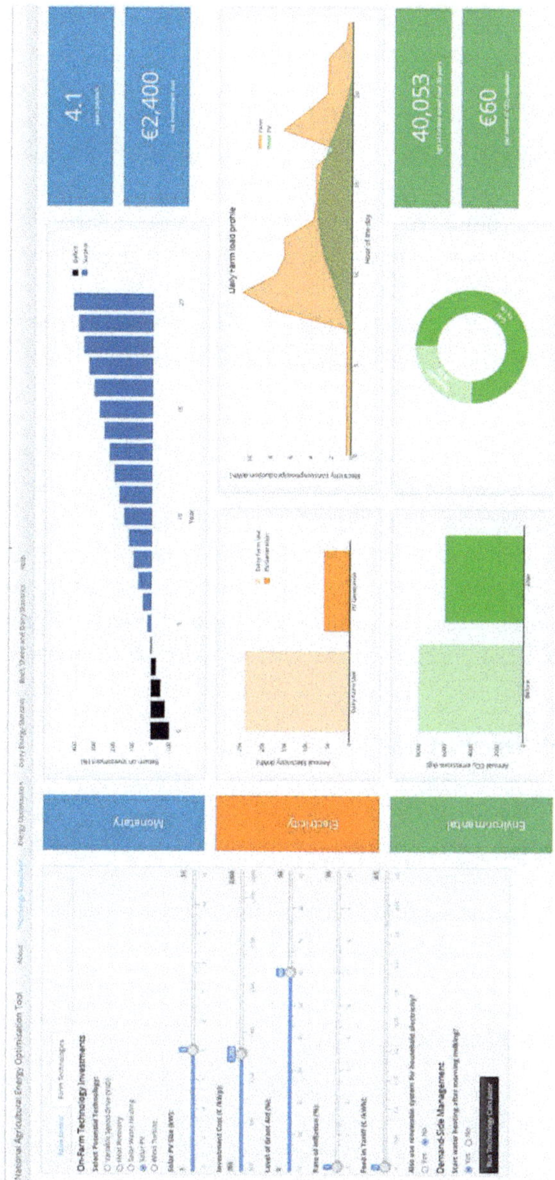

Figure 2 The dashboard of the Dairy Energy Decision Support Tool.

on developing predictive models for specific on-farm energy demands. These models incorporate mechanistic, regression and machine learning methodologies that can predict liquid fuel use, heat energy demand and electricity use. Recent studies have combined mechanistic modelling of electricity consumption with renewable energy supply models, allowing studies regarding the applicability of these technologies on an individual farm-to-farm basis to be carried out.

Future studies could incorporate the development of machine-learning models for simulating dairy energy use, given their ability to quantify non-linearities and interactions between input variables. This would improve prediction accuracy compared to common multiple linear regression methods. It is important that these studies incorporate robust cross-validation methods and performance metrics to allow for the comparison of modelling techniques between cognate studies. These models can be used as part of life cycle analyses to generate electricity consumption values, thus reducing the necessity for expensive and time-consuming physical metering equipment, thereby allowing for a greater number of farms to be assessed. In addition, these models may be utilised for macro-level predictions of energy use on a national basis related to the production of milk.

To help maximise the impact, the prominence of software such as the Dairy Energy Decision Support Tool is likely to become more commonplace. These tools are becoming increasingly important mechanisms for packaging advanced scientific methods in an easily understandable format for the end-user, thus helping guide sustainable energy use and cost-efficient decision making.

7 Where to look for further information

7.1 Further reading

- Shine et al. (2020) also provides a comprehensive review of the dairy energy literature.
- Pagani et al. (2016) assessed differences in direct and indirect energy use between conventional and organic dairy farms in Italy.

7.2 Key journals/conferences

- Dairy energy research is regularly published in Applied Energy, Computers and Electronics in Agriculture, the Journal of Dairy Science, and Energies (open-source).
- Regular meetings are held by the European Association for Precision Livestock Farming (EA-PLF), the annual international meeting of American Society of Agricultural and Biological Engineers (ASABE), and the European Society for Agricultural Engineers (EurAgEng).

8 References

Aguirre-Villegas, H. A., Passos-Fonseca, T. H., Reinemann, D. J. and Larson, R. 2017. Grazing intensity affects the environmental impact of dairy systems. *Journal of Dairy Science* 100(8), 6804–6821. doi: 10.3168/jds.2016-12325.

Baldini, C., Gardoni, D. and Guarino, M. 2017. A critical review of the recent evolution of life cycle assessment applied to milk production. *Journal of Cleaner Production* 140, 421–435. doi: 10.1016/j.jclepro.2016.06.078.

Barnett, J. and Russell, J. 2010. Energy use on dairy farms. *Bulletin of the International Dairy Federation* 443(1), 23-32. Available at: http://www.fil-idf.org/EventsCalendar.htm.

Breen, M., Murphy, M. D. and Upton, J. 2015. Development and validation of photovoltaic and wind turbine models to assess the impacts of renewable generation on dairy farm electricity consumption. In: Engineers, A. S. of A. and B. (ed.) 2019 ASABE Annual International Meeting. New Orleans, Louisiana, pp. 1–11. doi: 10.13031/aim.20152189379.

Breen, M., Murphy, M. D. and Upton, J. 2019a. Development of a Dairy Multi-Objective Optimization (DAIRYMOO) method for economic and environmental optimization of dairy farms. *Applied Energy* 242, 1697–1711. doi: 10.1016/j.apenergy.2019.03.059.

Breen, M., Upton, J. and Murphy, M. D. 2019b. Development of a discrete infrastructure optimization model for economic assessment on dairy farms (DIOMOND). *Computers and Electronics in Agriculture* 156, 508–522. doi: 10.1016/j.compag.2018.11.018.

Breen, M., Upton, J. and Murphy, M. D. 2020. Photovoltaic systems on dairy farms: financial and renewable multi- objective optimization (FARMOO) analysis. *Applied Energy* 278. doi: 10.1016/j.apenergy.2020.115534.

Brøgger Rasmussen, J. and Pedersen, J. 2004. *Electricity and Water Consumption at Milking*. Danish Agricultural Advisory Service, Aarhus.

Bruinsma, J. and Alexandratos, N. 2012. World agriculture towards 2030/2050: the 2012 revision. Available at: http://www.fao.org/docrep/016/ap106e/ap106e.pdf.

Calcante, A., Tangorra, F. M. and Oberti, R. 2016. Analysis of electric energy consumption of automatic milking systems in different configurations and operative conditions. *Journal of Dairy Science* 99(5), 4043–4047. doi: 10.3168/jds.2015-10490.

Cederberg, C. and Mattsson, B. 2000. Life cycle assessment of Swedish milk production—a comparison of conventional farming. *Journal of Cleaner Production* 8(1), 49–60. doi: 10.1016/S0959-6526(99)00311-X.

Dunn, P., Butler, G., Bilsborrow, P. and Brough, D. 2010. Energy + efficiency—renewable energy and energy efficiency options for UK dairy farms. Available at: https://docplayer.net/6454694-Energy-efficiency-renewable-energy-and-energy-efficiency-options-for-uk-dairy-farms.html.

Edens, W. C., Pordesimo, L.O., Wilhelm, L.R. and Burns, R.T. 2003. Energy use analysis of major milking components at a dairy experiment station. *Applied Engineering in Agriculture* 19(6), 711–716. doi: 10.13031/2013.15659.

FAO 2018. Agricultural outlook 2018–2027 dairy and dairy products. Available at: http://www.agri-outlook.org/commodities/Agricultural-Outlook-2018-Dairy.pdf.

Finneran, E., Crosson, P., O'Kiely, P., Shalloo, L., Forristal, D., Wallace, M. 2010. Simulation modelling of the cost of producing and utilising feeds for ruminants on Irish farms. *Journal of Farm Management* 14(2), 95–116.

Fuentes-Pila, J., DeLorenzo, M. A., Beede, D. K., Staples, C. R. and Holter, J. B. 1996. Evaluation of equations based on animal factors to predict intake of lactating Holstein cows. *Journal of Dairy Science* 79(9), 1562–1571. doi: 10.3168/jds.S0022-0302(96)76518-9.

Hartman, K. and Sims, R. 2006. Saving energy on the dairy farm makes good sense. In: Proceedings of the 4th Dairy Conference. Centre for Professional Development and Conferences, Massey University, Palmerston North, New Zealand, pp. 11–21.

Hörndahl, T. 2008. *Energy Use in Farm Buildings–A Study of 16 Farms with Different Enterprises* (2nd ed.). Available at: https://pub.epsilon.slu.se/3396/1/Eng-rapport 145-v1.pdf.

Hospido, A., Moreira, M. T. and Feijoo, G. 2003. Simplified life cycle assessment of Galician milk production. *International Dairy Journal* 13(10), 783–796. doi: 10.1016/S0958-6946(03)00100-6.

IDF 2010. Bulletin of the International Dairy Federation–a common carbon footprint approach for dairy 445/2010. Available at: http://www.idf-lca-guide.org/Files/media/Documents/445-2010-A-common-carbon-footprint-approach-for-dairy.pdf.

Karlsson, A. E., Hörndahl, T. and Nordman, R. 2012. Energy recover from milk cooling. Report 401. Agriculture & Industry. JTI–Swedish Institute of Agricultural and Environmental Engineering, Uppsala, Sweden.

Kraatz, S. 2012. Energy intensity in livestock operations–modeling of dairy farming systems in Germany. *Agricultural Systems* 110, 90–106. doi: 10.1016/j.agsy.2012.03.007.

Lanigan, G., Donnellan, T., Hanrahan, K., Paul, C. and Shalloo, L. 2018. An analysis of abatement potential of Greenhouse Gas emissions in Irish agriculture 2021–2030. Available at: https://www.teagasc.ie/media/website/publications/2018/An-Analysis-of-Abatement-Potential-of-Greenhouse-Gas-Emissions-in-Irish-Agriculture-2021-2030.pdf.

Ludington, D. and Johnson, E. 2003. Dairy farm energy audit summary. Available at: https://www.nyserda.ny.gov/-/media/Files/Publications/Research/Energy-Audit-Reports/dairy-farm-energy.pdf.

Meul, M., Nevens, F., Reheul, D. and Hofman, G. 2007. Energy use efficiency of specialised dairy, arable and pig farms in Flanders. *Agriculture, Ecosystems and Environment* 119(1–2), 135–144. doi: 10.1016/j.agee.2006.07.002.

Mhundwa, R., Simon, M. and Tangwe, S.L. 2017. Modelling of an on-farm direct expansion bulk milk cooler to establish baseline energy consumption without milk pre-cooling: a case of fort hare dairy trust, South Africa. *African Journal of Science, Technology, Innovation and Development* 1338, 62–68. doi: 10.1080/20421338.2017.1385132.

Morison, K., Gregory, W. and Hooper, R. 2007. Improving dairy shed energy efficiency. Available at: https://ir.canterbury.ac.nz/bitstream/handle/10092/11588/Dairy_Technical_Report.pdf;sequence=1.

Murgia, L., Todde, G., Caria, M. and Pazzona, A. 2013. A partial life cycle assessment approach to evaluate the energy intensity and related greenhouse gas emission in dairy farms. *Journal of Agricultural Engineering* 44(2s), 186–190. doi: 10.4081/jae.2013.279.

Murphy, M., Shine, P., Breen, M. and Upton, J. 2018. DSSED: decision support system for energy use in dairy production. Available at: https://www.seai.ie/resources/publications/SEAI-DSSED-Final-Report.pdf.

Murphy, M. D., Upton, J. and O'Mahony, M. J. 2013. Rapid milk cooling control with varying water and energy consumption. *Biosystems Engineering* 116(1), 15–22. doi: 10.1016/j.biosystemseng.2013.05.007.

O'Brien, D., Shalloo, L., Patton, J., Buckley, F., Grainger, C. and Wallace, M. 2012. A life cycle assessment of seasonal grass-based and confinement dairy farms. *Agricultural Systems* 107, 33–46. doi: 10.1016/j.agsy.2011.11.004.

Pagani, M., Vittuari, M., Johnson, T. G. and De Menna, F. 2016. An assessment of the energy footprint of dairy farms in Missouri and Emilia-Romagna. *Agricultural Systems* 145, 116–126. doi: 10.1016/j.agsy.2016.03.009.

Rajaniemi, M., Jokiniemi, T., Alakukku, L. and Ahokas, J. 2015. Direct energy consumption and saving possibilities in milk production. Available at: https://agronomy.emu.ee/wp-content/uploads/2015/05/13_1_31_B5.pdf.

Rajaniemi, M., Jokiniemi, T., Alakukku, L. and Ahokas, J. 2017. Electric energy consumption of milking process on some Finnish dairy farms. *Agricultural and Food Science* 26(3), 160–172. doi: 10.23986/afsci.63275.

Sefeedpari, P., Rafiee, S. and Akram, A. 2013. Application of artificial neural network to model the energy output of dairy farms in Iran. *International Journal of Energy Technology and Policy* 9(1), 82. doi: 10.1504/IJETP.2013.055819.

Sefeedpari, P., Rafiee, S., Akram, A. and Komleh, S. H. P. 2014. Modeling output energy based on fossil fuels and electricity energy consumption on dairy farms of Iran: application of adaptive neural-fuzzy inference system technique. *Computers and Electronics in Agriculture* 109, 80–85. doi: 10.1016/j.compag.2014.09.010.

Sefeedpari, P., Rafiee, S., Akram, A., Chau, K. and Komleh, S. H. P. 2015. Modeling energy use in dairy cattle farms by applying Multi-Layered Adaptive Neuro-Fuzzy Inference System (MLANFIS). *International Journal of Dairy Science* 10(4), 173–185. doi: 10.3923/ijds.2015.173.185.

Shine, P., Scully, T., Upton, J., Shalloo, L. and Murphy, M. D. 2018a. Electricity & direct water consumption on Irish pasture based dairy farms: a statistical analysis. *Applied Energy* 210, 529–537. doi: 10.1016/j.apenergy.2017.07.029.

Shine, P., Murphy, M. D., Upton, J. and Scully, T. 2018b. Machine-learning algorithms for predicting on-farm direct water and electricity consumption on pasture based dairy farms. *Computers and Electronics in Agriculture* 150, 74–87. doi: 10.1016/j.compag.2018.03.023.

Shine, P., Scully, T., Upton, J. and Murphy, M. D. 2018c. Multiple linear regression modelling of on-farm direct water and electricity consumption on pasture based dairy farms. *Computers and Electronics in Agriculture* 148, 337–346. doi: 10.1016/j.compag.2018.02.020.

Shine, P., Breen, M., Upton, J., O'donovan, A. and Murphy, M. D. 2019a. A decision support and optimization platform for energy technology investments on dairy farms. In: American Society of Agricultural and Biological Engineers (ed.) 2019 ASABE Annual International Meeting. Boston, MA, pp. 1–8. doi: 10.13031/aim.201901055.

Shine, P., Scully, T., Upton, J. and Murphy, M. D. 2019b. Annual electricity consumption prediction and future expansion analysis on dairy farms using a support vector machine. *Applied Energy* 250(103), 1110–1119. doi: 10.1016/j.apenergy.2019.05.103.

Shine, P., Upton, J., Sefeedpari, P. and Murphy, M. D. 2020. Energy consumption on dairy farms : a review of monitoring , prediction modelling and analyses. *Energies* 13(5), 1288. doi: 10.3390/en13051288.

Shortall, J., Shalloo, L., Foley, C., Sleator, R. D. and O'Brien, B. 2016. Investment appraisal of automatic milking and conventional milking technologies in a pasture-based dairy system. *Journal of Dairy Science* 99(9), 7700–7713. doi: 10.3168/jds.2016-11256.

Shortall, J., O'Brien, B., Sleator, R. D. and Upton, J. 2018. Daily and seasonal trends of electricity and water use on pasture-based automatic milking dairy farms. *Journal of Dairy Science* 101(2), 1565–1578. doi: 10.3168/jds.2017-13407.

Sjaunja, L. O., Baevre, L., Junkkarinen, L., Pedersen, J. and Setälä, J. 1990. A Nordic proposal for an energy corrected milk (ECM) formula. In: *27th Session of the International Commission for Breeding and Productivity of Milk Animals*. Paris, France, pp. 156–157.

Smith, P., Martino, D., Cai, Z., Gwary, D., Janzen, H., Kumar, P., McCarl, B., Ogle, S., O'Mara, F. and Rice, C. 2007. Policy and technological constraints to implementation of greenhouse gas mitigation options in agriculture. *Agriculture, Ecosystems and Environment* 118(1–4), 6–28. doi: 10.1016/j.agee.2006.06.006.

Thomassen, M. A., van Calker, K. J., Smits, M. C. J., Iepema, G. L. and de Boer, I. J. M. 2008. Life cycle assessment of conventional and organic milk production in the Netherlands. *Agricultural Systems* 96(1–3), 95–107. doi: 10.1016/j.agsy.2007.06.001.

Todde, G., Murgia, L., Caria, M. and Pazzona, A. 2017. Dairy Energy Prediction (DEP) model: a tool for predicting energy use and related emissions and costs in dairy farms. *Computers and Electronics in Agriculture* 135, 216–221. doi: 10.1016/j.compag.2017.02.014.

Todde, G., Murgia, L., Caria, M. and Pazzona, A. 2018a. A comprehensive energy analysis and related carbon footprint of dairy farms, Part 1: direct energy requirements. *Energies* 11(2), 1–14. doi: 10.3390/en11020451.

Todde, G., Murgia, L., Caria, M. and Pazzona, A. 2018b. A Comprehensive energy analysis and related carbon footprint of dairy farms, Part 2: investigation and modeling of indirect energy requirements. *Energies* 11(2), 1–13. doi: 10.3390/en11020463.

Upton, J., Humphreys, J., Groot Koerkamp, P. W., French, P., Dillon, P. and De Boer, I. J. 2013. Energy demand on dairy farms in Ireland. *Journal of Dairy Science* 96(10), 6489–6498. doi: 10.3168/jds.2013-6874.

Upton, J., Murphy, M., Shalloo, L., Groot Koerkamp, P. W. and De Boer, I. J. 2014. A mechanistic model for electricity consumption on dairy farms: definition, validation, and demonstration. *Journal of Dairy Science* 97(8), 4973–4984. doi: 10.3168/jds.2014-8015.

Upton, J., Murphy, M., De Boer, I. J., Groot Koerkamp, P. W., Berentsen, P. B. and Shalloo, L. 2015. Investment appraisal of technology innovations on dairy farm electricity consumption. *Journal of Dairy Science* 98(2), 898–909. doi: 10.3168/jds.2014-8383.

von Keyserlingk, M. A. G., Martin, N. P., Kebreab, E., Knowlton, K. F., Grant, R. J., Stephenson, M., Sniffen, C. J., Harner, J. P., Wright, A. D. and Smith, S. I. 2013. Invited review: sustainability of the US dairy industry. *Journal of Dairy Science* 96(9), 5405–5425. doi: 10.3168/jds.2012-6354.

Wells, C. 2001. Total energy indicators of agricultural sustainability: dairy farming case study. Technical Paper. Ministry of Agriculture and Forestry, Wellington. ISBN: 0-478-07968-0.

Chapter 10

Energy-smart pig farming

Lee J. Johnston and Kelsey L. Hammers, West Central Research and Outreach Center, University of Minnesota, USA

1 Introduction

The primary reason for the existence of any pig farming system is to produce pigs that will provide pork products for human consumption. Throughout human civilization, a myriad of systems and their permutations have been employed to achieve this primary goal. Regardless of the production system used, energy in various forms is required to produce pork. In the past few decades, pig farmers and society, in general, have become increasingly aware of, and concerned about, the environmental footprint of food production, including pork. Energy use is one portion of the environmental footprint of pork production but concerns about energy use in pork production are not new. In the early 1980s, researchers acknowledged the need to improve the energy efficiency of pig farming (Fredeen and Harmon, 1983).

In the context of this chapter, 'energy-smart pig farming' refers to the efficient use of energy and reduced reliance on non-renewable energy sources in the production and housing of pigs that ultimately will be processed into a range of pork products.

The energy footprint of pork production refers to impacts resulting from energy use to produce consumer-ready pork. These impacts begin upstream of the pig farm and include the energy required to generate and process feed

http://dx.doi.org/10.19103/AS.2022.0100.17

ingredients for pigs, together with the manufacture of building materials and equipment that house and manage the pigs. Impacts of direct energy used on farms where pigs are raised, such as electricity and vehicle fuel, are also part of the energy footprint. Finally, the impacts of energy used downstream of the pig farm for transporting and harvesting pigs, processing carcasses, and packaging pork products are included in the overall energy footprint. In this chapter, we will focus primarily on the energy used on the pig farm ('behind the farm gate'), with some discussion of upstream energy demand for feed production.

Energy flows through a pork production system along different pathways. The energy balance of pig production on farms includes all the necessary energy inputs and the resulting outputs (Fig. 1). On-farm energy represents a major portion of the total energy inputs of pork production.

The overarching goal of a pork production system is to maximize carcass outputs, with lean meat more than fat, and minimize the outputs of manure, heat and gaseous products. This approach optimizes energy output in all the consumable products for a given energy input. Regardless of the system of pork production used, there are obligatory losses of energy, though some of these losses can be re-captured to improve the energy efficiency of pig farming. For example, manure produced by pigs represents a loss of energy from the system, but it can be digested anaerobically to re-capture the energy as biogas that can then be used for heating or electricity generation (Nardin and Mazzetto, 2014; Grando et al., 2017). Capturing lost energy from manure can be part of an energy-smart pig farming system.

Multiple disciplines in animal and veterinary sciences, including nutrition, genetics, reproduction, health and engineering, focus on improving the efficiency of biological performance in modern pork production systems.

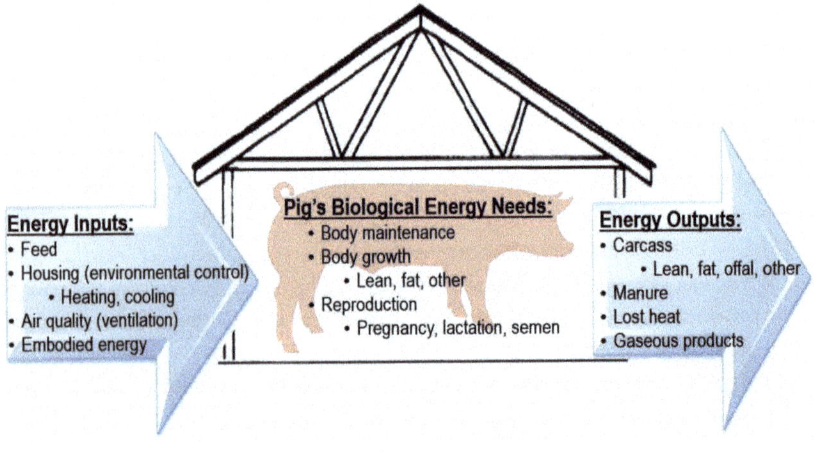

Figure 1 Energy flows through a pig production system.

Enhancing biological performance improves the overall energy efficiency of the system. Carcass outputs can be maximized by precisely satisfying the pig's biological energy needs for optimal performance (Fig. 1). Rapid and efficient lean growth translates into greater carcass weight for the feed consumed and fewer days to reach market weight compared with slow-growing, less-efficient pigs. Similarly, increasing reproductive rates realized by proper feeding and care of sows increases the number of piglets produced for the energy inputs used for feed, housing and ventilation.

Several direct and indirect energy inputs merge to satisfy a pig's biological energy needs (Fig. 1). Feed is by far the largest energy input in any pork production system (Lammers et al., 2010a, 2012; Tallaksen et al., 2020). Pigs extract the energy supplied by feed ingredients when they consume and digest the nutrients. The energy inputs required to produce and process feed ingredients, and complete feed and deliver it, are part of the energy footprint of a pig farm. Improving upstream energy-use efficiency in feed production will reduce the energy footprint of pork production.

The housing system is designed to provide an environment for pigs that is conducive to optimal biological performance. The housing system might be very complicated and include environmentally controlled barns, or be rather simple, such as an outdoor system with open-sided sheds or sunshades. Depending on the level of sophistication, the housing system requires varying quantities of energy to maintain the ideal temperature and air quality for pigs. Energy embodied in materials and construction of a building also contributes to the energy footprint of pig farming. Less sophisticated housing systems that reduce energy inputs and result in poorer environmental control will require increased energy inputs from feed to satisfy the biological energy needs of pigs.

Energy use in pork production is influenced by a myriad of factors that individually have small influences on the overall energy footprint. Energy-smart pig production will result from combinations of many small changes that add up to significant improvements in energy efficiency. This chapter does not discuss all the points that could improve energy efficiency but presents ideas for improving energy efficiency based on pig farms that have been evaluated under controlled research or commercial conditions and so have broad practical applications on modern pig farms.

2 Energy generation on pig farms

Pig farming is an energy-intensive industry that relies on electricity and heating fuels to provide a thermo-neutral environment for pigs in large-scale commercial production. In most parts of the world, electricity and heating fuels are derived from fossil fuels. Rising fossil fuel cost, the imminent depletion of non-renewable energy sources and consumer pressure to reduce the environmental footprint of pig production have stimulated interest in the use of renewable energy technologies. Renewable energy is defined as energy that is derived from natural sources that can be replaced within a few years from initial use (Lund, 2010). Fossil fuels are the conventional energy source used in the swine industry; however, renewable energy sources such as solar, geothermal and biomass can be used to meet the energy demands of swine production and simultaneously offer economic and environmental benefits (Ogunlade et al., 2018). Renewable energy technologies can be deployed on farms or can be a component of the commercial utility companies that are increasingly generating renewable electricity and heating fuels.

Solar energy generation systems have been labeled the most prominent method to produce on-farm renewable energy (USDA, 2011). Solar energy is an increasingly available energy source due to the steadily developing solar industry (Kannan and Vakeesan, 2016) and the abundance of energy from the sun (Panwar et al., 2011). The high initial investment costs, subsidies, tax breaks and other financial incentives have directly influenced the adoption of this technology and will continue to increase implementation (Devabhaktuni et al., 2013). Solar photovoltaic panels have been increasingly incorporated on pig building roofs or on nearby land where available, to provide electricity and help displace fossil fuels (Resch et al., 2008). Apprehension about converting land from cultivation to carry solar systems has driven the development of agrivoltaics, a method to integrate solar systems into agricultural production while providing synergistic benefits to both (Barron-Gafford et al., 2019). Photovoltaic panels mounted on raised structures can provide shade to animals and reduce heat stress (Sharpe et al., 2020), reduce drought stress to the plants growing below (Dinesh and Pearce, 2016) and maintain a more desirable soil/water balance (Amaducci et al., 2018).

Solar energy can also be harvested and used to directly heat pig buildings. This approach requires some form of heat storage to account for the lack of solar heat at night. Godbout et al. (2004) used a solar wall to pre-heat incoming air for a 1000-pig nursery building in Canada. They observed a 23% to 31% reduction in heating costs over six months compared with using LPG (liquid petroleum gas, mainly propane) and reduced CO_2 emissions. Similarly, Mun et al. (2015) used a solar space-heating system in a pig nursery to reduce both electricity use and CO_2 emissions by 15% compared with conventional heating.

The amount of solar energy available for use is highly dependent on latitude, climate and time of year (Brooks, 2014). The location of a solar array influences the amount of energy captured and electricity produced. This amount can be predicted using various tools such as PVWatts (NREL, 2021a) or the System Advisor Model (NREL, 2021b). Extensive planning is needed to determine the optimal size and number of panels, the optimal placing of the array, governmental ordinances, the tax credits available and the current energy or electricity demands (CERT, 2021a).

Geothermal heating and cooling offer possible benefits for optimizing the performance of pig production (Dominguez and Predicala, 2011; Belyakov, 2019). Geothermal heating relies on the capture and movement of heat from below the Earth's surface. Reduction of greenhouse gas (GHG) generation can be realized compared with traditional heating systems.

Ground-source heat pumps can utilize heat present in subsoil and groundwater to heat a wean-to-finish pig barn and reduce fossil fuel consumption, electricity costs and GHG emissions compared to an identical, conventionally heated barn (Mun et al., 2020). Ground-source heat pumps can also provide cooling in summer. Sow farms that require heat for optimal growth and comfort of piglets can benefit from well-suited and efficient geothermal heating systems (Licharz et al., 2020). However, widespread adoption of ground-source heat pumps is limited by the high initial investment and the need for specialized equipment (Krommweh et al., 2014; Yahyaoui, 2017).

For cooling, Shah et al. (2017) evaluated an earth-to-water heat exchanger for cooling finishing pigs. They observed greater than 50% reduction in electricity use compared to pigs cooled through the conventional evaporative means with fans and sprinklers. Similarly, Bostami Rubayet et al. (2016) reduced energy expenditure by 28% and emitted 37% less CO_2 compared to a conventional cooling system for growing pigs. Krommweh et al. (2014) used geothermal-tempered inlet air to significantly reduce electricity and heating fuel demand for environmental control of a pig-finishing barn. Geothermal heating and cooling systems can be implemented with very little influence on pig performance (Bostami Rubayet et al., 2016; Shah et al., 2017).

The manure produced by pigs can be considered an obligatory energy loss in a pig production system. However, in an energy-smart production system, energy embedded in the manure can be captured and offensive odors avoided. The anaerobic digestion of pig manure to produce biogas is an increasingly popular method to reduce negative environmental impacts (Adair et al., 2016; Hijazi et al., 2020). In modern pig barns, manure management systems are often liquid- or slurry-based, and therefore are conducive to the implementation of an anaerobic digestion system (Lusk, 1998).[1] The by-product is an odorless effluent that can be applied to crops as a value-added fertilizer (Massé et al., 2010). Anaerobic digestion has been widely adopted in Europe (Nardin and Mazzetto, 2014; Grando et al., 2017; González-García et al., 2020), and to a lesser degree in the U.S. (Cheng et al., 2004). The capital and operating costs of an in-ground anaerobic digester are relatively high, but Adair et al. (2016) concluded that

1 In theory, manure can also be converted into other 'green fuels' such as biodiesel (Kim et al., 2020) and biomethane (Worawimut et al., 2019).

in countries with renewable energy credits and carbon offsets, revenue can be generated to substantially reduce these costs. Acceptance of biogas production and utilization will continue to expand as pig farmers pay increasing attention to the environmental impacts of their operations and governments implement legislation and policies designed to improve the environmental footprint of swine farms through renewable energy programs (Zering et al., 2005).

While there are benefits to utilizing renewable resources, the initial investments in technologies are still relatively large and carry a high financial risk. Costs continue to decline however, so there is little doubt that renewable energy generation will continue to grow, with great potential to reduce the environmental footprint of pig farms.

3 Feed energy related to pig production

The energy needed to produce, process and transport feed for a pig farming enterprise cannot be overstated (Barber et al., 1989). It includes the energy used to till the soil, to plant and harvest crops and to produce the pest control products, fertilizers and equipment used in crop production. Energy used in the manufacturing and transportation of feed is also included in the energy contribution of feed to pig farming.

Life cycle assessments (LCAs) determine the energy use and environmental impacts associated with the development of a specific product throughout its life cycle (van Middelaar et al., 2019). LCAs conducted for pork production agree that large amounts of energy are required to provide feed, which is influenced heavily by cropping sequences, diet formulation and ingredient choice (Lammers et al., 2010b; Pelletier et al., 2010; Kebreab et al., 2016). Approximately 70% of the energy used in pig production is attributed to growing, harvesting, processing and manufacturing feed (Nguyen et al., 2012; Shurson, 2020). This also greatly contributes to total GHG emissions per kilogram of pork produced (Zhou et al., 2018).

By improving the efficiency of feed utilization, energy-smart pig farms can reduce GHG emissions because they require less feed-related energy to produce the same or greater amount of pork (van der Werf et al., 2005; Nguyen et al., 2010). Changes in diet formulation for pigs can influence the environmental impacts of pig farming. Kebreab et al. (2016) suggested that the increased use of crystalline amino acids and phytase in diets could reduce global warming potential and other environmental impacts but would have no meaningful effect on primary energy demand. In contrast, Schmitt et al. (2018) found that the use of crystalline amino acids in corn-based diets increased GHG emissions, but phytase supplementation reduced or had no effect on GHG emissions, depending on the region of the U.S. Clearly, improved efficiency of feed utilization plays an important role in energy-smart pig farming. The nuanced effects of diet formulation on the energy demands of pig farming will become clearer as this topic is studied more thoroughly in the future.

4 Building design to improve energy efficiency

The environment where an animal is housed is the most important factor affecting animal productivity (Scott, 1984). Construction of pig barns in the United States has undergone minimal changes regarding energy efficiency in recent years due to the availability of relatively inexpensive heating fuels and electricity. Pig barns have been constructed with an eye toward reducing capital costs – resulting in the construction of tunnel-ventilated barns and deep-pit, curtain-sided barns where over 80% of growing-finishing pigs are housed (Jacobson et al., 2011). These designs typically have insufficient insulation of the structure, making it difficult to provide a constant thermo-neutral environment for pigs without the high use of fossil fuels. If temperatures in the pig space are cooler than desired, pigs will increase feed intake to compensate for the cooler environment. Farmers and designers must decide whether to rely more heavily on building design or feed intake to best meet the pigs' energy needs. Currently, pig farmers rely more heavily on feed but will likely increase their reliance on energy-efficient building design and operation in the future.

Increasing societal pressure for more sustainable food systems has spurred some in the swine industry to investigate 'greener' housing methods that reduce the environmental impacts and energy consumption of pig farming. Jacobson et al. (2011) studied a 'green pig barn' concept by placing emphasis on the incorporation of energy-saving technologies to provide optimal environmental conditions for pigs that will support biologically efficient pig growth. The goal was to design a building for 2400 pigs from weaning to market weight, with shallow pits and under-floor scrapers for manure removal that would reduce energy consumption and air emissions by 50% compared to a standard tunnel-ventilated finishing barn.

Four variations of the basic design were proposed and modeled to evaluate various design aspects that could improve energy efficiency while providing a near-ideal environment for pigs. Features of the various designs included partially or fully slatted floors, a ground-source heat pump for heating and cooling solid floor sections, geothermal cooling, evaporative cooling pads, and enhanced insulation. All four versions reduced air emissions due to the shallow pits and cooler room temperatures and saved energy in winter months due to increased insulation. These designs were expected to cost between 1.3 and 2 times more than typical construction of a similar-sized tunnel-ventilated barn with deep manure storage pits. However, to offset these costs, the average daily weight gain of pigs was predicted to improve by 3% to 7% and feed efficiency (feed:body weight gain) was expected to decline by 5% to 10% (Jacobson et al., 2011). The authors estimated that improved pig performance would pay for the added construction costs in 6 to 12.8 years. Of course, long-term prices for feed and energy will significantly influence the economic outcomes of these designs when actually built. However, theoretically, they could contribute meaningfully to energy-smart pig farming.

While this 'green pig barn' might be a standard to strive for in the future, additional technologies may be designed today into newly constructed barns to help minimize energy consumption. Maintaining air quality and thermal comfort of pigs are key factors in the environmental management of barns for optimal pig performance. Electricity is used to operate fans, air inlets and controllers of ventilation systems. Controlling electricity demand for specific motors can reduce energy expenditure. Teitel et al. (2008) demonstrated a 75% reduction in electricity used in poultry buildings when a variable frequency drive (VFD) was installed to modulate fan speed, compared to a conventional on/off system, with no negative impact on bird performance. Evaluations of VFDs on ventilation fans in commercial pig farms have not been reported; however, similar savings would be expected.

Thermal conditions in pig buildings are often difficult to manage because the thermal requirements of pigs change with age. Maintaining a consistent barn temperature that is conducive to optimal pig performance is difficult to achieve because energy is lost in the air exhausted through the ventilation system. A typical on/off heater modified for variable heat output can improve temperature stability but does not reduce energy consumption (van Utrecht et al., 2002). Riva et al. (2000) used a reversible air-to-air heat pump for heating and cooling in farrowing rooms and reduced heating energy costs by 11% compared to a standard gas boiler. Heat pumps may find more prominent uses in future pig barns because of their high efficiency and low GHG emissions, depending on the source of electricity.

5 Practices to improve energy efficiency of existing buildings

To avoid the high cost of new building investments, farmers can improve their existing buildings to make them more energy-smart. Improving energy efficiency usually is more economical than building new barns or installing equipment to generate renewable energy on-site (Hayter and Kandt, 2011; Chwieduk, 2017). Approaches to improve the energy efficiency of existing pig production facilities are discussed in the remainder of this chapter.

5.1 Energy use in pig barns

A first step toward optimizing energy use in pig barns is to know how much energy is being consumed. Over a two-year period, Sharpe et al. (2018)

monitored the use of electricity and LPG heating fuel in six commercial pig barns in the Upper Midwest region of the United States: two were farrowing (breed-to-wean), two nursery (weaning at 21 days to about 10 weeks of age), and two finishing (10 weeks of age to market weight of about 127 kg). All were environmentally controlled using heaters and mechanical ventilation, with the exception of 'Finishing B' that relied on natural ventilation during summer. The electricity and heating fuel inputs to produce one pig within each stage of production were fairly consistent between barns, except for discrepancies in the finishing phase (Table 1). Finishing A was tunnel-ventilated, with several large exhaust fans that increased the consumption of electricity by 3.5 times compared to Finishing B, which was a curtain-sided barn. This difference in construction likely explained the 43% increase in the use of LPG noted in Finishing B compared with A. Hanna et al. (2016) similarly reported that tunnel-ventilated barns required 43% more electricity per pig space than curtain-sided barns.

Electricity and heating fuel in pig enterprises will vary based on the stage of production, building design, weather conditions and the farmer's management of the ventilation and heating systems. Harmon and Schweitzer (2016) reported that electricity and LPG use averaged 240 kWh and 23 L per sow space, respectively, on sow farms. Assuming 30 pigs were weaned per sow annually, this equates to 8 kWh electricity and 0.8 L of LPG per weaner produced. Jacobson et al. (2011) used energy invoices and pig performance data supplied by farmers to estimate energy use in eight growing-finishing barns of various designs located in Iowa and Minnesota. Assuming a market pig weighed 118 kg, 10.6 kWh electricity and 3.5 L LPG were required to produce one market pig. Harmon and Schweitzer (2016) reported that in curtain-sided and tunnel-ventilated finishing buildings, respectively, electricity use was 22.6 and 25.9 kWh per pig space (9.41 and 10.79 kWh/market pig, assuming

Table 1 Total average annual electricity and LPG inputs per pig produced in buildings housing pigs of various ages[a]

Barn	Pig class produced[b]	Pigs produced annually	Annual electricity usage (kWh/pig)	Annual LPG usage (L/pig)
Breed-to-wean A	Weaner	58,419	11.36	1.29
Breed-to-wean B	Weaner	87,672	11.91	1.17
Nursery A	Feeder	19,103	2.38	1.63
Nursery B	Feeder	71,650	2.10	1.55
Finishing A	Market	6,328	14.40	1.29
Finishing B	Market	2,813	4.12	1.85

[a] Sharpe et al. (2018).
[b] Weaner = 21 days old; feeder = about 10 weeks old; market = slaughter-ready.

2.4 groups were produced annually). LPG use averaged about 2.5 L per pig space (1.04 L/market pig) in 30 finishing barns located in Iowa, Minnesota and Missouri. Hanna et al. (2016) measured 11.0 kWh of annual electricity use (4.7 kWh/market pig assuming 2.4 groups produced annually) and 17 L of LPG use per pig space (7.1 L/market pig) in a curtain-sided finishing barn in Iowa. These authors indicated that LPG use was very high, probably due to malfunctioning heaters.

Pig farmers generally know how much electricity and heating fuel they are consuming in their barns, based on the invoices provided by utility companies. However, pig farmers usually do not know the amount of energy being used for specific functions. Knowing which activities consume the most energy in a barn can help identify areas where the greatest impact on energy savings can be realized. Sharpe et al. (2018) converted the electricity and heating fuel consumed to a common basis, for comparison of energy used for specific functions (Fig. 2). In the breed-to-wean barns, LPG accounted for 45% of the total energy used, with the electricity for heat lamps used to warm suckling pigs following closely at 36%. Heat lamps used about 60% of all the electricity consumed in the breed-to-wean barns. Similarly, APL (2014) reported that heat lamps for suckling pigs represented 77% of the total energy consumed in Australian farrowing barns.

In the two nursery barns, heating of the pig space consumed 88% of the energy used and the ventilation system consumed about 50% of the electricity. The design of the two finishing barns greatly influenced how the energy was used. In the tunnel-ventilated barn, heating and ventilation accounted for 42% and 47% of total energy consumption, respectively, but in the curtain-sided design, heating and ventilation represented 76% and 20% of total energy

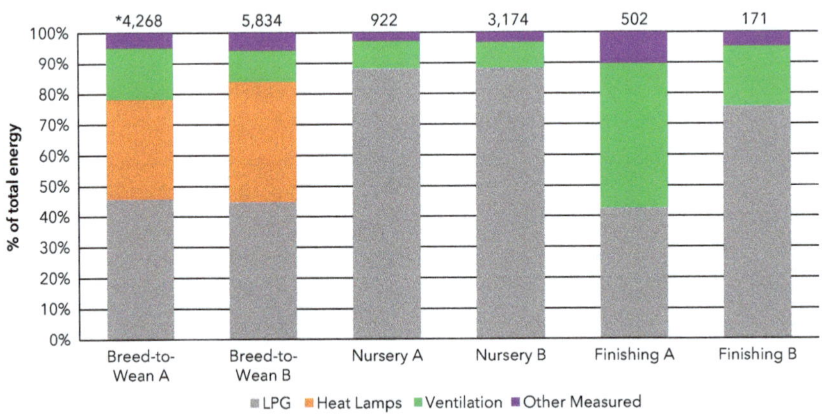

Figure 2 Total annual electrical and thermal energy use (GJ) for buildings producing pigs at different stages. *Total annual energy demand (GJ) for each building. From Sharpe et al. (2018). Used with permission.

consumption, respectively. Regardless of design, ventilation accounted for 72% to 81% of the total electricity used in the finishing phase.

This study of environmentally controlled pig barns under commercial production conditions showed that the largest users of energy were for space heating in breed-to-wean and nursery barns, providing zone heating for suckling pigs (breed-to-wean) and operating ventilation fans across all building types. These functions are ripe for interventions that could lead to meaningful reductions in energy use within pig barns.

5.2 Optimal operation of temperature control and ventilation systems

Systems installed to control temperature and maintain desirable air quality within the pig space can be significant users of energy. Each class of pig has a thermo-neutral temperature zone in which they are most comfortable (NRC, 1981; Curtis, 1985). Maintaining room temperature within this zone usually results in optimal pig performance. Temperatures below the thermo-neutral zone cause decreased efficiency of body weight gain because increased feed intake is required to maintain body temperature and growth rates (Lopez et al., 1991). In contrast, heat-stressed pigs reduce feed intake, which reduces growth performance (Renaudeau et al., 2008; Pearce et al., 2013) and causes them to use energy to dissipate excess body heat, which does not contribute directly to producing pork.

The ventilation system plays a key role in exhausting noxious gases, dust, humidity and, sometimes, excess heat. Build-up of noxious gases and dust in the pig space can compromise pig health and/or performance (Martin and Willoughby, 1972; Carpenter et al., 1986; Carpenter, 1986). When pig performance is compromised, there is a good possibility that the energy efficiency of the pork production system is compromised.

Temperature control and ventilation systems are designed to work in concert to optimize environmental conditions. However, in practice, they are often not managed properly or may even work against each other (Harmon et al., 2012). Such systems have become increasingly complex over time, making proper operation of them difficult for many on-farm users. Over-ventilation by exhausting more air than necessary is a common error. Ventilation rates as little as 10% over that required can increase LPG use for heating by 27% and 40% over requirements can double LPG use in wean-to-finish buildings (Harmon et al., 2010). Allowing room heaters to shut off at temperatures too close to the room setpoint causes a condition called 'heater run-by.' This causes room temperature to exceed the setpoint so that exhaust fans speed up to drive room temperature back down to the setpoint. This causes the heat just added to the room to be exhausted, which wastes energy (Harmon et al., 2012).

Negative pressure ventilation systems are common in pig buildings. Throttling air entering the pig space through inlets will increase static pressure in the room and require more power to achieve the desired airflows, causing elevated electrical usage (Harmon et al., 2012).

Routine maintenance of temperature control and ventilation systems is also necessary to reduce energy use. Dust exhausted from pig barns can build up on exhaust fans, shutters and blades and reduce air delivery by as much as 40%. Dirty discharge cones on exhaust fans can reduce fan efficiency by 15%, and loose drive belts on fans may reduce efficiency by 50% (Hanna et al., 2010). Reduced efficiency means that fans must run longer or more fans are needed to achieve the desired air exchange rate, and this will increase the use of electricity.

Proper settings and maintenance of temperature control and ventilation systems are relatively easy to implement at minimal costs. Therefore, they can be considered 'low-hanging fruit' related to improving energy efficiency. Pig farmers simply need to become adept at the proper operation of their systems and continuously monitor the effectiveness of their systems.

5.3 Structural deficiencies

Structural deficiencies can compromise the proper operation of the temperature control and ventilation systems, leading to increased energy use. Common structural deficiencies include air infiltration through unplanned inlets and faulty insulation. These deficiencies may also be present in new buildings if poorly designed but generally become more significant avenues of energy waste as the building ages. Air infiltration was greater in pig-finishing buildings that were over 13 years old compared with newer barns (Jadhav et al., 2018a).

Commonly used negative pressure ventilation systems rely on fans to exhaust air from the pig space, which creates a vacuum. This vacuum draws air into the space through planned inlets as well as by leakage through unplanned cracks in the building envelope (Zhang and Barber 1995; Jadhav et al., 2015). Such leakage compromises air distribution, which can lead to non-uniform air pockets, create drafts that compromise pig comfort and performance and contribute to more rapid deterioration of the building envelope (Masse et al., 1994). Air infiltration can account for 47% of the total air entering the building (Zhang and Barber, 1995), which will likely increase energy usage.

Sealing unplanned inlets will reduce air infiltration and improve the effectiveness of the ventilation system. Jadhav et al. (2018b) suggested that aggressive use of foaming insulation in the attic of a simulated pig barn could reduce air infiltration by 80%.

Having no insulation or deteriorating insulation can result in heat energy loss from pig barns. However, the diagnosis of this problem can be difficult in existing buildings because these problems are hidden inside walls, above ceilings and around fixtures such as fans, windows and doors in the building envelope.

Infrared thermography has been an increasingly useful, non-destructive tool when undertaking energy audits of residential and commercial buildings (Lucchi, 2018). Commercial cameras detect infrared radiation from objects and convert it to visual images that enable identification of heat loss through the building envelope, the sites of air infiltration, moisture migration into the walls and other structural defects (Fox et al., 2014; Lucchi, 2018). Energy audits using thermal cameras can be conducted by walking through or around the outside of a building and capturing thermal images of the building envelope and fixtures. Interpretation of these images with regard to improving energy efficiency must be conducted with careful consideration of the many structural and climatic conditions that can influence thermal images (Lehmann et al., 2013; Lucchi, 2018). The scientific literature is practically devoid of studies on infrared applications in agricultural structures such as pig barns.

Advances in consumer electronics have brought infrared thermography to the ubiquitous smartphone. Thermal camera attachments and supporting apps are available that allow practically anyone to capture thermal images (FLIR, 2021). However, untrained users must recognize that many factors can influence thermographic images (Lehmann et al., 2013; Fox et al., 2014). Consequently, training and professional experience are needed to properly interpret images when undertaking energy audits of buildings. Mauriello et al. (2017) provided novice users with smartphone thermal cameras for private energy auditing and found that they lacked the confidence to interpret thermal images and understand the severity of energy losses that they identified. Smartphone infrared thermography can probably help pig farmers undertake preliminary on-farm energy audits. However, more in-depth analysis of potential energy efficiencies will require input from professionals trained in performing energy audits.

5.4 Energy-saving practices

5.4.1 Piglet heating

Piglets are born with a limited ability to regulate their body temperature because they have little hair, a small mass relative to body surface area, low body fat content, and limited glycogen reserves (Mersmann, 1974). Excessive chilling can often lead to piglet death. Therefore, supplemental heat is required for them. From a practical perspective, providing supplemental heat for piglets is challenging because the ideal environmental temperature for newborn piglets is about 32°C, while sows are most comfortable at 16°C (Holden and

Ensminger, 2006). To address this challenge, pig farmers operate the farrowing room at temperatures comfortable for sows and provide zone heating for piglets. Traditionally, zone heating for piglets has been provided by electric heat lamps suspended over the creep area of the farrowing stall or pen (Xin et al., 1997).

Pig farmers have used heat lamps extensively because they are easy to install, relatively inexpensive and easily moved (Zhu et al., 2020). However, heat lamps have a relatively short lifespan because they are easily broken. Heat lamps provide radiant heat from above the piglets, with spatial variation across the heated area, which may limit the useful space for the piglets (Zhang and Xin, 2001) and may contribute unwanted heat to the sow's space. Consequently, heat lamps are an energy-inefficient method of providing supplemental heat, although different designs and wattages of heat lamps can improve performance. Xin et al. (1997) found that an energy-efficient, 175 W radiant heat lamp improved piglet survival by 1.2 percentage units and decreased electricity consumption by 21% compared with a 250 W infrared heat lamp.

Electric heat mats provide more consistent heating over a larger area compared with heat lamps (Zhang and Xin, 2001). Although they have much higher capital and installation costs, their functional life can last for up to 10 years (Lane et al., 2020). Electric heat mats provide heat to piglets through conduction, as piglets lay on the mats located in the creep area of the farrowing stall or pen. Heat mats consume only 27% to 63% of the electricity required by heat lamps, without any effects on piglet mortality or growth rate (Beshada et al., 2014; Stinn and Xin, 2014; Johnston et al., 2019b; Lane et al., 2020; Zhu et al., 2020). In the first few days of life, piglets prefer heat lamps to heat mats (Zhang and Xin, 2001; Lane et al., 2020) but this preference declines as piglets age (Zhu et al., 2020). In early life, piglets seek the warmth rather than the light provided by the heat lamps (Larsen et al., 2017). With no advantage in piglet survival or growth rate, the higher capital cost of heat mats must be offset by savings in electricity before pig farmers will use electric heat mats. Lane et al. (2020) calculated a payback period of 11.7 months in an intensive farrowing system when comparing electric heat mats to a 125 W heat lamp by assuming a 72% reduction in electricity consumption. Based on operational costs alone, electric heat mats are preferred to heat lamps, but the economic choice is less clear when including initial purchase and installation costs (Zhu et al., 2020).

Smith et al. (2019) studied a semi-enclosed heated micro-climate (SEHM) provided for piglets in the creep area of farrowing stalls. The SEHM consisted of a parabolic-shaped shield that directed heat from a horizontal infrared heating element toward the piglets. Clear vinyl curtains hung from the edge of the shield helped contain the heat, creating a micro-climate for piglets. Piglet mortality was not influenced by SEHM, but the incidence of crushed piglets was reduced compared with using 125 W heat lamps. Average electricity use was reduced from 6 kWh/day for heat lamps to 3.25 kWh/day for SEHM. The authors calculated a payback period of 6.1 years based on this 59% reduction in electricity use and assuming an initial capital cost of USD400 per SEHM unit.

5.4.2 Reduced nocturnal temperature

The largest use of fossil fuels in pig nursery buildings is for heating the rooms that house pigs (Fig. 2). Nursery pigs will select a cooler night-time room temperature compared to daytime temperature when allowed to choose (Bench and Gonyou, 2007). Therefore, one way to reduce heating fuel use is to reduce the nocturnal temperature (RNT). In practice, night-time room temperature (from around 1900 to 0700 hours) can be reduced by 6°C to 8°C from the daytime setpoint temperature by reducing the activation temperature for the room heater without changing exhaust fan settings. This approach allows the room to cool down gradually so that recently added heat is not abruptly exhausted when the RNT protocol initiates each night.

In the 1980s, RNT reduced LPG and electricity use in Nebraska, in the range of 15.5% to 32.4% (Brumm et al., 1985; Brumm and Shelton, 1988). This was concurrent with increases in the daily feed intake and weight gain of pigs. In a subsequent study with a more aggressive RNT of 10°C, Brumm and Shelton (1991) reported a 21% and 27% reduction in LPG and electricity use respectively, with no effect on pig performance. Similarly, Swinkels et al. (1988) reported a 35% reduction in the use of natural gas to heat nursery rooms when RNT was implemented, with no effect on growth performance but a slight increase in the incidence of diarrhea.

Despite the apparent benefits of reducing energy use, RNT regimens have not been adopted widely in the commercial swine industry for a variety of reasons, including a lack of sophisticated heating and ventilation controllers that could automatically implement different room temperatures during the day and at night. However, recent advances in controller technology have overcome this problem. A more recent analysis of RNT in pig nursery barns (Johnston et al., 2013) reduced target night-time room temperature by 6°C, 7 days after weaned pigs were placed in the nursery (experiment 1), and by 8.3°C after 5 days (experiment 2). RNT began at 1900 h and ended at 0700 h each day. Temperature reductions did not affect the performance or health of pigs during the 5- to 6-week nursery period (Table 2). In experiment 1, consumption of heating fuel and electricity declined by 18% and 9%, respectively, in the RNT rooms compared with control rooms, but these differences were not statistically significant. In experiment 2, the more aggressive temperature reduction lowered heating fuel and electricity consumption by 30% and 20%, respectively, which was statistically and practically significant.

Table 2 Effect of reduced nocturnal temperature (RNT) on nursery pig performance and energy use in the US[1]

	Experiment 1			Experiment 2		
Trait	Control	RNT	SE[2]	CON	RNT	SE
No. of rooms	6	6		9	10	--
Initial pig weight (kg)	6.2	6.2	0.13	6.2	6.2	0.03
Final pig weight (kg)	23.7	23.7	1.04	21.8	21.5	0.64
Average daily gain (kg)	0.43	0.43	0.022	0.45	0.44	0.016
Average daily feed intake (kg)	0.62	0.62	0.025	0.61	0.60	0.019
Gain:feed intake efficiency	0.69	0.69	0.011	0.75	0.75	0.012
Heating fuel used (Btu/pig/day)	9658	7958	1662	10 019[a]	7061[b]	1467
Electricity used (kWh/pig/day)	0.138	0.125	0.014	0.026[a]	0.021[b]	0.004

[a,b] Within a row and experiment, means with different superscripts differ ($P < 0.05$).
[1] Johnston et al. (2013).
[2] Pooled standard error.

RNT regimens appear to have consistent benefits in reducing fossil fuel use in pig nurseries where heating demands are high, without compromising pig performance and health. Farmers may be able to realize similar benefits in the early part of the growing-finishing period when supplemental heat is required in some climates. Benefits in later stages are less likely. Lemay et al. (2001) used an RNT regimen to increase air exchange rates at night in an attempt to mitigate the heat stress of growing-finishing pigs in Saskatchewan, Canada. They reported small numerical improvements in pig performance due to RNT but increased electricity consumption to power the ventilation fans.

5.4.3 Lighting

Typically, lighting in swine barns is provided by incandescent bulbs that produce light by electrical resistance, which also generates significant heat (Harmon and Petersen, 2011). This decreases energy efficiency. Several other types of light bulbs are available for livestock facilities, with varying advantages and disadvantages. Recent improvements in light-emitting diodes (LEDs) have garnered the increasing attention of livestock producers. The electricity consumption of LED bulbs is about 15% to 20% of that used for incandescent bulbs to produce the same quantity of light (CERT, 2021b). LED bulbs emit directional light rather than the 360 degree lighting produced by incandescent bulbs. This characteristic of LED bulbs may dictate that more LED bulbs need to be installed to achieve adequate illumination of farm buildings compared to incandescent bulbs. The increased number of bulbs and higher initial cost of LEDs results in greater initial investment. However, a far longer life is expected, together with reduced electricity consumption, compared with incandescent bulbs.

Evaluations of LED lighting in poultry barns determined the energy use, durability and quality of light produced. While not exactly the same, the harsh environment of a poultry barn is similar to that of a pig barn with regard to airborne dust, humidity and ventilation systems. Four poultry barns with existing compact fluorescent bulbs (CFL) and four poultry barns with incandescent bulbs in Minnesota were studied. Annual electricity savings of 8% to 48% were achieved by replacing CFL bulbs with LEDs, and 84% to 88% when compared to incandescent bulbs (Ebinger, 2014). Electricity savings realized in the switch from CFL to LED bulbs did not cover the initial costs of the LED system, but when switching incandescent bulbs to LEDs, electricity savings gave a 1- to 2-year payback period. Evaluation of LED lighting systems in modern commercial pig barns has not been undertaken to date and is needed.

5.4.4 Cooling

Heat stress of livestock occurs when the environmental temperature exceeds the animal's upper critical temperature such that evaporative heat loss from sweating and/or panting occurs (NRC, 1981). Heat stress reduces the performance of both sows (Black et al., 1993) and growing pigs (Ross et al., 2015). The occurrence of heat stress has and will increase further in temperate regions due to climate change (Mikovits et al., 2019). Effective cooling of pigs improves their biological efficiency and allows greater output to be realized from the energy-intensive inputs of feed and electricity. Therefore, cooling the pigs can be an important part of energy-smart pig farming.

Cooling approaches can require little to no additional energy input or can require substantial energy input, to operate effectively. In either case, improved comfort and performance of pigs must be realized, to contribute positively to energy-efficiency benefits. Schauberger et al. (2020) divided cooling options into building-related or animal-level practices. Building-related designs such as evaporative cooling of inlet air, earth-tempered cooling of inlet air, and geothermal cooling of inlet air by groundwater, were deemed to be the most effective cooling techniques but required additional capital investment, although they needed minimal additional energy to operate. Animal-level cooling practices, such as increased velocity of ventilation air, cooled drinking water, cooled flooring in lying areas and sprinkling or fogging animals with water, did provide some measure of cooling but were less effective than building-level designs.

In some cases, animal-level interventions require substantial energy inputs to operate. For instance, Johnston et al. (2019a) studied a system that circulated cooled water under the floor and provided cooled drinking water to heat-stressed lactating sows housed in farrowing stalls. Water was cooled by an air-source heat pump that was powered by a solar photovoltaic array. They reported a significant decline in heat stress signs in sows but no improvement in sow or litter performance. The farrowing room equipped with the cooling system required 1.6 to 2.6 times more electricity to operate than the control room without the cooling system installed. In this instance, additional energy inputs were required to operate the cooling system with no additional output. Hence, the energy efficiency of pig production was not improved by cooling the sows.

In contrast, other researchers reported improved sow and litter performance as a result of mitigating heat stress when lactating sows were housed on cooled floors (Silva et al., 2006, 2009; Cabezon et al., 2017). The small energy input of 27.2 kWh/sow/lactation required to cool sows (Silva et al., 2006) resulted in increased output of pork, thus improving the energy efficiency of production. Other less complicated cooling systems such as drip coolers, sprinklers, snout cooler, and foggers can improve sow and growing pig performance. However, quantitative evaluations of the energy required to operate these systems compared with improved pig performance have not yet been undertaken.

6 Conclusion and future trends

Energy-smart pig farming will become increasingly important in the foreseeable future as society demands food with less negative environmental impacts. Pig farmers can contribute to a more environmentally friendly food production system by improving energy efficiency. The conservation of energy in existing pig barns, implemented with minimal to no new capital investment, is the first step to improving energy efficiency. Proper maintenance and operation of heating and ventilation systems can be implemented on every pig farm using minimal labor inputs. Any costs avoided from reducing LPG heating fuel or electricity inputs can be used to pay for moderate capital investments in heating mats for suckling pigs and computerized controllers needed for RNT regimens in pig nurseries. Optimizing the performance of all classes of pigs will enhance the overall energy footprint of pig farming by maximizing the output of pork from energy inputs. Genetic improvements in pig performance, proper nutritional programs, health maintenance practices and advanced reproductive management approaches will all contribute to optimizing pig performance. Management approaches that both conserve energy and optimize pig performance can be implemented in existing enterprises to achieve energy-smart pig farming.

Pig barns in the future will probably incorporate more effective, energy-efficient designs and include on-farm generation of electricity from solar PV and possibly heat from geothermal systems. They will have more robust wall, roof and floor insulation, assuming that this will become more economically viable for farmers to achieve. Improvements in the energy efficiency of feed production will also contribute to energy-smart pig farming with a lower energy footprint in the future.

7 Where to look for further information

As mentioned in the introduction to this chapter, discussion of and thinking about energy efficiency of pig farming is not new. However, researchers and commercial pig farmers historically have not dedicated much effort to improving energy efficiency of pig farming probably because the cost of energy has represented a small portion of the total costs required to produce a market pig. Consequently, there are not expansive texts or comprehensive references on the topic available currently. Recent technological advances that allow improvements in energy efficiency and increasing societal pressures that encourage pig farmers to scrutinize and reduce their energy and environmental footprints have renewed scientific and practical interest in improving energy efficiency of pig farming. These renewed interests have prompted university, governmental, and industry groups to assemble current information that is updated regularly on websites. Readers interested in gaining additional perspectives on energy-smart pig farming are directed to the following sources:

- University of Minnesota's Clean Energy Resource Teams (CERTS) at: https://www.cleanenergyresourceteams.org/.
- United States Department of Energy's National Renewable Energy Laboratory (NREL) at: https://www.nrel.gov/.
- United States Environmental Protection Agency's AgSTAR: Biogas Recovery in the Agriculture Sector at: http://epa.gov/agstar.
- Australian Pork Limited at: https://australianpork.com.au/sites/default/files/2021-06/2017-2212.pdf.

8 Acknowledgement

The authors appreciate the critical review of this manuscript provided by Dr. E. Cortus, University of Minnesota Department of Bioproducts and Biosystems Engineering.

9 References

Adair, C. W., Xu, J., Elliott, J. S., Simmons, W. G., Cavanaugh, M., Vujic, T. and Deshusses, M. A. 2016. Design and assessment of an innovative swine waste to renewable energy system. *Trans. Amer. Soc. Agric. Biol. Eng.* 59:1009–1018. http://doi:10.13031/trans.59.11731.

Amaducci, S., Yin, X. and Colauzzi, M. 2018. Agrivoltaic systems to optimise land use for electric energy production. *Appl. Energy* 220:545–561. http://doi:10.1016/j.apenergy.2018.03.081.

APL (Australian Pork Limited) 2014. Fact sheet: piggery total energy usage. Energy Efficiency Information Fact Sheet Series. Project 2012.2407. Available at: https://australianpork.infoservices.com.au/items/2012-2407REPORT.

Barber, E. M., Classen, H. L. and Thacker, P. A. 1989. Energy use in the production and housing of poultry and swine - an overview. *Can. J. Anim. Sci.* 69(1):7–21.

Barron-Gafford, G. A., Pavao-Zucherman, M. A., Minor, R. L., Sutter, L. F., Barnett-Moreno, I., Blackett, D. T., Thompson, M., Dimond, K., Gerlak, A. K., Nabhan, G. P. and Machnick, J. E. 2019. Agrivoltaics provide mutual benefits across the food-energy-water nexus in drylands. *Nat. Sustain.* 2(9):848–855.

Belyakov, N. 2019. *Sustainable Power Generation: Current Status, Future Challenges, and Perspectives*. Academic Press, Cambridge, MA. pg. 475–500.

Bench, C. J. and Gonyou, H. W. 2007. Temperature preference in piglets weaned at 12 – 14 days of age. *Can. J. Anim. Sci.* 87(3):299–302.

Beshada, E., Zhang, Q. and Boris, R. 2014. Energy consumption of heat pads and heat lamps and aerial environment in a commercial swine farrowing facility. *Can. Biosyst. Eng.* 56(1):5.1–5.6. http://doi.org/10.7451/CBE.2014.56.5.1.

Black, J. L., Mullan, B. P., Lorschy, M. L. and Gilles, L. R. 1993. Lactation in the sow during heat stress. *Livest. Prod. Sci.* 35(1–2):153–170.

Bostami Rubayet, A. B. M., Mun, H. S., Islam, M. M. and Yang, C. J. 2016. Potential of geothermal and ground channel system house on reduction of energy consumption and CO_2 emissions with maintenance of performance of growing pigs. *Int. J. Recent Sci. Res.* 7:8974–8979.

Brooks, A. E. 2014. *Future Energy: Improved, Sustainable and Clean Options for Our Planet*. Elsevier Sci., Amsterdam, Netherlands. pg. 383–404.

Brumm, M. C. and Shelton, D. P. 1988. A modified reduced nocturnal temperature regimen for early-weaned pigs. *J. Anim. Sci.* 66(5):1067–1072.

Brumm, M. C. and Shelton, D. P. 1991. Two reduced nocturnal temperature regimens for early-weaned pigs. *J. Anim. Sci.* 69(4):1379–1388.

Brumm, M. C., Shelton, D. P. and Johnson, R. K. 1985. Reduced nocturnal temperatures for early weaned pigs. *J. Anim. Sci.* 61(3):552–558.

Cabezon, F. A., Schinckel, A. P., Marchant-Forde, J. N., Johnson, J. S. and Stwalley, R. M. 2017. Effect of floor cooling on late lactation sows under acute heat stress. *Livest. Sci.* 206:113–120. http://doi.org/10.1016/j.livsci.2017.10.017.

Carpenter, G. A. 1986. Dust in livestock buildings - review of some aspects. *J. Agric. Eng. Res.* 33(4):227–241.

Carpenter, G. A., Cooper, A. W. and Wheeler, G. E. 1986. The effect of air filtration on air hygiene and pig performance in early-weaner accommodation. *Anim. Sci.* 43(3):505–515.

CERT (Clean Energy Resource Teams) 2021a. Simple steps to solar. Available at: https://www.cleanenergyresourceteams.org/simple-steps-solar.

CERT (Clean Energy Resource Teams) 2021b. Find energy efficient lights. Available at: https://www.cleanenergyresourceteams.org/lighting.

Cheng, J., Shearin, T. E., Peet, M. M. and Willits, D. H. 2004. Utilization of treated swine wastewater for greenhouse tomato production. *Water Sci. Technol.* 50(2):77–82. http://doi:10.2166/wst.2004.0093.

Chwieduk, D. A. 2017. Towards modern options of energy conservation in buildings. *Renew. Energy* 101:1194–1202. http://doi.org/10.1016/j.renene.2016.09.061.

Curtis, S. E. 1985. *Stress Physiology in Livestock* (vol. II). CRC Press Inc, Boca Raton, FL.

Devabhaktuni, V., Alam, M., Depuru, S. S. R., Green, R. C., Nims, D. and Near, C. 2013. Solar energy: trends and enabling technologies. *Renew. Sustain. Energy Rev.* 19:555–564. http://doi:10.1016/j.rser.2012.11.024.

Dinesh, H. and Pearce, J. M. 2016. The potential of agrivoltaic systems. *Renew. Sustain. Energy Rev.* 54:299–308. http://doi:10.1016/j.rser.2015.10.024.

Dominguez, L. T. and Predicala, B. 2011. Evaluation of the use of heat recovery ventilator and ground source heat pump in swine grow-finish rooms. *Proceedings of the CSBE/SCGAB 2011 Annual Conference*, Winnipeg, Manitoba.

Ebinger, F. 2014. LEDs: energy savings and replicability in MN livestock facilities. Conservation Appl. Res. Dev., Minnesota Depart. Commerce, St. Paul, MN. Available at: https://www.cleanenergyresourceteams.org/sites/default/files/poultry-report.pdf.

FLIR 2021. Pro-grade thermal cameras for smartphones. Available at: https://www.flir.com/products/flir-one-pro-lt/.

Fox, M., Coley, D., Goodhew, S. and de Wilde, P. 2014. Thermography methodologies for detecting energy related building defects. *Renew. Sustain. Energy Rev.* 40:296–310. http://doi.org/10.1016/j.rser.2014.07.188.

Fredeen, H. T. and Harmon, B. G. 1983. The swine industry: changes and challenges. *J. Anim. Sci.* 57 (Suppl. 2):100–118.

Godbout, S., Pouliot, F., Lachance, I., Guimont, H., Leblanc, R., Pelletier, F. and Hamelin, L. 2004. Feasibility and energy recovery of a solar wall in pig nursery. *Proc. 2004 CSBE/SCGAB Amer. Soc. Agric. Biol. Eng.*, Edmonton, Canada.

González-García, I., Riaño, B., Molinuevo-Salces, B., Vanotti, M. B. and García-González, M. C. 2020. Improved anaerobic digestion of swine manure by simultaneous ammonia recovery using gas-permeable membranes. *Water Res.* 190. http://doi:10.1016/j.watres.2020.116789.

Grando, R. L., de Souza Antune, A. M., da Fonseca, F. V., Sánchez, A., Barrena, R. and Font, X. 2017. Technology overview of biogas production in anaerobic digestion plants: a European evaluation of research and development. *Renew. Sustain. Energ. Rev.* 80:44–53. http://doi:10.1016/j.rser.2017.05.079.

Hanna, H. M., Harmon, J. D. and Petersen, D. 2010. Farm energy: energy efficient fans for swine production. *Agric. Environ. Ext. Pub.* 25. Available at: http://lib.dr.iastate.edu/extension_ag_pubs/25.

Hanna, H. M., Harmon, J. D. and Schweitzer, D. D. 2016. Energy use for field operations, crop drying, and swine housing on university farms. *Appl. Eng. Agric.* 32(6):769–781. http://doi.org/10.13031/aea.32.11720.

Harmon, J. D., Brumm, M. C., Jacobson, L. D., Pohl, S. H., Stender, D. R. and Stowell, R. R. 2012. Field performance evaluation of a ventilation system: a swine case study. *Appl. Eng. Agric.* 28(2):251–257.

Harmon, J. D., Hanna, H. M. and Petersen, D. 2010. Farm energy: sizing minimum ventilation to save heating energy in swine housing. *Agric. Environ. Ext. Pub.* 28. IA State Univ. Available at: http://lib.dr.iastate.edu/extension_ag_pubs/28.

Harmon, J. D. and Petersen, D. 2011. Farm energy: indoor lighting for livestock, poultry, and farm shop facilities. *Agric. Environ. Ext. Pubs.* 32. Available at: http://lib.dr.iastate.edu/extension_ag_pubs/32.

Harmon, J. and Schweitzer, D. 2016. Farm energy: case studies / benchmarking energy usage for swine producers. Iowa State Univ. PM 3063E. Available at: https://store.extension.iastate.edu/product/14499.

Hayter, S. J. and Kandt, A. 2011. Renewable energy applications for existing buildings: Preprint. Natl. Renew. Energy Lab, United States. Paper presented at the 48th AiCARR International Conf., Baven-Lago Maggiore, Italy. Available at: https://www.osti.gov/servlets/purl/1022419.

Hijazi, O., Abdelsalam, E., Samer, M., Amer, B. M. A., Yacoub, I. H., Moselhy, M. A., Attia, Y. A. and Bernhardt, H. 2020. Environmental impacts concerning the addition of trace metals in the process of biogas production from anaerobic digestion of slurry. *J. Clean. Prod.* 243:118593. http://doi:10.16/j.clepro.2019.118593.

Holden, P. J. and Ensminger, M. E. 2006. Buildings and equipment. In: *Swine Science* (7th edn.). Pearson Prentice Hall publishers. pg. 391.

Jacobson, L. D., Schmidt, D. R., Lazarus, W. F. and Koehler, R. 2011. Reducing the environmental footprint of pig finishing barns. ASABE Annual International Mtg:Paper no. 1110589.

Jadhav, H. T., Hoff, S. J., Harmon, J. D., Alvarez, I., Andersen, D. S. and Passe, U. 2018a. Swine finishing room air infiltration: Part 1. Quantification and prediction. *Appl. Eng. Agric.* 34(2):413–424. http://doi.org/10.13031/aea.12396.

Jadhav, H. T., Hoff, S. J., Harmon, J. D. and Andersen, D. S. 2018b. Swine finishing room air infiltration: Part 2. Infiltration as affected by room characteristics. *Appl. Eng. Agric.* 34(4):735–745. http://doi.org/10.13031/aea.12559.

Jadhav, H. T., Hoff, S. J., Harmon, J. D., Jacobson, L. D. and Hetchler, B. P. 2015. Infiltration characteristics of swine finishing and gestation buildings: review and quantification. *Amer. Soc. Agric. Biol. Eng. Mtg. presentation*. Paper number 152190046. http://doi.org/10.13031/aim.20152190046.

Johnston, L. J., Brumm, M. C., Moeller, S. J., Pohl, S., Shannon, M. C. and Thaler, R. C. 2013. Effects of reduced nocturnal temperature on pig performance and energy consumption in swine nursery rooms. *J. Anim. Sci.* 91(7):3429–3435. https://doi.org/10.2527/jas.2012-5824.

Johnston, L., Reese, M., Buchanan, E., Li, Y., Hilbrands, A., Janni, K., Hetchler, B. and Cortus, E. 2019a. Effects of a solar cooling system on sow performance. Final rep. Minnesota Legislature, Legislative-Citizen Comm. MN Resources. Available at: https://www.lccmr.leg.mn/projects/2016/finals/2016_07e_rpt_sow_performance_25pgs.pdf.

Johnston, L. J., Sharpe, K. T., Reese, M. H., Buchanan, E. S., Tallaksen, J. E., Hilbrands, A. M., Janni, K. A., Hetchler, B. and Li, Y. 2019b. Quantifying use of electricity to heat piglet creep areas in farrowing rooms. *Adv. Anim. Biosci.* 10 (Suppl.1):s22.

Kannan, N. and Vakeesan, D. 2016. Solar energy for future world: a review. *Renew. Sustain. Energy Rev.* 62:1092–1105. http://doi:10.1016/j.rser.2016.05.022.

Kebreab, E., Liedke, A., Caro, D., Deimling, S., Binder, M. and Finkbeiner, M. 2016. Environmental impact of using specialty feed ingredients in swine and poultry production: a life cycle assessment. *J. Anim. Sci.* 94(6):2664–2681. http://doi:10.2527/jas.2015-9036.

Kim, M., Jung, S., Lee, D. J., Lin, K. A., Jeon, Y. J., Rinklebe, J., Klinghoffer, N. B. and Kwon, E. E. 2020. Biodiesel synthesis from swine manure. *Bioresour. Technol.* 317. http://doi:10.1016/j.biortech.2020.124032.

Krommweh, M. S., Rösmann, P. and Büscher, W. 2014. Investigation of heating and cooling potential of a modular housing system for fattening pigs with integrated geothermal heat exchanger. *Biosyst. Eng.* 121:118–129. http://doi:10.1016/j.biosystemseng.2014.02.008.

Lammers, P. J., Honeyman, M. S., Harmon, J. D. and Helmers, M. J. 2010a. Energy and carbon inventory of Iowa swine production facilities. *Agric. Syst.* 103(8):551–561. http://doi:10.1016/j.agsy.2010.06.003.

Lammers, P. J., Kenealy, M. D., Kliebenstein, J. B., Harmon, J. D., Helmers, M. J. and Honeyman, M. S. 2010b. Nonsolar energy use and one-hundred-year global warming potential of Iowa swine feedstuffs and feeding strategies. *J. Anim. Sci.* 88(3):1204–1212. http://doi:10.2527/jas.2009-2371.

Lammers, P. J., Kenealy, M. D., Kliebenstein, J. B., Harmon, J. D., Helmers, M. J. and Honeyman, M. S. 2012. Energy use in pig production: an examination of current Iowa systems. *J. Anim. Sci.* 90(3):1056–1068.

Lane, K. J., Johnson, A. K., Stilwill, C. E. J., Karriker, L. A., Harmon, J. D. and Stalder, K. J. 2020. Comparison of heat lamps and heat mats in the farrowing house: effect on piglet production, energy use, and piglet and sow behavior through live observation. *J. Swine Health Prod.* 28:205–201.

Larsen, M. L. V., Thodberg, K. and Pedersen, L. J. 2017. Radiant heat increases piglets' use of the heated creep area on the critical days after birth. *Livest. Sci.* 201:74–77. http://doi.org/10.1016/j.livsci.2017.05.008.

Lehmann, B., Ghazi Wakili, K., Frank, Th., Vera Collado, B. and Tanner, Ch. 2013. Effects of individual climatic parameters on the infrared thermography of buildings. *Appl. Energy* 110:29–43. http://doi.org/10.1016/j.apenergy.2013.03.066.

Lemay, S. P., Guo, H., Barber, E. M. and Chenard, L. 2001. Performance and carcass quality of growing-finishing pigs submitted to reduced nocturnal temperature. *Trans. Amer. Soc. Agric. Eng.* 44:957–965.

Licharz, H., Rösmann, P., Krommweh, M. S., Mostafa, E. and Büscher, W. 2020. Energy efficiency of a heat pump system: case study in two pig houses. *Energies* 13(3):662. http://doi:10.3390/en13030662.

Lopez, J., Jesse, G. W., Becker, B. A. and Ellersieck, M. R. 1991. Effects of temperature on the performance of finishing swine: II. Effects of a cold, diurnal temperature on average daily gain, feed intake, and feed efficiency. *J. Anim. Sci.* 69(5):1850–1855.

Lucchi, E. 2018. Applications of the infrared thermography in the energy audit of buildings: a review. *Renew. Sustain. Energy Rev.* 82:3077–3090. http://doi.org/10.1016/j.rser.2017.10.031.

Lund, H. 2010. Introduction. *Renewable Energy Systems*. Academic Press, Cambridge, MA. pg. 1–14.

Lusk, P. 1998. *Methane Recovery from Animal Manures: The Current Opportunities Casebook*. 3rd edn. NREL/SR-25145. Golden, CO: National Renewable Energy Laboratory, Washington, DC. https://www.nrel.gov/docs/fy99osti/25145.pdf. Accessed 12 February 202.

Martin, S. W. and Willoughby, R. A. 1972. Organic dusts, sulfur dioxide, and the respiratory tract of swine. *Arch. Environ. Health* 25(3):158–165. https://doi.org/10.1080/00039896.1972.10666155.

Massé, D. I., Masse, L., Xia, Y. and Gilbert, Y. 2010. Potential of low-temperature anaerobic digestion to address current environmental concerns on swine production. *J. Anim. Sci.* 88(13) (Suppl.):E112-E120. http://doi:10.2527/jas.2009-2432.

Masse, D. I., Munroe, J. A. and Jackson, H. A. 1994. Air leakage through farm building envelopes in Eastern Ontario. *Can. Agric. Eng.* 36:159-163.

Mauriello, M. L., Saha, M., Brown, E. and Froehlich, J. E. 2017. Exploring novice approaches to smartphone-based thermographic energy auditing: a field study. *Proceedings of the ACM SIGCHI Conference on Human Factors in Computing Systems*. pg. 1768-1780. http://doi.org/10.1145/3025453.3025471.

Mersmann, H. J. 1974. Metabolic patterns in the neonatal swine. *J. Anim. Sci.* 38(5):1022-1030. https://doi.org/10.2527/jas1974.3851022x.

Mikovits, C., Zollitsch, W., Hortenhuber, S. J., Baumgartner, J., Niebuhr, K., Piringer, M., Anders, I., Andre, K., Hennig-Pauka, I., Schonhart, M. and Schauberger, G. 2019. Impacts of global warming on confined livestock systems for growing-fattening pigs: simulation of heat stress for 1981 to 2017 in central Europe. *Int. J. Biometeorol.* 63(2):221-230. https://doi.org/10.1007/s00484-018-01655-0.

Mun, H. S., Ahmed, S. T., Islam, M. M., Park, K. J. and Yang, C. J. 2015. Retrofitting of a pig nursery with solar heating system to evaluate its ability to save energy and reduce environmental pollution. *Eng. Agric. Environ. Food* 8(4):235-240. (Abstr). http://doi:10.1016/j.eaef.2015.02.001.

Mun, H. S., Dilawar, M. A., Jeong, M. G., Rathnayake, D., Won, J. S., Park, K. W., Lee, S. R., Ryu, S. B. and Yang, C. J. 2020. Effect of a heating system using a ground source geothermal heat pump on production performance, energy-saving and housing environment of pigs. *Animals (Basel)* 10(11):2075. http://doi:10.3390/ani10112075.

Nardin, F. and Mazzetto, F. 2014. Mapping of biomass fluxes: a method for optimizing biogas-refinery of livestock effluents. *Sustainability* 6(9):5920-5940. http://doi:10.3390/su6095920.

Nguyen, T. L. T., Hermansen, J. E. and Mogensen, L. 2010. Fossil energy and GHG saving potentials of pig farming in the EU. *Energy Policy* 38(5):2561-2571. http://doi:10.1016/j.enpol.2009.12.051.

Nguyen, T. L. T., Hermansen, J. E. and Mogensen, L. 2012. Environmental costs of meat production: the case of typical EU pork production. *J. Clean. Prod.* 28:168-176. http://doi:10.1016/j.clepro.2011.08.018.

NRC (National Research Council) 1981. *Effect of Environment on Nutrient Requirements of Domestic Animals*. National Academy Press, Washington, DC.

NREL (National Renewable Energy Laboratory) 2021a. PVWatts® calculator. Available at: https://pvwatts.nrel.gov/index.php.

NREL (National Renewable Energy Laboratory) 2021b. System Advisor Model (SAM). Available at: https://sam.nrel.gov/.

Ogunlade, C. A., Akinyele, O. A., Oladmeji, S. T. and Oladipo, O. O. 2018. Renewable energy usage for agricultural practices: a review. International Conference of the Nigerian Institution of Agricultural Engineers, Oshodi, Lagos, Nigeria. Pg. 58-69.

Panwar, N. L., Kaushik, S. C. and Kothari, S. 2011. Role of renewable energy sources in environmental protection: a review. *Renew. Sustain. Energy Rev.* 15(3):1513-1524. http://doi:10.1016/j.rser.2010.11.037.

Pearce, S. C., Gabler, N. K., Ross, J. W., Escobar, J., Patience, J. F., Rhoads, R. P. and Baumgard, L. H. 2013. The effects of heat stress and plane of nutrition on metabolism of growing pigs. *J. Anim. Sci.* 91(5):2108–2118. https://doi.org/10.2527/jas.2012-5738.

Pelletier, N., Lammers, P., Stender, D. and Pirog, R. 2010. Life cycle assessment of high- and low-profitability commodity and deep-bedded niche swine production systems in the upper midwestern United States. *Agric. Syst.* 103(9):599–608. http://doi:10.1016/j.agsy.2010.07.001.

Renaudeau, D., Kerdoncuff, M., Anais, C. and Gourdine, J. L. 2008. Effect of temperature level on thermal acclimation in Large White growing pigs. *Animal* 2(11):1619–1626. http://doi:10.1017/S1751731108002814.

Resch, G., Held, A., Faber, T., Panzer, C., Toro, F. and Haas, R. 2008. Potentials and prospects for renewable energies at global scale. *Energy Policy* 36(11):4048–4056. http://doi:10.1016/j.enpol.2008.06.029.

Riva, G., Foppa Pedretti, E. and Fabbri, C. 2000. Utilization of a heat pump in pig breeding for energy saving and climate and ammonia control. *J. Agri. Eng. Res.* 77:449–455. http://doi:10.1006/jaer.2000.0624.

Ross, J. W., Hale, B. J., Gabler, N. K., Rhoads, R. P., Keating, A. F. and Baumgard, L. H. 2015. Physiological consequences of heat stress in pigs. *Anim. Prod. Sci.* 55(12):1381-1390. http://doi.org/10.1071/AN15267.

Schauberger, G., Hennig-Pauka, I., Zollitsch, W., Hortenhuber, S. J., Baumgartner, J., Niebuhr, K., Piringer, M., Knauder, W., Anders, I., Andre, K. and Schonhart, M. 2020. Efficacy of adaptation measures to alleviate heat stress in confined livestock buildings in temperate climate zones. *Biosys. Eng.* 200:157–175. http://doi.org/10.1016/j.biosystemseng.2020.09.010.

Schmitt, J., Urriola, P. and Shurson, G. 2018. Carbon, water, and land use for pork production when modifying type and regional sourcing of feed ingredients. *Res. Rep. NPB #17-128*. Available at: https://pork.org/research/carbon-water-land-use-pork-production-modifying-type-regional-sourcing-feed-ingredients/. National Pork Board, Des Moines, IA.

Scott, N. R. 1984. Livestock buildings and equipment: a review. *J. Agric. Eng. Res.* 29(2):93–114.

Shah, S. B., Lentz, Z. A., van Heugten, E., Currin, R. D., Jr. and Singletary, I. 2017. Tempering ventilation air in a swine finishing barn with a low-cost earth-to-water heat exchanger. *J. Renew. Sustain. Energy* 9(2):023901. http://doi:10.1063/1.4979359.

Sharpe, K. T., Heins, B. J., Buchanan, E. S. and Reese, M. H. 2020. Evaluation of solar photovoltaic systems to shade cows in a pasture-based dairy herd. *J. Dairy Sci.* 104(3):2794–2806. http://doi:10.3168/jds.2020-18821.

Sharpe, K. T., Reese, M. H., Buchanan, E. S., Tallaksen, J. E., Janni, K. A. and Johnston, L. J. 2018. Electrical and thermal energy consumption in Midwest commercial swine facilities. *Appl. Eng. Agric.* 34(5):857–864. http://doi.org/10.13031/aea.12771.

Shurson, G. C. 2020. "What a waste" – can we improve sustainability of food animal production systems by recycling food waste streams into animal feed in an era of health, climate, and economic crises? *Sustainability* 12:7071. http://doi:10.3390/su12177071.

Silva, B. A. N., Oliveira, R. F. M., Donzele, J. L., Fernandes, H. C., Abreu, M. L. T., Noblet, J. and Nunes, C. G. V. 2006. Effect of floor cooling on performance of lactating sows during summer. *Livest. Sci.* 105(1–3):176–184. http://doi:10.1016/j.livsci.2006.06.007.

Silva, B. A. N., Oliveira, R. F. M., Donzele, J. L., Fernandes, H. C., Lima, A. L., Renaudeau, D. and Noblet, J. 2009. Effect of floor cooling and dietary amino acids content on performance and behavior of lactating primiparous sows during summer. *Livest. Sci.* 120(1-2):25-34. http://doi:10.1016/j.livsci.2008.04.015.

Smith, B. C., Ramirez, B. C., Hoff, S. J. and Greiner, L. L. 2019. Pilot-scale assessment of a novel farrowing creep area supplementary heat source. *Animals (Basel)* 9(11):996. https://doi.org/10.3390/ani9110996.

Stinn, J. P. and Xin, H. 2014. Heat lamp vs. heat mat as localized heat source in swine farrowing crate. *Anim. Ind. Rep. AS660. ASL* (vol. R2931). https://doi.org/10.31274/ans_air-180814-1213.

Swinkels, J. W. G. M., Kornegay, E. T. and Verstegen, M. W. A. 1988. The effect of reduced nocturnal air temperature and feed additives on the performance, immune response and scouring index of weanling pigs. *J. Anim. Physiol. Nutr.* 60(1-5):137-145.

Tallaksen, J., Johnston, L., Sharpe, K., Reese, M. and Buchanan, E. 2020. Reducing life cycle fossil energy and greenhouse gas emissions for Midwest swine production systems. *J. Cleaner Prod.* 246. https://doi.org/10.1016/j.jclepro.2019.118998:118998.

Teitel, M., Levi, A., Zhao, Y., Barak, M. Bar-lev, E. and Shmuel, D. 2008. Energy saving in agricultural buildings through fan motor control by variable frequency drives. *Energy Build.* 40(6):953-960. http://doi:10.1016/j.enbuild.2007.07.010.

USDA 2011. USDA's 2009 On-farm energy production service. *Natl. Agric. Stat. Serv.* http://www.agcensus.usda.gov/Publications/2007/Online_Highlights/OnFarm_Energy_Production/index.asp (accessed 1 February 2021).

van der Werf, H. M. G., Petit, J. and Sanders, J. 2005. The environmental impacts of the production of concentrated feed: the case of pig feed in Bretagne. *Agric. Syst.* 83(2):153-177. http://doi:10.1016/j.agsy.2004.03.005.

van Middelaar, C. E., van Zanten, H. H. E. and de Boer, I. J. M. 2019. *Poultry and Pig Nutrition: Challenges of the 21st Century*. Wageningen Academic Publishers, Wageningen, Netherlands. pg. 307-314.

van Utrecht, D. M., Hoff, S. J. and Harmon, J. D. 2002. Variable rate heater control for livestock space heating. *Appl. Eng. Agric.* 18(2):245-253.

Worawimut, C., Vivanpatarakij, S., Watanapa, A., Wiyaratn, W. and Assabumrungrat, S. 2019. Performance evaluation of biogas upgrading systems from swine farm to biomethane production for renewable hydrogen source. *Int. J. Hydrog. Energy* 44(41):23135-23148. http://doi:10.1016/j.ijhydene.2019.07.072.

Xin, H., Zhou, H. and Bundy, D. S. 1997. Comparison of energy use and piglet performance between conventional and energy-efficient heat lamps. *Appl. Eng. Agric.* 13(1):95-99.

Yahyaoui, I. 2017. *Specifications of Photovoltaic Pumping Systems in Agriculture: Sizing, Fuzzy Energy Management and Economic Sensitivity Analysis*. Elsevier Science, Amsterdam, Netherlands. Pg. 1-14.

Zering, K., Atkinson, A., Chvosta, J., Marra, M., Norwood, F. B., Renkow, M. and Wossink, A. 2005. Cost and returns analysis of manure management systems evaluated in 2004 under the North Carolina attorney general agreements with Smithfield Foods, Premium Standard Farms, and Front Line Farmers. Res. Rep. Depart. Agric. Res. Econ. North Carolina State Univ., Raleigh, North Carolina. http://citeseerx.ist.psu.edu/viewdoc/download?doi=10.1.1.593.1014&rep=rep1&type=pdf.

Zhang, Q. and Xin, H. 2001. Responses of piglets to creep heat type and location in farrowing crate. *Appl. Eng. Agric.* 17(4):515-519.

Zhang, Y. and Barber, E. M. 1995. Air leakage and ventilation effectiveness for confinement livestock housing. *Trans. ASAE* 38:1501-1504.

Zhou, Y., Dong, H., Xin, H., Zhu, Z., Huang, W. and Wang, Y. 2018. Carbon footprint assessment of a large-scale pig production system in northern China: a case study. *Trans. ASABE* 61(3):1121–1131. http://doi: 10.13031/.

Zhu, Y., Li, Y., Reese, M., Buchanan, E., Tallaksen, J. and Johnston, L. 2020. Behavior and performance of suckling piglets provided three supplemental heat sources. *Animals (Basel)* 10(7):1155. https://doi.org/10.3390/ani10071155.

Chapter 11

Energy-smart poultry farming

Yi Liang, University of Arkansas, USA

1 Introduction

Poultry meat and eggs are the primary sources of high-quality animal proteins, minerals, and vitamins to balance the human diet. Broilers, bred specifically for meat due to their low feed-meat conversion ratio, are mainly produced in modern integrated buildings. Original breeds of hen lay around 60 eggs per year but due to natural selection and cross-breeding, commercially bred layers can lay over 300 eggs. Turkeys, ducks, and geese are farmed outdoors or in intensive livestock facilities for meat and eggs. Manure is a by-product of the poultry industry with its high soil nutrient value that can also be used as a feedstock for biogas production with the effluent then applied to the land. Down and feathers also have a value for insulation and warm clothing.

The worldwide broiler chicken industry has grown continuously over the last 50 years and produced over 100 million tons of live weight in 2020. The United States is the world's largest poultry meat producer (18%), followed by China and Brazil (FAO, 2021). In 2018, close to 9 billion broilers were raised in the United States, yielding over 16 million live-weight tons. In the last three decades, world egg production has increased by more than 150%. China is by far the world's largest egg producer, with 42% of global production, followed by the United States (7%) and India (6%). In 2017, chickens accounted for about 92% of the world's poultry population, followed by ducks (5%), and turkeys (2%). The development and transfer of feed, slaughter, and processing technologies have led the poultry industry and the associated feed industry to scale up rapidly and to integrate vertically with increased safety and efficiency.

http://dx.doi.org/10.19103/AS.2022.0100.18

Energy is one of the critical inputs to modern poultry farming, and its overall use in poultry production and processing is constantly growing (Pelletier, 2008). In the United States, the energy cost has been the third-largest production cost after feed inputs and capital investment payment of production buildings for broiler farmers (Cunningham, 2006). Fossil fuel-based energy use produces emissions of greenhouse gases, contributing to climate change. The volatility of energy prices and uncertainties with respect to long-term fossil fuel availabilities are major concerns of the industry as a whole, and a risk factor to sustaining the ever-growing demand for protein by the global population. Therefore, quantifying energy consumption for different products and different components is important to enable comparison between products, and to look for ways to reduce cost and improve efficiencies (Ramirez et al., 2006; Pishgar-Komleh et al., 2017).

This chapter describes how energy is used in intensive poultry production in buildings today and discusses the factors that have driven the growth of energy use. It then identifies opportunities toward energy-smart production including the uptake of existing technologies and management practices that can lead to significant energy conservation and improved energy efficiency. Technologies that are emerging in certain regions with a high potential of wider adoption are presented as is the current status of research and development of renewable energy systems applicable to poultry production.

2 Energy use

Energy consumption is a combined effect of climate region, housing type, and the level of mechanization of production operation and control of equipment. Intensive poultry production occurs inside buildings where energy is consumed for internal climate management involved with heating, cooling, ventilation, lighting, and for the operation of equipment (feeding, manure management, egg collection, etc.). Energy consumption profiles differ among the major poultry products, namely broilers, laying hens, turkeys, and ducks. Heating consumes a large share of energy in raising broilers and turkey poults but tends to be minimal in laying hen operations, or even zero in the case of battery-caged hen facilities in the United States. Note that in other countries there is a strong movement away from producing hens in cages for animal welfare reasons. The birds that are produced 'free-range' can have lower energy demands per unit of protein.

2.1 Broiler and chicken meat production

Broiler houses have undergone a steady change toward energy-intensive, environmentally controlled houses over the past several decades, concurrent with an increase of production evidenced by the growth of about 5% per year from 1960 to 2000 (Ollinger et al., 2000). In the 1970s, in the United States, over 57% of the broilers produced were in relatively open houses, over 19% were in environmentally controlled houses, and the remainder were in semi-environmental houses (Rogers et al., 1974). US broiler production has since moved away from open houses toward enclosed environmentally controlled houses with a higher level of heating, ventilating, and cooling capability thanks to industry specialization and the economies of scale (Wheeler et al., 2006).

The shift to environmentally controlled broiler houses and enclosed sidewall, tunnel ventilated houses replacing curtain-side window houses (Xin et al., 1993; Liang et al., 2009) is largely a result of a higher demand of the thermal environment to support the continuous improved genetic potential of broiler strains. This results in increased daily weight gain and feed conversion efficiency (Havenstein et al., 2001; Liang et al., 2013) and a higher tendency toward heat stress due to market demand for birds of heavy weights in the United States (Bottcher et al., 1994; Dozier et al., 2006). Additionally, building energy consumption is also impacted by the adoption of advanced technologies such as radiant heating replacing unvented gas heaters with pilot lights or convective furnace-type heating equipment (Chastain, 2008), higher efficiency lighting (Tabler et al., 2015), and more efficient ventilation fan designs (Chastain et al., 2017).

In a typical commercial broiler house, electricity is used for feed and water delivery, lighting, ventilation, and cooling, while fossil fuels are used for heating – though electric heating is an option. Energy demand within environmental-controlled houses fluctuates throughout the year, predominately due to varying heating and ventilation demand associated with flock age. The relative proportions between heating fuel and electricity use depend largely on the local climate and bird age. Higher fuel demand for heat occurs during cold seasons with younger birds, and higher electricity demand for cooling occurs during warm seasons with older birds.

An energy intensity index can be determined as the ratio of energy input and production output. Energy inputs include direct and indirect energy. Direct energy inputs are used for poultry production, such as fuel energy, electricity energy, and human labor energy. Indirect energy inputs include the manufacture of machinery and feedstock. In this chapter, only direct energy inputs as fuel and electricity are considered, although there is literature that includes indirect energy.

The market live weight per bird (kg) (instead of slaughtered meat) was selected as production output for broilers, and the weight of eggs used for laying hens for their nutritional values. Energy intensity has also been reported as the ratio of annual energy input and annual production number of broiler chickens. The normalization of energy intensity over the number of chickens was avoided due to the wide range of market weights of broilers that can vary from 1.8 kg to 4.1 kg across regions of the world.

Many EU studies frequently report the energy intensity indices using primary or final energy inputs, while those in the United States are reported as on-farm energy. In this chapter, the primary energy reported was converted back to the amount of on-farm energy by dividing the conversion factors of the reported energy sources. The choice of on-farm energy is deemed easier to compare when energy savings are discussed later.

On-farm energy consumption and energy use efficiency have been reported for broiler houses in North America (Rogers et al., 1974; Lomax et al., 1984; Xin et al.,1993; Wheeler et al., 2000; Overhults, 2008, 2009; Liang et al., 2020), Eastern European countries (Baxevanou et al., 2017; Kilic, 2016), and Northern European countries (Hörndahl, 2008; Rajaniemi et al., 2013). On-farm electricity and heating fuel consumption for broiler production varies around the world (Table 1). Despite a large variation, heat energy dominates the total direct energy in almost all regions. For similar types of housing and production management practices, heating degree days are a good statistical measure of the difference in the heating fuel use, exemplified by a three-fold difference between lowland and mountainous regions in Greece (Baxevanou et al., 2017).

2.2 Laying hen production

Egg production has evolved considerably over the recent three decades by incorporating new technologies and production practices, with the implication of natural resource usage including energy demand and consumption. The most significant change is the introduction of welfare-based housing systems since 1980 (Leinonen et al., 2012; Pelletier et al., 2014). Variations of alternative hen housing systems exist in different countries, with their legal status changing over the years. European legislation sets down rules for three cage-free alternative rearing systems, namely barn, free-range, and organic (Eurogroup for Animals, n.d.). Barn rearing systems have to meet requirements such as stocking density, littered area, adequate nest boxes, and a dustbathing area with litter, etc. and can be either single-tier or multi-tier housing system. Aviary houses is an example of multi-tier barn rearing system where hens are allowed to access littered areas. Furnished cages have been designed and adopted by EU countries to overcome some of the welfare concerns of battery cages whilst retaining their economic and husbandry advantages, and also provide some of the welfare advantages over non-cage systems. Furnished cages, also called an enriched colony, is another alternative housing system with multi-tier enriched colonies containing perches, nestbox, and scratch pads (60 hens per colony at a stocking density of 752 cm^2/hen) adopted by the Humane Society of the United States and the United Egg Producers (Zhao et al., 2013).

Table 1 On-farm electricity and heating fuel demands for broiler production

References	Country	Heating fuel, MJ/kg live-weight	Electricity, MJ/kg live-weight	Total energy, MJ/kg	Market live-weight, kg	Relevant information
Rogers et al. (1974)	US	2.09	0.33			
Lomax et al. (1984)	US	1.48				
Overhault et al. (2008)	US	1.32	0.277		1.8	
Overhault et al. (2009)	US	1.10	0.349		2.8	
Liang et al. (2009)	US	1.87	0.288		1.7–3.7	Sidewalls with curtains
		1.61	0.367			Solid sidewalls
Liang et al. (2020)	US	1.1	0.4			
Hörndahl (2008)	Sweden	1.80				
Rajaniemi et al. (2013)	Finland	4.82			1.8	
Baxevanou et al. (2017)	Greece	0.504	0.149		2.4	Lowland
		1.476	0.112		2.4	Mountainous
Kilic (2016)	Turkey			0.88	1.8	Western Turkey
MacKinnon et al. (2018)	Canada	0.147	0.658		2.0	

A free-range egg system in Australia requires regular access to an outdoor range during daylight hours, as well as indoor nest boxes, perches, and automatic feeding and watering systems. Wiedemann and McGahan (2011) reported approximately 63.5% of Australia's egg production from battery cage systems and 26.6% from free-range systems. While comparing the performance using sustainability indicators of egg production, Mollenhorst et al. (2006) presented four of the most common housing systems in the Netherlands of about equal share: the battery-cage system, the deep-litter system with and without an outdoor run, and the aviary system with outdoor run.

EU countries have been in the transition towards alternative hen housing systems for over 20 years. Based on a study of 10 countries the average electricity consumption per kg egg production increased 20% cost when moving from conventional cage housing to cage-free housing (AgraCEAS Consulting, 2004). With a change in the production system, the electricity used by different components also changed. Stout (1984) partitioned energy use for egg production in conventional houses with incandescent lighting but without manure cleaning belts as 64% for mechanical ventilation, 17% for lighting, 9% for operation of egg coolers, 5% for feeder operation, and 5% miscellaneous. By way of comparison some years later, Hayes et al. (2014) partitioned electricity use for two identical aviary houses (without outdoor run) with compact fluorescent lighting and manure belt drying as 54% for manure belt blowers, 15% for mechanical ventilation, 6% for lighting, 3% for feed augers, and 25% miscellaneous.

On-farm electricity and heating fuel consumption for egg production reported globally varies widely (Table 2). The electricity demand for aviary hen houses was 0.33 kWh/kg egg being almost double that of conventional cage houses at 0.18 kWh/kg egg, with low heating fuel consumption in both types of houses (Hayes et al., 2014). Zhao et al. (2013) simulated the effect of supplemental heat requirement based on ventilation rates and housing characteristics during winter weather conditions in mid-west United States in aviary and enriched colony versus conventional cage systems. Supplemental heat cost for the alternative housing systems accounted for less than 0.5% of the total production cost.

3 Energy conservation and energy efficiency opportunities

The most effective pathway toward an energy-smart poultry industry should be a tiered approach with energy conservation being the first step, for example, turning off loads not required in the first place. Improved energy efficiency measures and management when maintaining and/or upgrading equipment should follow. Adoption of renewable energy sources and low carbon technologies can then be more economically viable (MacKinnon et al., 2018).

Table 2 Examples of the range of on-farm electricity and heating fuel consumption for laying hen production

References/Year	Geographic Region	Heating Fuel, MJ/ ton egg	Mean Electricity, MJ/ton egg	Range of Electricity, MJ/ton	Type of Houses
Hayes et al. (2014)	US	14.8	1184		Aviary
Sonesson et al. (2009)	EU		594		Furnished cage houses
Sonesson et al. (2009)	EU			630–1620	Barn floor houses

The biggest barrier to making energy efficiency improvements and renewable energy investments is the internal competition for capital. Farmers usually have limited capital available for end-use efficiency projects and frequently look for very short payback periods of one to three years. Government incentives in some counties have been a major driver for energy-related poultry housing retrofits and equipment replacement with the aim of conserving natural resources and reducing emissions of greenhouse gases.

3.1 Chick rearing

Brooding of young chicks is the most energy-intensive activity in broiler production. The recommended brooding temperatures are 29–31°C during week 1 and 26.7°C during week 2 (Aviagen Inc., 2018; Deaton et al., 1996). After measuring on-site heating fuel demand in a commercial broiler house over 4 years, Flood et al. (1994) reported that on average, 55.8% of the total annual heating fuel consumption occurred during the first week, while 91.2% occurred during the first 3 weeks.

Partial house brooding is a common practice where houses are partitioned into two sections and the chicks are delivered into a pre-heated brooding section. Young birds are allowed into the entire house at the end of the brooding period, which varies from 7 to 14 days in the United States. For whole-house brooding, Wheeler et al. (2000) measured an average demand for LPG[1] of 9000 L per flock per house (with a dimension of 15.2 m × 122 m) and around 5350 L where partial-house brooding was used over three winter flocks on three Pennsylvanian broiler farms. Partial-house brooding saved around one-third of fuel demand using a similar amount of air exchange as in a whole house because better air mixing alleviated temperature stratification problems and also maintains a more appropriate floor level temperature.

1 Liquid petroleum gas that is mainly propane.

3.1.1 Lighting

Traditional lamps used within poultry houses have been dominated by incandescent, fluorescent, and mercury vapor lamps. Electricity demand for lighting has decreased in recent years due to new, affordable technologies with the same lumen level but at a lower wattage. Light-emitting diodes (LEDs) are the latest lighting innovation, offering high-efficiency and extended life. Lamp replacement to LED in broiler houses is relatively straightforward and, with the recent market penetration, can result in electricity savings of approximately 80% over incandescent lighting.

3.1.2 Ventilation fans

Ventilation fans are key components of mechanically-ventilated poultry houses They remove air contaminants, including moisture generated inside the building, as well as excess heat thereby controlling interior temperature rise levels above ambient during periods of high outdoor temperature. Ventilation fans account for 40–80% of the total amount of electricity demand in mechanically ventilated poultry buildings (Chastain et. al, 2017).

The performance and efficiency of ventilation fans used in agricultural facilities have steadily improved due to design and development by both equipment manufacturers and the research community (Li and Heber, 1992; Heber and Li, 1995). The performance of the fan systems is affected by many engineering modifications of the fan motor, impeller design factors, and the complex effects of housing dimensions and associated inner attachments on aerodynamic losses (Li and Heber, 1992). Additional factors such as the speed of the fan motor and blades and fan housing design also influence the energy efficiency of a ventilation fan.

Ventilation in livestock buildings usually involves low static pressures of 12–30 Pa. Ventilation efficiency ratios (VER) are measured by the amount of air moved per unit of electrical energy input to the fan motor ($m^3/h/W$) at a static pressure drop of 24 Pa. The airflow ratio (AFR) is an index that represents the flatness of the fan curve. It is defined as the ratio of airflow rate at 48 Pa to that at 12 Pa. A wide range of ventilation efficiency ratios have been published in a database by Bioenvironmental and Structural Systems Laboratory, University of Illinois, a qualified testing laboratory (Chastain et al., 2017). There is an obvious benefit from reducing electricity operating costs by choosing more energy-efficient fans, especially during new house construction. The recommendation is to select ventilation fans with efficiencies greater than about 34 $m^3/h/W$ at 25 Pa (Czarick and Lacy, 1999).

When installed in poultry houses, fans are often fitted with one or more accessories such as safety guards, shutters, and discharge cones. The first two usually reduce the airflow and fan efficiency, whereas discharge cones improve airflow. Shutters, mostly installed on the suction side inside the houses, reduce airflow significantly while preventing backflow through non-operating fans when the other fans aren't running. Simmons and Lott (1997) reported that shutters on a 91 cm diameter ventilation fan in a broiler house reduced fan airflow rates by either 10.9% under clean condition, up to 16.3% after dirt had accumulated over a 42-day growth cycle. The authors recommended the use of a shutter motor designed to pull open the vanes to improve the airflow rate. Casey et al. (2008) reported reduced resistance to airflow of plastic shutters in comparison with aluminum shutters. Discharge cones increase airflow rates and the VER significantly for fans with a diameter of more than 250 mm but do not provide a consistent improvement in airflow ratios (Chastain et al., 2017).

Regular maintenance of fans, such as checking belt tension, replacing worn belts, and cleaning off dirt from the blades and shutters ensure fans perform close to their specification. Laboratory test data are available for new fans but those operating in a livestock house for several years decline below the published fan test data (Bottcher et al., 1996; Casey et al., 2008).

Most fans used in US poultry houses are single-phase 230 V, axial flow fans whereas in some European countries many are three-phase 380 V. These fans usually operate under ON–OFF control so start when either the temperature or the humidity within the house exceeds a certain set point and turn off when the micro-climate in the house reaches the desired conditions. Alternatively, they operate for a period of time on a predetermined timer based on a minimum ventilation program.

Variable frequency drives (VFD) allow fans to operate over a longer period of time with the reduced rotational speed of the motors. Variable-speed controls automatically and smoothly adjust airflow (Q) according to the difference between internal building and set-point temperatures. As a result, the specific energy consumption (power required to deliver a certain airflow rate) can be reduced based on the fan law (power consumed by the fans is proportional to N^3, whereas the flow rate Q is proportional to rotational speed (N)), hence electricity savings.

Ford et al. (1998) carried out experiments to evaluate the VFD as a method for controlling fan speeds in agricultural buildings. Tests between VFD and triac speed controllers showed that the VFD-controlled fans gave a more stable airflow as static pressure varied and also had higher ventilating energy efficiency at reduced speeds. They calculated that the higher initial cost of the VFD system was recovered in energy savings over 4.3 years. Teitel et al. (2008) compared the electricity consumption and micro-climate between ON–OFF and VFD systems for poultry houses. VFD units were installed on fans driven by three-phase 380 V/50 Hz electric motors, one group rated 0.75 kW and another group rated 1.1 kW. The average electricity consumption with the VFD control system was about 75% of that with an ON-OFF system. Variable speed technology on ventilation fans not only has a large potential to reduce electricity consumption but also reduces the amplitude variations in temperature and humidity inside the poultry facilities.

3.2 Housing construction

Construction standards and quality of materials have long been recognized to directly impact the energy consumed in poultry houses. Careful selection of insulation material and degree of insulation for buildings with low infiltration rates is crucial to maintain the optimum temperature and humidity levels required to raise modern broilers (Koon et al., 1993; Berry and Miller, 1989). Solid-wall houses are more energy-efficient than conventional poultry houses with curtains because they allow for wall insulation and lack the air leaks that come with curtains.

Berry and Miller (1989) demonstrated the economic advantages of various amounts of wall and ceiling insulation levels on fuel consumption by conducting energy balance analysis and simulation in commercial broiler houses. The choice and degree of insulation are based on the prevailing climate, ease of installation, durability including resistance to compression, high moisture, degradation, cost which is generally related to durability and effectiveness, and the mode of heat transfer. Bulk insulators are most useful in cold conditions where significant convective/conductive losses occur but less useful in hot conditions where solar radiation is the main source of heat gain.

Another common issue of poultry houses is air leakage due to insect and rodent damage over the years of service so that they lose tightness resulting in higher infiltration rates and energy consumption. Proper detection and regular building maintenance go a long way towards conserving energy during production.

3.3 Heat recovery ventilation (HRV)

Energy requirements for supplemental heating of poultry houses can be lowered by reclaiming the heat lost through ventilation to pre-heat incoming fresh air. Efficient heating of a poultry building makes it more affordable to bring more fresh air into the building and maintain a better air quality which benefits the birds' health and wellbeing. Assuming newer or retrofitted buildings have lower conductive heat loss through the building envelope, ventilation represents a greater portion of the total heating load. Recovering waste heat from the warm exhaust air is one of the most economical ways to reduce heating fuel costs (Kennedy et al., 1991; Manbeck et al., 1987; Petersen, 1990; Liang et al., 2011; Han et al., 2013).

Sokhansaj et al. (1980) developed a method for sizing counterflow heat exchangers and an economic criterion for theoretical evaluation of different sizes of heat exchangers in a turkey-production building located in Minnesota. Ice formation in cold climates and filtration of air outflows to prevent dust accumulation are two major issues that need to be properly addressed by equipment developers before their adoption by poultry growers.

A side-by-side comparison of actual fuel savings over three consecutive bird growing cycles was conducted when effective ventilation rates were kept similar (Liang et al., 2011), resulting in a fuel saving of 40-60%. A feature of HRV that is unique to livestock and poultry facilities was the high humidity level in the outgoing air compared with commercial buildings. Some of the latent heat of condensation in the outgoing airstream is picked up as sensible heat by the incoming air stream, providing a large reduction in heating system operating hours, fuel consumption, and cost (Liang et al., 2011). Han et al. (2013) reported a 55% reduction in total heating energy consumption, as well as improved temperature uniformity in a house with the HRV and energy recovery ventilation (ERV), due to the continuous recirculation of indoor air.

Waste heat recovery ventilators in poultry farms are now in use in many parts of the world (Xu et al., 2015; Bokkers et al., 2010). Heat recovery is generally the first stage of ventilation when supplemental heating is needed. Newer products have incorporated variable speed drive (VFD) technology on both exhaust and intake fan motors so that they can be operated continuously at reduced speed to match the ventilation demands when birds are young. This can result in significant electricity savings.

Liang and Costello (2015) simulated the amount of potential heat recovery using a spreadsheet broiler energy model (Berry and Miller, 1989). Two levels of efficiency were simulated for the HRV system, and for two insulation levels of a turkey brood building and a broiler building, both with industry-standard stocking densities and ventilation management. Partial house brooding was assumed on broiler farms, while the whole house was assumed to raise poults in the turkey brood house. The influence of building insulation and HRV system efficiency upon annual recovered heat and savings due to the installation of a heat recovery system was calculated (Table 3). An installed ventilation capacity of 0.85 m^3/h/bird was assumed for the heat recovery system. A decision to install a larger ventilation capacity would incur higher capital investment with reduced returns.

Heat exchangers with a thermal efficiency between 0.5 and 0.8 will save around 40–60% of the supplemental heating costs than without heat recovery. With similar heat recovery system capacities installed, higher efficiency units will recover more heat through ventilation.

Heat recovery ventilators can be operated to allow more fresh air into the building and maintain better air quality, but with less energy savings. Among 25

Table 3 Projected annual recovered heat (GJ), heat demand with heat recovery, and percentage savings in poultry buildings with low or high insulation levels and low or high heat exchanger efficiencies in NW Arkansas

		Broiler[b]			Turkey Brooder[c]		
		Recovered heat	Heat needed (with heat recovery)	Saving	Recovered heat	Heat needed (with heat recovery)	Saving
Insulation level[a]	Heat exchanger efficiency	GJ		%	GJ		%
Low	0.5	274	390	41	327	559	37%
	0.8	390	274	59	475	411	54%
High	0.5	211	285	43	232	295	44%
	0.8	306	190	62	327	200	62%

[a]Low insulation level consists of wall R-value of 1.2 m^{2}°C/W, ceiling R-value of 1.9 m^{2} °C/W, while high insulation level consists of wall R-value of 1.9 m^{2} °C/W, ceiling R-value of 3.3 m^{2} °C/W.
[b]Assume 20 000 birds per house with partial house brooding; 5 1/2 growout cycles each year with 6.5-week each cycle.
[c]Assume 18 000 poults per house with whole-house brooding; 4 brood cycles each year with 5.5-week each cycle.

Netherlands farmers surveyed, Bokkers et al. (2010) reported that the use of gas for heating was reduced by 38% after installing a heat exchanger (HE) based on data from 13 farms. Lower fuel consumption as a result of heat recovery means less combustion-related by-products, thereby improving air quality and potentially drier litter. Three farms had similar total energy use before and after installation of a HE, partially due to an increase in fan electricity. Higher daily weight gains of broilers were reported, 42% of the farmers observed a decrease in respiratory problems, and 54% of the farmers reported a decrease in contact dermatitis.

In summary, heat recovery systems are most attractive for poultry houses that are not sufficient in heat production and for buildings that have some combination of (a) high room temperatures, (b) high ventilation rates, (c) low winter temperatures, and (d) high energy costs. Economic feasibility greatly depends on the potential fuel savings, the cost of the HRV system, and the price of fuel.

4 Renewable energy in poultry applications

Renewable energy applications have penetrated all sectors of the economy due to becoming more competitive and with the potential to reduce greenhouse gases. Cui et al. (2020) provided a systematic review of different types of renewable and sustainable technologies with the potential of application in poultry houses. Solar is the leading renewable energy source for poultry farms at present as discussed below, but biomass (such as wood pellet boilers and biogas) (Overhults, 2017), geothermal (including ground source heat pumps) (Choi et al., 2012) and wind (for electricity generation and water pumping) also have a place.

Solar radiation is transformed into direct current (DC) electricity through photovoltaic (PV) cells which are converted into alternating current (AC) through the use of inverters. Solar PV applications have seen significant growth in the past two decades, largely due to the continuous improvement of PV module efficiency, a steady decrease of PV module price due to market penetration, more trained installers, and policy support and financial incentives at both federal, state and local levels in many countries. The application of solar PV to generate electricity on-site is suitable for poultry houses due to its modular design and easy integration with the existing power supply (utility grid). Investment can be favorable especially in areas with good solar radiation levels, high electricity tariffs, and net-metering policies.

An early feasibility study of solar PV systems for Delaware's poultry farms (Byrne and Glover, 2005) was followed by numerous economic analyses (Bazen and Brown, 2009). Today broiler farms around the world have adopted solar PV technology. Cui et al. (2020) summarized production characteristics of recently installed PV solar systems on commercial broiler farms. Most systems are grid-connected and sized to generate enough electricity to save between 30% to 85% of utility bills. Hybrid systems, such as solar-wind-diesel, have also been deployed to provide a small, isolated power system in areas with less reliable power supply or experiencing an energy crisis (Karki and Billinton, 2001).

4.1 Solar thermal technologies

The design of efficient solar heating systems for brooding broiler chickens is more complicated than solar electricity because of the uneven time distribution of heat requirements over different seasons. Flood et al. (1979) presented large differences between heat requirements for brooding during winter months and summer months, as well as the minimal heat requirement for the last 3 weeks of grow cycles in spring and summer months. The challenge is to design solar heating systems that can be utilized efficiently when there are such pronounced differences in the magnitude of heat requirements with time. Solar water heating can be economically feasible, especially where there is an electric or gas backup system to ensure hot water supply even during winter periods (Lomax et al., 1984).

Early studies of solar thermal technologies focused on solar thermal collectors for space heating in order to reduce heating load during the day and during the night by adding thermal storage (Mirzaee-Ghaleh et al., 2013). For example, Rokeby et al. (1983) evaluated the thermal performance of a solar heating system with flat plate collectors and rock storage. The system provided from 42% to 72% of the supplemental heat required for brooding in winter and the efficiency of the system over a 56-day brooding period ranged up to 28.5%.

The global challenge to reduce environmental impacts has renewed the interest of the research community in developing cost-effective solar heating systems. El Zanaty (2015) conducted an engineering design and economic analysis of a solar heating system that consists of evacuated tubes, water storage tanks, and fan coil units to meet the heating demand of a broiler poultry house with a capacity of 24,000 birds located in Egypt. The best design reduced 77% of the heat demand of the house. A completely green heating solution was proposed by integrating biogas methane as an auxiliary source of heat produced from bio-digesters using the organic wastes and manure of the birds.

4.2 Solar collectors

Transpired solar walls (TSW) have been used in commercial and industrial buildings in Canada and EU countries for several years (Kozubal et al., 2008). The TSW consists of a dark-colored, unglazed perforated metal wall installed on south-facing building surfaces (in the northern hemisphere). When heated by solar radiation the air inside the wall heats, rises up, and flows through into the buildings, hence reducing the use of heating fossil fuel and also reducing CO_2 emissions.

The efficiency of a TSW depends on its design influencing surface airflow distribution. Application in livestock and poultry buildings is challenging, partly because building air ventilation varies with the season and age of the animals housed, and having distributed inlets across sidewalls. The TSW concept has been evaluated in temperate climate conditions such as for Canadian livestock and poultry farms. Godbout et al. (2004) evaluated the impact of a TSW on a 1000-head pig nursery in Quebec, Canada, from November 2002 to April 2003, and reported the LPG energy savings ranged from 23 to 31% of annual demand. Deutsch reported that solar walls reduced LPG use by at least 25% in two pig nurseries in Quebec with a simple payback period of less than 5 years.

Cordeau and Barrington (2010) evaluated TSW installed on two broiler buildings. The unglazed solar air pre-heaters consisted of a wooden box frame built over the fresh air inlets. It is unprotected facing of black corrugated metal sheeting had small perforations covering 1% of its surface through which fresh ventilation air entered the building. The solar air heaters were installed on the southeast face of the building and could be bypassed by opening the bottom portion of the wooden box frame. Overall energy recovery efficiency ranged from 25% to 63% and was negatively correlated to ambient wind velocity. The solar air pre-heaters were able to recover CAN\$14.80/m^2 during the cold season assuming an energy cost of CAN\$ 0.08/kWh, representing an annual return on investment of 4.7% (Cordeau and Barrington, 2010, 2011). Shah et al. (2010) evaluated TSW installed on a pig nursery. LPG use was 25% lower in the test nursery compared with a conventional nursery. It was concluded that TSC was technically feasible but not economical, because of an unacceptable long payback period.

Research efforts on developing a low-cost solar collector were reported (Poole et al., 2018; Yu et al., 2018). Yu et al. (2018) evaluated a landscape fabric-based TSW for heating air. Livestock buildings experience an increase of heat output as animals grow and mature, to certain ages making overheating a possibility. Therefore, a by-pass mechanism has to be integrated into the environmental control system, posing significant challenges. Further research and development of the solar thermal collector concept are necessary before it can be widely adopted by intensive livestock farmers.

5 Conclusion and future trends in research

Energy is one of the critical inputs for modern poultry production because the demand for the controlled environment is relatively high. Due to the continuous improvement of poultry strains, nutrition, and management practices, recent advances in quality housing construction, energy-efficient lighting, radiant heating, ventilation system with heat recovery ventilators, etc. have already contributed to more efficient poultry production. Research and development of economic solutions for further improving energy efficiency and developing renewable energy applications will promise a greener and more sustainable future for poultry production.

6 Where to look for further information

- Bioenvironmental and Structural Systems Laboratory: http://bess.illinois.edu/index2.htm.
- Egg Industry Center: https://www.eggindustrycenter.org/.
- Food and Agricultural Organisation of the United Nations (FAO): http://www.fao.org/poultry-production-products/en/.
- International Poultry Council: https://internationalpoultrycouncil.org/.
- This is European Poultry! High-quality poultry with European guarantee: https://en.eu-poultry.eu/eu-poultry-sector/.
- UGA Poultry House Environmental Management and Energy Conservation: https://www.poultryventilation.com/.
- University of Arkansas Poultry Science: https://poultry-science.uark.edu/.

7 References

Agra CEAS Consulting. (2004). *Study on the Socio-Economic Implications of the Various Systems to Keep Laying Hens*. The European Commission.

Aviagen Inc. (2018). *Broiler Management Manual*. Huntsville, AL: Aviagen Inc.

Baxevanou, C., Fidaros, D., Bartzanas, T. and Kittas, C. (2017). Energy consumption and energy saving measures in poultry. *Energy and Environmental Engineering* 5(2):29–36. DOI: 10.13189/eee.2017.050201.

Bazen and Brown (2009). Feasibility of solar technology (photovoltaic) adoption: a case study on Tennessee's poultry industry. *Renewable and Sustainable Energy Reviews* 16(5):2926–2939.

Berry, I. L. and Miller, J. K. (1989). Spreadsheet model of broiler house environments. *Transactions of the ASAE* 5(2):237–244.

Bokkers, E. A. M., van Zanten, H. H. E. and van den Brand, H. (2010). Field study on effects of a heat exchanger on broiler performance, energy use, and calculated carbon dioxide emission at commercial broiler farms, and the experiences of farmers using a heat exchanger. *Poultry Science* 89(12):2743–2750. DOI: 10.3382/ps.2010-00902.

Bottcher, R. W., Bisesi, P. S., Brake, J., Pardue, S. L. and Etheredge, A. M. (1994). Reducing mixing fan thermostat set-points in naturally ventilated broiler housing during hot weather. *Journal of Applied Poultry Research* 3(3):289–296.

Bottcher, R. W., Baughman, G. R. and Madura, J. T. (1996). Field measurements of fan speed and power use in poultry houses. *Journal of Applied Poultry Research* 5(1):56–62.

Byrne, J. and Glover, L. (2005). *The Potential of Solar Electric Applications for Delaware's Poultry Farms, Center for Energy and Environmental Policy*. USA: University of Delaware. Available at: https://www.yumpu.com/en/document/read/30779506/potential-of-solar-electric-applications-for-delawarespoultry-farms (last accessed February 2021).

Casey, K. D., Gates, R. S., Wheeler, E. F., Xin, H., Liang, Y., Pescatore, A. J. and Ford, M. J. (2008). On-farm ventilation fan performance evaluations and implications. *Journal of Applied Poultry Research* 17(2):283–295.

Chastain, J. P. (2008). Fuel savings estimates for conversion from pilot ignition to spark ignition brooders for broiler production. *ASABE Annual Meeting Providence*, Rhode Island.

Chastain, J. P., Anderson, P. T. and Vassalos, M. (2017). Analysis of available efficiency and performance data for axial flow agricultural ventilation fans. *Paper #1700116*. Spokane, WA: ASABE.

Choi, H. C., Salim, H. M., Akter, N., Na, J. C., Kang, H. K., Kim, M. J., Kim, D. W., Bang, H. T., Chae, H. S. and Suh, O. S. (2012). Effect of heating system using a geothermal heat pump on the production performance and housing environment of broiler chickens. *Poultry Science* 91(2):275–281.

Cordeau, S. and Barrington, S. (2010). Heat balance for two commercial broiler barns with solar preheated ventilation air. *Biosystems Engineering* 107(3):232–241.

Cordeau, S. and Barrington, S. (2011). Performance of unglazed solar ventilation air pre-heaters for broiler barns. *Solar Energy* 85(7):1418–1429.

Cui, Y., Theo, E., Gurler, T., Su, Y. and Saffa, R. (2020). A comprehensive review on renewable and sustainable heating systems for poultry farming. *International Journal of Low-Carbon Technologies* 15(1):121–142.

Cunningham, D. L. (2006). 2005–2006 broiler production systems in Georgia: costs and return analysis. *B1240*. The University of Georgia Cooperative Extension Service, Athens, GA.

Czarick, M. and Lacy, M. P. (1999). Exhaust fan performance factors. *Poultry Housing Tips* 11(3).

Deaton, J. W., Branton, S. L., Simmons, J. D. and Lott, B. D. (1996). The effect of brooding temperature on broiler performance. *Poultry Science* 75(10):1217–1220.

Dozier, W. A., Purswell, J. L. and Branton, S. L. (2006). Growth responses of male broilers subjected to high air velocity for either twelve or twenty-four hours from thirty-seven to fifty-one days of age. *Journal of Applied Poultry Research* 15(3):362–366.

El Zanaty, H. (2015). A techno-economic study for heating poultry houses using renewable energy. MSc Thesis. The American University in Cairo.

Eurogroup for Animals. n.d. Optimizing laying hen welfare in cage-free systems. Available at: https://www.eurogroupforanimals.org/sites/eurogroup/files/2020-03/E4A-Optimising_Laying_Hens_Welfare.pdf (last accessed April 2021).

Flood Jr., C. A., Koon, J. L. and Brewer, R. N. (1979). Solar heating of poultry houses. 1. *Poultry Science* 58(2):314–318.

Flood Jr., C. A., Koon, J. L., Trumbull, R. D. and Brewer, R. N. (1994). Weekly heat usage for broiler chickens. *Applied Engineering in Agriculture* 10(2):283–284.

Food and Agriculture Organization (2021). Gateway to poultry production and products. Available at: http://www.fao.org/poultry-production-products/en/ (last accessed April 2021).

Ford, S. E., Riskowski, G. L., Funk, T. L. and Zhivov, A. M. (1998). Evaluation of variable drives for agricultural ventilation fan speed control. *Paper no. 984041. ASAE Annual International Meeting*, Orlando, FL, 8 pp. St. Joseph, MI: ASABE.

Godbout, S., Pouliot, F., Lachance, H., Guimont, H., Leblanc, R., Pelletier, F. and Hamelin, L. (2004). Feasibility and energy recovery of a solar wall in pig nursery. ASAE Paper No. 044140. St. Joseph, Mich.: ASAE.

Han, H., Kim, K., Jang, K. J., Han, G.-S. and Lee, I. B. (2013). Energy consumption and indoor environment of broiler houses with energy recovery ventilators. *Applied Engineering in Agriculture* 29(5):751–759.

Havenstein, G. B., Ferket, P. R. and Qureshi, M. A. (2001). Growth, livability, and feed conversion of 1957 versus 2001 broilers when fed representative 1957 and 2001 broiler diets. *Poultry Science* 82(10):1500–1508.

Hayes, M. D., Xin, H., Li, H., Shepherd, T. A. and Stinn, J. P. (2014). Electricity and fuel use of aviary-laying hen houses in the mid-western United States. *Applied Engineering in Agriculture* 30(2): 259–266.

Heber, A. J. and Li, J. (1995). Predicting variable-speed fan performance from motor and impeller characteristics. Part I. using complete motor data. *Transactions of the ASAE* 38(4):1247–1252.

Hörndahl, T. (2008). Energy use in farm buildings – a study of 16 farms with different enterprises. Revised and translated second edition. Swedish University of Agricultural Sciences, Faculty of Landscape Planning, Horticulture and Agricultural Science. Report 2008:8. Available at: http://pub.epsilon.slu.se/3396/1/Engrapport145-v1.pdf (last accessed February 2021).

Karki, R. and Billinton, R. (2001). Reliability/cost implications of PV and wind energy utilization in small isolated power systems. *IEEE Transactions on Energy Conversion* 16(4):368–373.

Kennedy, D. A., Leonard, J. J. and Feddes, J. J. (1991). Field test of a PVC plate heat exchanger for animal housing. *Applied Engineering in Agriculture* 7(4):457–460.

Kilic, I. (2016). Analysis of the energy efficiency of poultry houses in the Bursa region of Turkey. *Journal of Applied Animal Research* 44(1):165–172.

Koon, J. L., Flood Jr, C. A., Trumbull, R. D. and Brewer, R. N. (1993). Energy savings with use of double sidewall curtains on broiler houses. *Transactions of the ASAE* 36(6):1901–1905.

Kozubal, E., Deru, M., Slayzak, S., and Norton, P. (2008). Evaluating the performance and economics of transpired solar collectors for commercial applications. 2008 ACEEE Summer Study on Energy Efficiency in Buildings, Pacific Grove, California. https://www.nrel.gov/docs/fy08osti/43305.pdf

Leinonen, I., Williams, A. G., Wiseman, J., Guy, J. and Kyriazakis, I. (2012). Predicting the environmental impacts of chicken systems in the United Kingdom through a life cycle assessment: egg production systems. *Poultry Science* 91(1):26–40.

Li, J. and Heber, A. J. (1992). Optimizing performance of variable-speed fans. *Transactions of the ASAE* 35(5):1737–1744.

Liang, Y., Tabler, G. T., Watkins, S. W., Xin, H. and Berry, I. L. (2009). Energy use analysis of open-curtain and enclosed systems at a commercial-scale broiler research farm. *Applied Engineering in Agriculture* 25(4):577–584.

Liang, Y., Tabler, G. T., Reynolds, W., Starkweather, R. and Jensen, K. N. (2011). Heat recovery ventilators in a broiler house to reduce energy use. *ASABE Annual Meeting, Paper number 1110595*, St. Joseph, MI: ASABE.

Liang, Y., Kidd, M. T., Watkins, S. E. and Tabler, G. T. (2013). Effect of commercial broiler house retrofit: a 4-year study of live performance. *Journal of Applied Poultry Research* 22(2):211–216.

Liang, Y. and Costello, T. A. (2015). Saving energy with ventilation heat recovery in poultry barns. *Fact sheet FSA 1077*. University of Arkansas Division of Agriculture.

Liang, Y., Costello, T. A. and Bautista, R. (2020). Energy consumption and conservation opportunity for Arkansas broiler production. *Factsheet 1099*. University of Arkansas Division of Agriculture.

Lomax, K. M., Collins, N. E., Jr., Beston, T. A. and McCarthy, G. S. (1984). Off-peak electric heat for broiler houses. *Transactions of the ASAE* 27:1131–1134.

MacKinnon, I., MacDonald, R., Joseph, T. and Ritchie, S. (2018). Net zero energy study for Broiler Barn. *Canadian Society of Bioengineering Annual Conference. Paper No. CSBE18-149*. The Canadian Society for Bioengineering: Guelph, ON.

Manbeck, H.B., Scheidemann, R.E., Roush, W.B. (1987). Simulatig ventilation heat recovery for broiler and swine facilities. *Transactions of the American Society of Agricultural Engineers* 30(3):746–752.

Mirzaee-Ghaleh, E., Omid, M., Keyhani, A. and Javadikia, P. (2013). Forecasting the thermal load for implementing solar energy in a model poultry house. *Journal of Agricultural Engineering and Biotechnology* 1(2):30–36.

Mollenhorst, H., Berentsen, P. B. M. and De Boer, I. J. M. (2006). On-farm quantification of sustainability indicators: an application to egg production systems. *British Poultry Science* 47(4):405–417. DOI: 10.1080/00071660600829282.

Ollinger, M., MacDonald, J. and Madison, M. (2000). Structural change in U.S. chicken and turkey slaughter/AER-787 USDA.

Overhults, D. G., Pescatore, A., Gates, R. S., Jacob, J. P., Miller, M. and Earnest Jr. (2009). House characteristics and energy utilization in poultry houses raising large broilers. *Poultry Science* 88(Supp. 1):87.

Overhults, D. G., Pescatore, A., Gates, R. S., Jacob, J. P., Miller, M. and Earnest Jr. (2009). House characteristics and energy utilization in poultry houses raising small broilers. *Poultry Science* 88(Supp. 1):192.

Overhults, D. G. (2017). A comparison of wood pellet and natural gas heating systems for broiler houses. *ASABE Annual International Meeting, Paper Number: 1700343*. Spokane, Washington, St. Joseph, MI: ASABE.

Pelletier, N. (2008). Environmental performance in the US broiler poultry sector: life cycle energy use and greenhouse gas, ozone depleting, acidifying and eutrophying emissions. *Agricultural Systems* 98(2):67–73.

Pelletier, N., Ibarburu, M. and Xin, H. (2014). Comparison of the environmental footprint of the egg industry in the United States in 1960 and 2010. *Poultry Science* 93(2):241–255.

Petersen, S. (1990). Chapter 2. Energy conservation in animal houses in Demark. In Pedersen (Ed.), Energy Conservation in Animal Houses. Food and Agriculture Organization of the United States.

Pishgar-Komleh, S. H., Akram, A., Keyhani, A. and Aelm, R. (2017). Life cycle energy use, costs, and greenhouse gas emissions of broiler farms in different production systems in Iran: a case study of Alborz province. *Environmental Science and Pollution Research International* 24(19):16041–16049. DOI: 10.1007/s11356-017-9255-3.

Poole, M. R., Shah, S. B., Boyette, M. D., Grimes, J. L. and Stikeleather, L. F. (2018). Evaluation of landscape fabric as a solar air heater. *Renewable Energy* 127:998 –1003.

Rajaniemi, M., Tuure, J. and Ahokas, J. (2013). Heating energy consumption, heat losses and heating energy saving possibilities in case broiler house in Finland. *ASABE Paper. # 131610092*. Kansas City, MO. St. Joseph, MI: ASABE.

Ramirez, C. A., Patel, M. and Blok, K. (2006). How much energy to process one pound of meat? A comparison of energy use and specific energy consumption in the meat industry of four European countries. *Energy* 31(12):2047–2063.

Rogers, G. B., Benson, V. W. and Van Dyne, D. L. (1974). Energy Use and Conservation in the Poultry and Egg Industry. *Agricultural Economic Report No. 354*. USDA Economic Research Service.

Rokeby, R. C., Redfern, J. M. and Pitts, D. J. (1983). Solar heating for a commercial broiler house: a further evaluation. *Transactions of the ASAE* 26(2):0507–0511. DOI: 10.13031/2013.33967.

Shah, S. B., Marshall, T. K. and Matthis, S. (2010). Long-term monitoring of a transpired solar wall on a pig nursery for its ability to reduce heat energy use and improve pig performance. *Final Report.*

Simmons, J. D. and Lott, B. D. (1997). Reduction of poultry ventilation fan output due to shutters. *Applied Engineering in Agriculture* 13(5):671–673.

Sokhansanj, S., Jordan, K. A., Jacobson, L. A. and Messer, G. L. (1980). Economic feasibility of using heat exchangers in ventilation of animal buildings. *Transactions of the American Society of Agricultural Engineers* 23(6):1525–1528.

Sonesson, U., Cederberg, C. and Bergland, M. (2009). Greenhouse gas emissions in egg production. *Report 2009:7 Klimatmärkning För Mat.*

Stout, B. A. (1984). *Energy Use and Management in Agriculture*. North Scituate, Mass.: Breton Publishers.

Tabler, T., Farnell, M. and Wells, J. (2015). LED bulbs: much to offer the poultry industry. *Publication 2894*. Mississippi State University Extension.

Teitel, M., Levi, A., Zhao, Y., Barak, M., Bar-lev, E. and Shmuel, D. (2008). Energy saving in agricultural buildings through fan motor control by variable frequency drives. *Energy and Buildings* 40(6):953–960.

Wheeler, E. F., Zajaczkowski, J. S., Weiss, R. W. J. and Hulet, R. M. (2000). Temperature stratification and fuel use during winter in three Pennsylvania broiler houses. *Journal of Applied Poultry Research* 9(4):551–563.

Wheeler, E., Casey, K., Gates, R., Xin, H., Zajaczkowski, J., Topper, P., Liang, Y. and Pescatore, A. (2006). Ammonia emissions from twelve US broiler chicken houses. *Transactions of the ASABE* 49(5):1495–1512.

Wiedemann, S. G. and McGahan, E. J. (2011). *Environmental Assessment of an Egg Production Supply Chain Using Life Cycle Assessment*. North Sydney, NSW: Australian Egg Corporation Limited.

Xin, H., Berry, I. L., Barton, T. L. and Tabler, G. T. (1993). Sidewall effects on energy use in broiler houses. *Journal of Applied Poultry Research* 2(2):176–183.

Xu, S. Y., Xu, T. and Reed, R. (2015). Propane saving in poultry farm through waste heat recovery system. *Midwest Poultry Federation Annual Meeting*. Available at: http://midwestpoultry.com/wp-content/uploads/Xu-Yun-Sheng-Propane-Saving-in-Poultry-Farm-Through-Waste-Heat-Recovery.pdf (last accessed 14 Febrauary 2021).

Yu, L., Shah, S. B., Knauer, M. T. and Boyette, M. D. (2018). Evaluation of a landscape fabric based solar air heater in a pig nursery. *ASABE Annual Meeting #1801405*. Detroit, MI. St. Joseph, MI: ASABE.

Zhao, Y., Xin, H., Shepherd, T. A., Hayes, M. D. and Stinn, J. P. (2013). Modelling ventilation rate, balance temperature and supplemental heat need in alternative vs. conventional laying-hen housing systems. *Biosystems Engineering* 115(3):311–323.

Index

www.ingramcontent.com/pod-product-compliance
Ingram Content Group UK Ltd.
Pitfield, Milton Keynes, MK11 3LW, UK
UKHW021826100926
8556IPUK00003B/174